THE MANAGEMENT OF
TELECOMMUNICATIONS NETWORKS

ELLIS HORWOOD SERIES IN
ELECTRICAL AND ELECTRONIC ENGINEERING
Series Editor: D. R. SLOGGETT, Technical Director, Marcol Group Ltd, Woking, Surrey

P.R. Adby	**APPLIED CIRCUIT THEORY: Matrix and Computer Methods**
J. Beynon	**PRINCIPLES OF ELECTRONICS: A User-Friendly Approach**
R.L. Brewster	**TELECOMMUNICATIONS TECHNOLOGY**
R.L. Brewster	**COMMUNICATION SYSTEMS AND COMPUTER NETWORKS**
M.J. Buckingham	**NOISE IN ELECTRONIC DEVICES AND SYSTEMS**
A.G. Butkovskiy & L.M. Pustylnikov	**THE MOBILE CONTROL OF DISTRIBUTED SYSTEMS**
S.J. Cahill	**DIGITAL AND MICROPROCESSOR ENGINEERING**
S.J. Cahill & I. McCrum	**DIGITAL AND MICROPROCESSOR ENGINEERING: Second Edition**
R. Chatterjee	**ADVANCED MICROWAVE ENGINEERING: Special Advanced Topics**
R. Chatterjee	**ELEMENTS OF MICROWAVE ENGINEERING**
P.G. Ducksbury	**PARALLEL ARRAY PROCESSING**
J.-F. Eloy	**POWER LASERS**
J. Jordan, P. Bishop & B. Kiani	**CORRELATION-BASED MEASUREMENT SYSTEMS**
F. Kouril & K. Vrba	**THEORY OF NON-LINEAR AND PARAMETRIC CIRCUITS**
P.G. McLaren	**ELEMENTARY ELECTRIC POWER AND MACHINES**
J.L. Min & J.J. Schrage	**DESIGNING ANALOG AND DIGITAL CONTROL SYSTEMS**
P. Naish & P. Bishop	**DESIGNING ASICS**
J.R. Oswald	**DIACRITICAL ANALYSIS OF SYSTEMS: A Treatise on Information Theory**
M. Ramamoorty	**COMPUTER-AIDED DESIGN OF ELECTRICAL EQUIPMENT**
J. Richardson & G. Reader	**ANALOGUE ELECTRONICS CIRCUIT ANALYSIS**
J. Richardson & G. Reader	**DIGITAL ELECTRONICS CIRCUIT ANALYSIS**
J. Richardson & G. Reader	**ELECTRICAL CIRCUIT ANALYSIS**
J. Seidler	**PRINCIPLES OF COMPUTER COMMUNICATION NETWORK DESIGN**
P. Sinha	**MICROPROCESSORS FOR ENGINEERS: Interfacing for Real Time Applications**
J.N. Slater	**CABLE TELEVISION TECHNOLOGY**
J.N. Slater & L.A. Trinogga	**SATELLITE BROADCASTING SYSTEMS: Planning and Design**
R. Smith, E.H. Mamdani, S. Callaghan	**THE MANAGEMENT OF TELECOMMUNICATIONS NETWORKS**
E. Thornton	**ELECTRICAL INTERFERENCE AND PROTECTION**
L.A. Trinogga, K.Z. Guo & I.C. Hunter	**PRACTICAL MICROSTRIP CIRCUIT DESIGN**
J.G. Wade	**SIGNAL CODING AND PROCESSING: An Introduction Based on Video Systems**
R.E. Webb	**ELECTRONICS FOR SCIENTISTS**
J.E. Whitehouse	**PRINCIPLES OF NETWORK ANALYSIS**
Wen Xun Zhang	**ENGINEERING ELECTROMAGNETISM: Functional Methods**
A.M. Zikic	**DIGITAL CONTROL**

THE MANAGEMENT OF TELECOMMUNICATIONS NETWORKS

Editors:
ROBIN SMITH
Intelligent Systems Research Section
BT Laboratories, UK

E.H. MAMDANI
Department of Electronic Engineering
Queen Mary and Westfield College
University of London

JAMES CALLAGHAN
Intelligent Systems Research Section
BT Laboratories, UK

ELLIS HORWOOD
NEW YORK   LONDON   TORONTO   SYDNEY   TOKYO   SINGAPORE

First published in 1992
and reprinted in 1993 by
ELLIS HORWOOD LIMITED
Market Cross House, Cooper Street,
Chichester, West Sussex, PO19 1EB, England

A division of
Simon & Schuster International Group
A Paramount Communications Company

© Ellis Horwood Limited, 1992

All rights reserved. No part of this publication may be reproduced, stored in a retrieval system, or transmitted, in any form, or by any means, electronic, mechanical, photocopying, recording or otherwise, without the prior permission, in writing, of the publisher

Printed and bound in Great Britain
by Hartnolls, Bodmin

British Library Cataloguing in Publication Data

A catalogue record for this book is available from the British Library

ISBN 0–13–015942–5

Library of Congress Cataloging-in-Publication Data

Available from the Publisher

Table of contents

FOREWORD	vii
INTRODUCTION	ix
STRUCTURE OF THE BOOK	xii

I - Logical Framework for TMN

Telecommunications Management Conceptual Models	3
Functional Description of Network Management	13
TMN Architecture	23
TMN Interoperable Interfaces	37

II - The Evolving TMN

TMN Evolution	51
TMN Reference Configuration Case Study Results	61
Towards Integrated TMNs - The Global Conceptual Schema	71

III - Modelling Aspects of TMN

Object Oriented Modelling in RACE TMN	85
Service and Network Model Implementation	97
Experience of Modelling and Implementing a Quality of Service Management System	109
Quality of Service Mappings	121

IV - Implementing the TMN

An Architecture and other Key Results of Experimental Development of Network and Customer Administration Systems	135
Viewpoints on Traffic and Quality of Service Management in Telecommunication Management Networks	147
Implementation of Management Applications for Network and Customer Administration Systems	159

V - Experimental Results

Virtual Path and Call Acceptance Management for ATM Networks	173
The Use of AIP Techniques in Traffic and Quality of Service Management Systems	185
A Generic Maintenance System for Telecommunication Networks	195
An Interconnected MANs Maintenance Prototype	213

VI - Recommendations

Experience Designing TMN Computing Platforms for Contrasting TMN Management Applications	225
Recommendations for the Use of AIP Techniques for Maintenance in Telecommunication Systems	241
HCI Considerations in TMN Systems	251

FOREWORD

Over the past few decades, telecommunications systems have been engaged in a quiet revolution. Transmission systems have matured rapidly from single circuit audio through multichannel frequency division multiplex to today's time division multiplexed digital systems. Switching systems have enjoyed a similar increase in power and functionality. The future systems currently being designed will continue the evolutionary process. These continuing increases in power and functionality have been achieved at the cost of increasing complexity. This complexity now requires radical new approaches to the management of the global telecommunications system.

In many regards the work presented in this book is of a pioneering nature. The previous five years or so have witnessed the emergence of telecommunications network management as an important engineering discipline in its own right. Gone are the days when each new transmission and switching system would be installed with its own unique monitoring and control system. Because telecommunications is central to the economic life of nations, customers rightly demand cost effective, reliable communications. The achievement of these twin goals is the responsibility of telecommunications equipment suppliers and the service suppliers. The efficient running of networks falls on the service suppliers and increasingly they rely on network management systems to augment the skills of their trained people.

Presented here are the results of over four years work on the definition and experimental studies in the area of network management for the future European Integrated Broadband Communications (IBC) network. The people who undertook this work are employed in companies and academic institutions spread across Europe. The co-ordination of their efforts and the common goals were set by the European Commission in their RACE I Programme. This coordinated research programme represents one of the largest undertakings of its kind in the network management field. But of course, the RACE Telecommunications Management Network (TMN) community is not inward looking. The maximum use is made of parallel activities being conducted across the world telecommunications community. Outputs from standards making bodies such as the ISO, CCITT, ISO/NMF, ANSI/T1M1 etc. have been taken fully into account where appropriate. And indeed, the RACE TMN community have made inputs to the appropriate standards making bodies.

The common objective shared by members of the project teams was to explore the application of modern computer based technologies to the difficult problem of managing future multiservice, multimedia broadband communications networks. Their task was especially difficult, since throughout the lifetime of the projects, the networks to be managed were themselves in their definition phase.

INTRODUCTION

Having briefly set the scene and introduced the importance of network management to modern communications, the remaining part of this editorial will provide a brief description of the RACE I programme, introduce the projects whose results are being presented and give a short introduction to the papers. A short account of a number of topics of common concern to the projects concludes the editorial.

THE RACE PROGRAMME

RACE is an acronym for "Research and Development in Advanced Communications in Europe". This programme of pre-competitive research and development was initiated by the European Commission following extensive consultation with the industrial sectors concerned, national governments and the European Community's institutions. Quoting from RACE'91 (DGXIII-F, Commission of European Communities):

> *The main objective of RACE is: To prepare for the "Introduction of Integrated Broadband Communications (IBC), taking into account the evolving ISDN and national introduction strategies, progressing to Community-wide services by 1995"'.*

The term IBC defines an evolutionary concept whose meaning can be understood from the following: Integrated refers to the integrity of telecommunications from narrowband voice through to fully interactive broadband services, embracing fixed and mobile networks; Broadband points to the expanding bandwidth requirements for future services, and Communications indicates the important need to embrace all aspects of telecommunications from high level network considerations through to user-friendly service provision.

RACE I

The first main phase of RACE began in January 1988 and will be completed in December 1992. The programme is organised into three concurrent parts:

  Part I   IBC Development and Implementation strategies
  Part II   IBC Technologies
  Part III Prenormative Functional Integration.

The work being carried out in these three areas is conducted by over 100 projects. Their activities cover the broad range of disciplines that characterise modern telecommunications. Along the technology dimension, these range from research into the physics of opto-electronic devices, bandwidth compression systems, transmission and switching technologies and customer premises equipment studies. Complementary work is being conducted into the economics of service provision, usability issues, approaches for verification etc. Embedded in this rich mix of R & D activity is the work on TMN which is the focus of this book.

Each project has its own goals, objectives and deliverables. But the real power of RACE I is that it is a concerted programme. The results of the projects are synthesised to produce a coherent overall statement. This consensus process is fundamental to success in the long run. The enduring output of RACE I will be:

- Common Functional Specifications for IBC
- Common Practices and Technology Recommendations
- Contributions to World Standards.

These documents will be the blueprint through which the various actors in Europe will build the integrated communications fabric essential for social and economic success in the years to come.

RACE I TMN

At the outset it is worth mentioning that the term Telecommunications Management Network conveys more than the notion of a network. A TMN is a complete network management system including the gathering of information from the managed network, storing information about the managed network, acting on the information and the mechanisms for controlling the managed network. There is some debate in the network management community whether the TMN is concerned only with automatic (as contrasted with manual) decision processes but that is of small concern here.

In this book the results of seven projects concerned with enhancing knowledge about TMN are presented. Two of the projects are Part I projects concerned with specification and evolution. Four are Part II projects investigating technology options for the future implementation of TMN and one is a Part III project concerned with verification of quality of service for IBC.

Diagrammatically the seven projects can be represented as shown below:

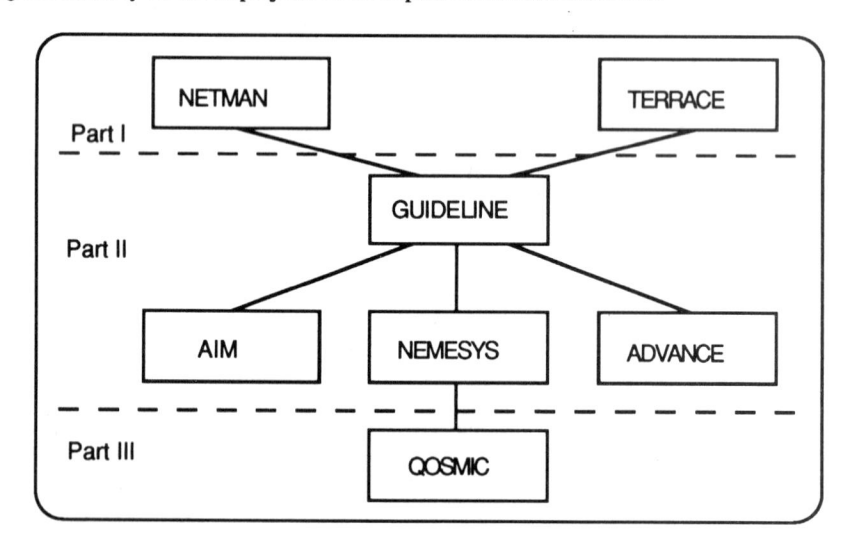

There are a number of other Part III projects which have an impact on the TMN area but their work is not presented in this book.

Brief Description of Project Responsibilities

The following provides a brief introduction to the work areas addressed by the seven projects. People interested in gaining a fuller account are recommended to read RACE'92, published by DGXIII - F, Commission of European Communities.

NETMAN (R1024) is charged with preparing a coherent set of Common Functional Specifications which describe the TMN in a technology independent manner. The work of this project has embraced case studies, modelling and consistency checking of developing specifications.

TERRACE (R1053) has the task of plotting evolution strategies which will guide the developers and installers of network management systems along technically and economically viable routes.

GUIDELINE (R1003) is charged with providing a coordination function for the Part II TMN technology projects, developing a common approach to TMN architectures and providing recommendations on Advanced Information Processing (AIP) technologies appropriate for TMN implementation. In the conduct of its coordination role, GUIDELINE has organised six TMN conferences. The papers in this book were presented at the Madeira Conference, 1st - 3rd September 1992.

AIM (R1006) is concerned with developing new approaches to the maintenance of telecommunications networks. Its principle target has been the definition of automatic network diagnostic and repair systems. Experimental studies have been a major part of the work.

NEMESYS (R1005) has the twin goals of investigating Traffic and QoS management for IBC. The traffic management investigations have involved a series of experiments on the application of AIP techniques to Asynchronous Transfer Mode (ATM) networks. These experiments have also been used to elaborate the QoS studies.

ADVANCE (R1009) is working on the production of recommendations for an architecture and implementation options for an integrated Network and Customer Administration System (NCAS) for IBC. This work has involved both theoretical studies and a set of linked experiments to test the validity of various AIP techniques.

QOSMIC (R1082) has the task of defining quality of service user requirements for IBC and the necessary verification methodologies and tools. The work is concerned with quality of service to network performance mappings and uses both theoretical studies and experimental platforms.

STRUCTURE OF THE BOOK

The twenty one papers presented in this book do not provide a complete definition of the TMN nor indeed a complete description of the work conducted in the RACE I TMN Programme. But what is presented is an insight into the depth and range of work undertaken. Each paper can be viewed as a pointer to the underlying continuum of work which stretches from specification through to experimental implementation. In order to provide a coherent structure to the material the papers are grouped in six themes.

The first theme addresses the logical architecture for TMN. Four papers provide information on the following topics:
- TMN conceptual models
- Functional description of network management
- TMN architecture
- TMN Interoperable interfaces

This is followed by three papers which give information on the evolutionary nature of TMN implementation:
- TMN evolution scenarios
- Case studies
- The role of the conceptual schema in TMN integration

The third theme addresses the important modelling aspects associated with describing the TMN. Four papers cover the following:
- Object oriented modelling
- Model based management
- Modelling QoS management
- QoS mappings

The next grouping provides experimental results of implementing TMN functions:
- Network & Customer Administration Systems (NCAS) architectures and key results
- Results from traffic and QoS management implementations
- Management application implementation results

The fifth theme continues on implementation aspects and presents experimental results in the following areas:
- Management aspects of ATM networks
- The use of AIP techniques in traffic & QoS management
- Maintenance prototype for MAN's
- Generic Maintenance System for Telecoms networks

The final theme concerns interim recommendations for TMN technology:
- TMN platforms
- AIP techniques for maintenance of telecoms systems
- Human Computer Interface (HCI) techniques

As would be expected from a concerted research programme such as RACE TMN there are a number of regularly occurring topics which are addressed in many of the papers. Important ones include:
- Architectures
- Methodologies
- Advanced Information Processing.

A brief explanation of these topics is given for those readers new to the area.

Architecture. This is a conceptual framework for the definition of systems. It provides the baseline specifications and the building bricks for a top level system design. An architectural definition will provide firm guidelines for the rigourous decomposition of the functions to be performed by the system under study. In essence an architecture assists the understanding of complex systems.

Methodologies. Formal approaches to the analysis and definition of systems is being given greater emphasis in modern systems engineering. This methodological approach permits the users of the material to comprehend the logic which supports the contents of the specifications, designs etc. This increasing rigour is considered to be an essential feature of the TMN design process.

Advanced Information Processing. AIP is a term used in RACE (and ESPRIT, a sister programme) to denote modern computer based information processing techniques. These include distributed processing, object oriented analysis and design, expert systems, distributed knowledge based systems etc. From the outset the RACE TMN programme embraced the AIP approach as the way to tame the scale and complexity of modern and future telecommunications management systems.

ACKNOWLEDGEMENTS

A list of all the people in the RACE I TMN programme who have contributed to the work of the seven projects reported here would be too long. Therefore, the editors will confine their acknowledgements to the authors of the papers for their efforts and pay tribute to the assistance given by the other members of the GUIDELINE management team (U. Apel, T. Turner, A. Galis and S Plagemann) in structuring the 6th RACE TMN conference and this book.

Robin Smith, BT Laboratories, UK
Professor Abe Mamdani, Queen Mary and Westfield College, London University, UK
James G. Callaghan, BT Laboratories, UK

I - Logical Framework for TMN

The Management of Telecommunications Networks
R. Smith, E. H. Mamdani, J. G. Callaghan (Editors)
© Ellis Horwood 1992

Telecommunications Management Conceptual Models

Stephen Plagemann, Terry Turner (Broadcom, Ireland)

ABSTRACT

Telecommunications Management (TM) conceptual models have a purpose to classify TM requirements and application functions in a top down fashion. This technical approach provides ease of system development and evolution when new requirements arise. The NETMAN CUBE conceptual model is based on an axiomatic definition of TM and general management theory and consists of three orthogonal parts: (1) a behavioural decision making process of Awareness, Decisions and Implementation (ADI); (2) a logical layering of organisational decisions making process into Business, Service, Network, and Network Element levels; (3) TM Application Functional Groupings into Customer Administration, IBCN Management and Pre-Service activities. A worked example is provided to demonstrate the utility of these conceptual models to structure logically a common problem of TM management of a customer complaint about a fault.

1. INTRODUCTION

Management theory based conceptual models classify and categorise Telecommunications Management (TM) requirements and application functions. Top-down decomposition allows ease of evolution for systems development purposes. The management conceptual models described are powerful enough to accommodate distinctions between deterministic routine management procedures such as customer billing and non-deterministic control loop governed management decisions like network planning, traffic management and equipment maintenance. The three orthogonal management models (collectively entitled the NETMAN CUBE model) are:

- The management based decision making conceptual model is characterised by three generic functions: Awareness Creation, Decision Making and Support and Decision Implementation (ADI).
- The Responsibility model segments decision making processes into four different logical layers of Business, Service, Network and Network Element Management.
- A system management lifecycle model classifies the complete range of management tasks found during requirements capture. Nine Telecommunications Management Functional Areas (TMFAs) resulted from the use of this model. A detailed description of these TMFAs may be found in [1].

A worked example demonstrates how the combination of the three management conceptual models logically structures a realistic functional specification concerning a customer complaint to a help desk relating to a fault report and its resolution when the network element is repaired. Finally, some reflections are presented on the role of a TM Functional Reference Model (FRM), ETSI management standards and future design processes specifying interworking between different management systems.

2. CURRENT CONCERNS IN OPERATION SYSTEM DESIGN

The future evolution of telecommunications management systems is threatened by the impact of an explosion in the data, the complexity, variety and distribution of Operations Systems (OS).

Traditionally each new type of overlay network and service contains a proprietary management system, with non-standard, non-portable management actions, data formats, goals and objectives. As long as telephony was the only service provided this situation was containable. Recently a host of new overlay networks and services were introduced by European service providers. A "spaghetti-syndrome" has arisen where bottom-up implementations of managing systems for new and old networks and services has created a number of tightly coupled proprietary OSs with data and functions allocated to each kind of application regardless of its common usefulness. This situation has caused the following types of problems:

- Lack of complete end-to-end telecommunications administration of public/private networks and services and customers
- OSs are connected in different ways to carry out their support functions according to no particular strategy. When this bottom-up policy is followed, sooner or later one system is dependant on more than one other system for the required data. This creates problems since OSs created for different purposes at different times will not have carefully specified interfunctional and information dependent relationships. When information exchanges are required between different OSs, they are carried out on an ad hoc basis using humans. To create new non-standard automatic interface linkages after the bottom-up development of OSs will require expenditure of considerable resources
- There is no way of reconfiguring or extending one system without affecting many of the others. This tight coupling of OSs means that changes reverberate around the management domain and require considerable resources to regenerate software and verify new system configurations
- Rapid change from one technology to another rapidly devalues the capital invested in technology dependent functional specifications, data structures and other OS components. Shorter product lifecycles of telecommunications equipment and dedicated management applications software means fewer versions of the same product before replacement with something new.

A different and more strategic approach to the specification and design of new telecommunications management systems has been explored in RACE. This more systematic approach offers a solution to this current crisis of the spaghetti syndrome of bottom up development of OSs. The creation of a relatively stable RACE Telecommunications Management (TM) Functional Reference Model (FRM), linking public and private management systems, should be seen as a primary objective of both vendors and service providers in the future. The TM FRM is only one of the many specifications and abstract conceptual models required as part of a systematic systems approach to problem solving needed to deal with the present crises of specification and development of OSs for the future. This conceptual model is a key starting point and stable focus for the less abstract conceptual models used at later stages of systems development.

3. MANAGEMENT CONCEPTUAL MODELS FOR THE RACE TM FRM

The foundation to RACE TM FRM conceptual models based on management theory is an axiomatic definition of telecommunications management:

"the human, automatic or combined decision making process which is triggered by network data, a request or information from a customer or from the business level of the organisation. This decision making process results in changes being made to the managed network or service, or a response to the customer or business level of the organisation. The decision making process has the objective of optimising the use of the network or service, satisfying the feature, quality or cost requirements of the customer as well as meeting the business objectives of the organisation."

The three combined orthogonal models of management that structure the TM FRM specifications are referred to as the NETMAN CUBE model and are shown in Figure 1.

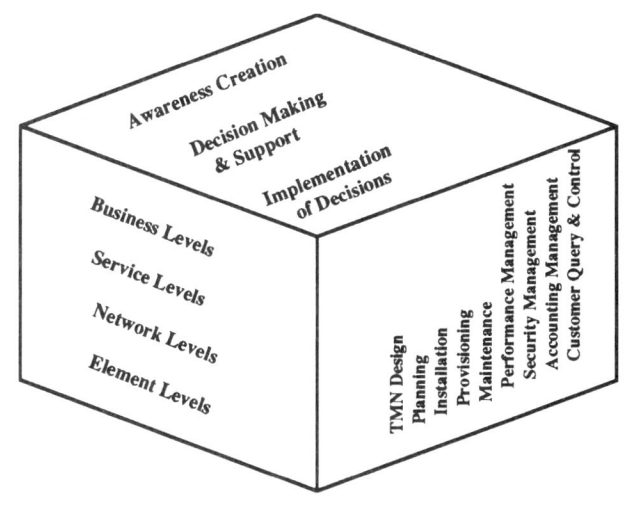

Figure 1 - The CUBE Model

When combined, these management based conceptual models show the following capabilities to:
- systematically categorise and classify the semantics of decision making within the framework of a TM FRM functional specification template
- organise or arrange the actions, procedures of the smallest components (functional entities) of the management systems according to explicitly stated theories which are widely accepted outside RACE
- support checking for completeness, correctness, consistency, and ease of revision
- reside within more general paradigms of stages of systems development and ISO/ODP distributed systems projections [2]
- enhance the network management application functional specifications with desirable features of ease of evolution and completeness
- provide a property of consistency that allows traceability between user requirements and TM functional specifications and vice versa.

3.1 The NETMAN ADI Model

ADI stands for Awareness Creation (A), Decision Making and Support (D) and Decision Implementation (I) - the process of performing management by a system such as a network management system. The concept is that such a system has a set of *A* performing functions, *D* performing functions and *I* performing functions. The processing passes logically from one function operation to another function as sequences of events (A to D to I). Many such processes may operate in parallel in the system. The ADI conceptual model is a development by the NETMAN project and was very successful in classification of the requirements and TM functions which were captured from a wide variety of sources. The degree of determinism of the decision making process is an important distinction contrasting between the management by objectives during which means can be chosen and the routine performance of management services in a fully prescribed manner. Telecommunications management needs both types of decision making.

Non-deterministic management processes are characterized by the presence of a control loop in which the outcome of a decision is not always predictable. Examples of non-routine behaviour are telecommunication system and equipment management processes such as IBCN Maintenance and Performance Management: repair of equipment components that in fact were not broken or the rerouting of traffic that mistakenly resulted in decreasing the QoS as perceived by the customer. The ADI model has been used to specify three generic functional modules for TMN prototypes of non-deterministic processes of broadband traffic management by the RACE NEMESYS project [3].

The ADI model has also successfully classified and categorised routine behaviour which forms a large part of the telecommunications management activity. Deterministic processes including routine customer administration activities like billing and provisioning, which do not usually deviate from prescribed processes, have been successfully modelled using the ADI model. The ADVANCE project have used the ADI model for prototyping routine deterministic processes of routine data processing in TMN prototypes of customer provisioning [4].

The activity sequence of generic decision making is an important property of this conceptual model. The order of generic functions A-D-I must be maintained whether the process is routine or non-deterministic. This property of the ADI conceptual model provides an inherent structure to the dynamic or behavioural aspect of functional specifications. In addition, omission of one or more of the generic functions A or D or I violates the basic axiom of the decision making process. The completion of the decision making process is a precondition of success. Decision Implementation in an active sense need not necessarily follow decision making, for to leave the situation as it is, may be a valid decision. The point is that the means to implement a decision must be provided by the specification. These properties of the ADI model have proved very useful as a check of the completeness of TM requirements and functional specifications.

"Pre-conditions" and "post-conditions" are modelling concepts for function performance. They are used to delineate the transitions that characterise the management activity. Pre-conditions are states of the system which must exist in order for the function to execute. Post-conditions are states which prevail after the function associated with the post condition has executed.

An Awareness Creation function acts as a sieve of information isolating the decision making process from a mass of external data most of which could be irrelevant. Awareness Creation events can take the form of alarms, customer requests for management services, messages or instructions from the organisation business level or other administrations. The network management system requires relevant information to monitor and detect certain external conditions which require action on the part of the decision maker (pre-conditions). It is also the responsibility of the Awareness Creation process to place a boundary on the problem to which it responds (post-condition). This boundary corresponds to the scope of action of the decision maker to be triggered. The preconditions for Awareness Creation are the changes to the states of the managed system or the customer or organisation being supervised by the function. The post-conditions for Awareness Creation include some of the following :

- The current status of the managed system has been determined from solicited or non-solicited information
- A comparison with design objectives has been made from which it has been decided if the managed system is meeting its objectives
- The problem cause cannot be identified to allow an decision making function to be invoked
- Validation of a customer request for a management service
- The firing of a valid awareness message to a decision making entity.

Decision Making and Support functions require both information and some form of intelligence. Decision making and decision support information are gathered from managed resources or from external sources via the Awareness Creation functions and directly to the sources. The IBC management system must as a minimum provide decision support of some kind. Intelligence is provided by human skill or expertise, perhaps in the form of sets of written procedures or guidelines. In the future, it is anticipated that expert systems or machine learning by neural networks will usurp that quality previously exclusive to the human brain. A real decision making IBC management system probably will rely on a combination of the above as well as decision support capabilities. The decision making process may request information from a Decision Support function. Decision support activity consists of the provision of timely, accurate, concise and relevant information to aid the decision making agent to reach a satisfactory decision. Examples of support functions are traffic count collection, QoS data collection and diagnostic procedures for telecommunications equipment.

The precondition for a Decision Making function execution is that an instability in the managed system is determined by the Awareness Creation function. Provision of support information required for the decision must also be part of the capability of the management system. The post-conditions for a Decision Making function include the following :

- If the managed system has not met the design objectives, a diagnosis of the cause has been made along with suggestions for improvement made to a planning process.
- If the managed system is not meeting its objectives, then a proposed cure or problem localisation action has been formulated as part of feedback control actions.
- If the managed system is meeting its objectives, but it is anticipated that it may not continue to do so, then appropriate feed forward control actions may be taken.
- If the managed system meets neither of the above conditions, then it may be decided to take no action.
- In the case of management services of a deterministic routine nature, a decision implementation function is triggered.
- Any exceptions to normal decision making and support situations are noted in logs and records.

A Decision Implementation function contains the final phase of generic functionality of management processes. This may occur as changes to real physical network resources, such as replacing network resources like faulty line cards in multiplexers or alteration of the value of data such as attribute values in a table or object. Changes may occur in stages and result in issuing written instructions for interpretation by operator staff, modification of databases or physical alteration of equipment or software. Decision Implementations may be summarised as generic activities such as local change translations and their issuing of verification or the production of new plans and schedules. The preconditions for a Decision Implementation function are instructions for corrective actions determined by a Decision Making function along with availability of procedures and models of managed system or procedures to be changed. Post-conditions for a Decision Implementation function are:

- Corrective actions have been transformed into control instructions in network elements which have been applied to change the status, state or behaviour of the managed system
- For management services of a routine nature, if the transaction has posed no exceptions, then action is completed and returned to control with closure message
- Verification of the correctness of actions taken
- Exceptional events associated with non-deterministic actions provide an awareness creation function for comparison with control models.

3.2 The Responsibility Model

The NETMAN Responsibility Model is based on original work by BT [6] and has been widely adopted within the RACE programme. The multi-layered nature of this management conceptual model, shown in Figure 2, was adopted by RACE TMN as it mapped to accepted principles of system behaviour and general management theory.

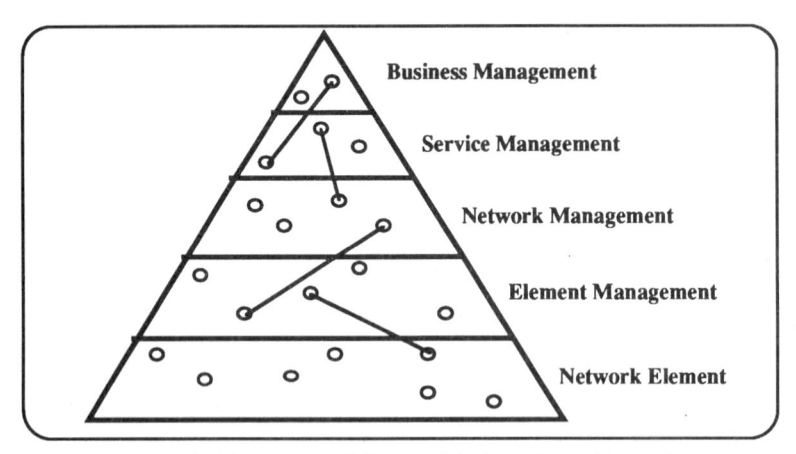

Figure 2 - The Responsibility Model : Inter-Layer Example

A corollary of the Responsibility Model, named the "Need to Know principle", provides guidelines structuring the generic ADI decision making functions. This is done by locating them at the logical layers or by defining relationships between layers of non-real-time management systems [6]. The "Need to Know principle" envisions decision making occurring at the logical layer at which sufficient information and expertise to perform the decision resides. This principle, when factored into the Responsibility Model's layering process of decision making in distributed systems, creates a powerful tool for specification and design of non-realtime systems that include human intervention in the management processes. The "Need to Know principle" provides the following guide-lines:

- A TM function shall be performed if sufficient information is received from a network element or other logical layer [EXECUTE]
- If additional information is required, the TM function shall be performed at the next highest logical layer where the required integration of information occurs [REFER then EXECUTE]
- Execution of a decision after REFERRAL requires control loops to prompt data transmission from other logical layers. Control messages are always downward towards lower logical layers
- Only that information strictly required to make a decision is passed upward in the logical hierarchy to higher layers.

Knowledge of the relationships between Responsibility model layers is the basis for management control and co-ordination in the management model. Responsibility for decision making concerns changing the state of abstract or real resources. These managing objects implement functions at different logical layers related to each other by certain roles [5].

The Business Management layer supervises the interaction between the providers of network and services and includes functions and information necessary for directing decisions concerning policies and strategies within the organisation that owns and operates the services (and possibly the networks). This layer has responsibility for a self contained part of the communications environment and has several principle roles of setting management or service policies or monitoring QoS and costs.

The **Service Management layer** is concerned with, and responsible for, the contractual aspects of services that are being provided to customers or available to potential customers. This layer includes functions and information to manage a telecommunications service.

The **Network Management layer** has responsibility for the management of all the network elements from a global system-wide viewpoint. It is not concerned with how an individual network element operates but may be aware, for alarm and inventory purposes, of some data concerning network elements. This layer is responsible for planning, design and operation of a network which satisfies the functional and financial requirements specified by the Service Management layer in the service specification.

The Network Management layer contains the functions and information addressing the management of a particular network as a complete entity. Higher or logical layers of networks support various services and manage the quality of interconnections and paths to meet the network performance objectives of the service. The physical transmission layer is concerned with the management functions of bit transport such as source policing parameter changes as well as installation of the physical infrastructure to support the logical networks.

This layer supports the Service Management layer in service provision activities such as connecting/removing the customer or changing the service. In addition the Network layer specifies, implements and manages a network to meet the service specification of the Service Management layer If the service specifications cannot be met, then the network manager notifies the service manager of this situation.

The **Network Element (NE) layer** is responsible for supporting network demands and facilities for all network elements. The NE layer includes the functions and information related to the management of the individual network elements. Repair of equipment faults, installation of cable in the local loop and the instantiation of particular equipment would be examples of NE functions. Information at the NE layer would include physical node and link repair data, records of building locations, ducts and poles, etc. Specific information on individual elements of hardware or software are included and only a very limited visibility of networks is allowed. Peer to peer interactions shall not be possible at this layer.

3.3 System Management Lifecycle Model

The System Management Lifecycle conceptual model has the purpose of classifying the different tasks that are part of the total scope of IBC telecommunications management of customers, services, networks and network elements past, present and future. Collections of management tasks into areas of concern are termed the TM Functional Areas.(TMFA). The IBC System Management lifecycle model categorises nine different TMFAs, as determined from requirements captured from standards, provider organisations, RACE research and literature surveys performed by NETMAN. The System Management Lifecycle model of pre-service and in-service TMFAs fully cater for these collected requirements for management of customers, services, networks and network elements. Further details of the TM application functions are provided in [1]. Future service changes to the TMN itself are for further study. Topics of Security Management are the responsibility of other projects in RACE and are outside the scope of this discussion. However a few security management functions have been specified using the TM functional template and these can be found in [1].

4. A WORKED EXAMPLE OF MANAGEMENT CONCEPTUAL MODEL'S INFLUENCE ON TM SPECIFICATIONS

A worked example shown in Figure 3 illustrates the influence of the combined NETMAN ADI and Responsibility models on the structure and logic of a RACE TM functional specification. The TMFAs chosen are Provisioning and Maintenance to illustrate mainly non-deterministic decision making processes that logically structure a problem faced everyday by provider organisations and their customers. The example chosen is deceptively simple and illustrates the fulfilment of the criteria of specification completeness and consistency as well as interfunctional interworking and use of control messages.

The logical specification of a non-realtime customer complaint system using the "Need to Know" principle allows for a consistent top-down approach to system design that would aid interworking between different provider administration's management systems. Several roles are presented which are part of the layer relationships in the Responsibility model, namely, level of authority, service and invoker of interaction.

Awareness Creation to the Operator is the customer reporting difficulties to the Help Desk, which is part of the Provisioning TMFA. After an exchange of dialogue to support the Decision Support process, the Help Desk staff decides on what entity within the managing system or organisation is optimally capable to solve the problem. The Decision Making and Support at the Service Management layer would select the appropriate functional area, say Maintenance, which is acquainted with the details of the difficulties. Implementation from the Help Desk at the Service Management layer then becomes Awareness at the Network Element (NE) Management layer. When the cause of the customer complaint has been rectified by the Decision Making and Implementation processes of the Maintenance TMFA at the NE Management layer, then the Provisioning TMFA at the Service Management layer is notified to respond to the customer query to close the control loop.

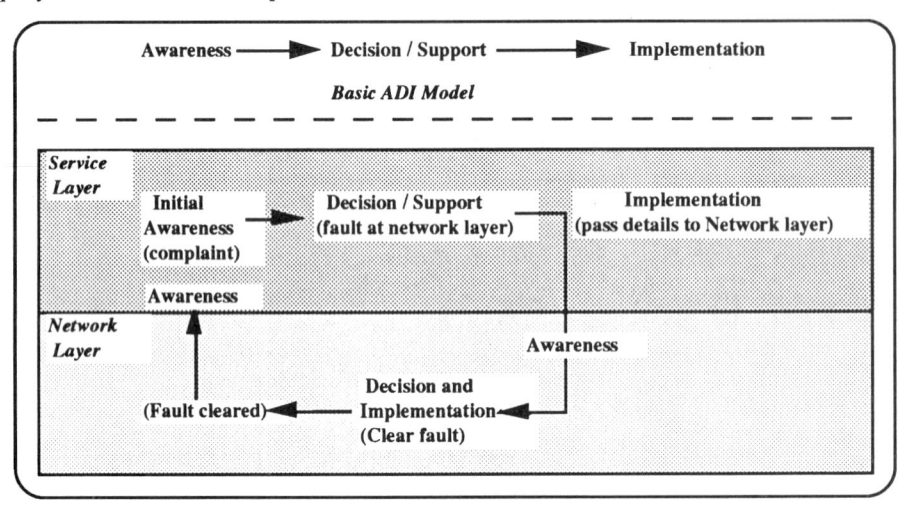

Figure 3 - Example of combination of ADI and Responsibility Models

The "Need to Know" principle and Responsibility model concern optimum placement of detailed information relating to a specific component failure for decision making at the NE Management layer. Maintenance decision making should not be required at layers in the Responsibility model which are not concerned with trouble diagnostics for specific hardware located in the network elements. Network or Service Management layer objects contain information about the logical organisation of networks and services, not about failure modes of telecommunications equipment. At a Business Management layer, a synopsis of trends in failures would be more appropriate for the formulation of maintenance strategies and costing of return on investment, rather than detailed repair procedures of specific telecommunications equipment.

The ADI model is the basis for the behavioural aspects of the TM FRM specification. If any of these abstract functional components of the decision making process outlined above is missing (A or D or I), then management decisions are not carried out. Imagine how far Maintenance in an exchange could proceed if the specifications omitted Decision Support information like repair instructions for failed equipment. This completeness factor allows rather easy but important validation checks of the functional specifications to be illustrated by this example.

Control aspects of management by objectives can be emphasized by considering the possible outcome if the ADI control loop in figure 3 was not closed. The customer complaint was recorded as Awareness Creation at the Service layer. If the NE Management layer fails to implement a report on the clearance of the failure to the Service Management layer, then the complaint remains active. The possibility of the customer being misinformed of the status of his complaint is very high. A secondary consideration is that after an elapsed time, the Service Management layer may make unnecessary enquiries (awareness messages that are false alarms) of the NE Management layer. If the failure has already been cleared, then confusion due to false alarms and fault status will exist concerning the exact status of the complaint. This lack of closure of a control loop is clearly an undesirable state of affairs in a management system.

5. CONCLUSIONS

The IBCN will require a distributed management system of great complexity. A current escalating crisis named as the "spaghetti syndrome" will prevent operators from taking full economic advantage of new demand for services, since geometric growth in management costs of building duplicated and overlapping OSs cannot be contained. RACE research has produced a more systematic cost effective approach to managing system development by creation and maintenance of the results of all three stages of systems development. Key to this new approach are conceptual models of various degrees of abstraction that help designers and customers to visualise the scope and nature of the next generation of IBC management systems (TMN). RACE research results will provide a sound basis for the European telecommunications industry to surmount the crises caused by the spaghetti syndrome. When RACE reference models are completely included in ETSI and CCITT standards, there will be a much broader international basis of understanding. Designers in the years to come can use the check-lists of TM functions and objects which are the modular components of any TMN system design.

The results of NETMAN show the need for and feasibility of a RACE TM FRM that is implementation independent, hierarchically decomposed and represents the Requirements Analysis stage of system development at the Enterprise and Information projections of distributed systems. Given a set of captured, classified and analyzed requirements, the TM FRM - an abstract conceptual model and associated application functional specifications - is considerably influenced by a set of management theory based conceptual models of decision making. The TM FRM is:

- an essential precursor to one of the key constituents of the TMN Reference Configurations
- a fundamental input to functional standards for IBC management systems
- a check list for top-down high level designs of IBC management systems.

The three management conceptual models (ADI, Responsibility, Lifecycle) either singly or in combination have proved to be of great value in classifying and categorising the TM requirements and functional specifications. The classification of both types of results allows traceability of functional specifications to requirements and provides ease of evolution for the inevitable user or designer changes during various stages of systems development. The management based ADI conceptual model is sufficiently universal in its application to have been validated in various TMN prototypes. The Responsibility Model (Business, Service, Network and Network Element layers) has provided new approaches to management system design, particularly for non-realtime data processing applications where separation of management decision making concerns is of importance. Interfunctional relationships have been preserved with the use of the Systems Management Lifecycle model to group management functions into eight different pre-Service and In-Service customer and IBCN management functional area (TMFAs) with Security management as a separate collection of TM functionality applicable to all stages of the lifecycle.

The main problem resolved by the RACE TM FRM concerns matching the specification paradigms, semantics and syntax of different types of conceptual models that relate the management functions and information which form the IBC managing system global conceptual schema. Designers can now create external schema with the assurance that relationships to other functions and information are known and accounted for by the TM FRM specifications. Cross transfer of information between different functional groupings has been anticipated in the specifications of the TMFAs contained in RACE CFS H 401-411 [1]. Each function must be aware of the status of other functions so that a proper response by the global telecommunication management system for correcting any problem situations can be made. These requests and responses can be adequately named (specified) by the TM information model aspect of the IBC managing system global conceptual schema. The TM objects named in the interworking specifications are specified in detail in [7]. The TM FRM approach to specification using conceptual models and a system wide approach to problem solving has successfully tackled the functional interworking problem left unsolved by the current "spaghetti syndrome" bottom-up approach to OSs development. Functional interworking is the "glue" that allows for communication between different proprietary managing systems. Provision is made in the RACE TM functional template for functional interworking and this information is summarised in [8].

6. ACKNOWLEDGEMENTS

This paper arises from work partly funded by the CEC RACE programme project, NETMAN and partly funded by LM Ericsson and Telecom Eireann.

7. REFERENCES

[1] Walles, A. : "Functional Descriptions of Network Management", Sixth RACE TMN Conference, Madeira, September 1992.

[2] ISO - International Standards Organisation : "Recommendation X.9yy : Basic Model of Open Distributed Processing, Part 2: Descriptive Model", ISO/IEC JTC1/SC21 N6079, August 1991.

[3] NEMESYS Deliverable 9 : "NEMESYS Experiment Case Study Description", 05/ICS/SS/DS/C/019/A1, September 1991.

[4] ADVANCE Deliverable : "Prototype Version Two Specifications", 09/BCM/RD3/DS/B/027/B1, December 1991

[5] BT : "CNA-M, Co-operative Networking Architecture - Management", 1990.

[6] NETMAN Deliverable 6 (Annex 3.4,) : "Draft Telecommunications Management Specifications", 24/BCM/RD2/DS/A/006/b2, September 1990

[7] RACE Common Functional Specification H550 : "Telecommunication Management Objects".

[8] RACE Common Functional Specification H 407 "Security Aspects of Interfunctional Information Exchanges".

The Management of Telecommunications Networks
R. Smith, E. H. Mamdani, J. G. Callaghan (Editors)
© Ellis Horwood 1992

Functional Descriptions of Network Management

Tony Walles (BT Laboratories, UK)

1. INTRODUCTION

The NETMAN Project has been charged with the task of defining an implementation independent view of the functions of the Telecommunication Management Network, (TMN) for the proposed Integrated Broadband Communications Network, (IBCN) [1]. In RACE the results are presented in a series of Common Functional Specifications (CFS). For the TMN these have been organised by Telecommunications Management Functional Areas, (TMFAs) which have been defined by the Project [2]. This paper sets out to explain the major areas of research which were undertaken by the project and how the results were synthesised to provide a realistic starting point for the Common Functional Specification preparation. The studies were carried out in the following areas.

- Requirements Capture [3]
- Modelling [3]
- Characterisation of the TMN Domain
- The Object-Oriented Approach
- Hypertext Representation of Results.

The life span of the IBCN will be extremely long compared to the technologies which will be used for its introduction and development. It follows therefore that the CFSs being produced by the project must be independent from, but cognisant of both the technology and the implementations which are presently used for telecommunications systems. Such an approach will allow the integration of technological advances into the IBCN while maintaining its overall integrity, so-called future proofing. It is contended that the results from this work on network management for the future Pan-European IBCN will be of direct relevance to today's network management problems.

2. CHARACTERISATION AND SPECIFICATION OF THE TMN DOMAIN

The initial process was, therefore, the establishment of what is understood by the term TMN. This interpretation is largely dependent upon, but not exclusive to, the point of view of the TMN user. In order to specify TMN functionality it is clearly essential to understand the requirements of the domain. For instance a Local Area Network will require management as part of a general office support environment. For example, access to the network, printers, servers etc. are all part of the LAN management which will probably be carried out by one person as part of their overall workload. On a larger scale, a company which has a number of different sites, possibly in various countries will have communications management problems of a more complex nature which will be handled by a communications manager. In both cases the communications system to be managed is secondary to and supportive of the main business of the company.

The telecommunications operating company has the management of its networks as its business. The company responsibilities are potentially to millions of customers and it has to manage the complete communications infrastructure from the customer connection to maintaining global services. From this short discourse it can be realised that the TMN field is particularly diverse and complex. For the TMN to function correctly an explicit understanding of its purpose is essential.

As mentioned previously, the TMN functionality which had been identified by the requirements capture exercise existed at a predominantly physical level. All that is specified about the IBCN is that it is a multi-media, multi-vendor and multi-service enterprise. Consequently a top-down approach has had to be adopted. As a parallel activity, the work of international standards bodies ISO and CCITT was also incorporated into developing the CFSs. The areas addressed by these bodies related to the in-service aspects of management, e.g. Fault, Configuration, Accounting, Performance and Security Management (FCAPS). It was considered that TMN functionality should additionally address pre-service and future service activities and some of the in-service aspects could be further refined. The result of this initiative was the expansion of the original five management areas to nine. Their applicability is illustrated by mapping them onto a simple life-cycle model, figure 1.

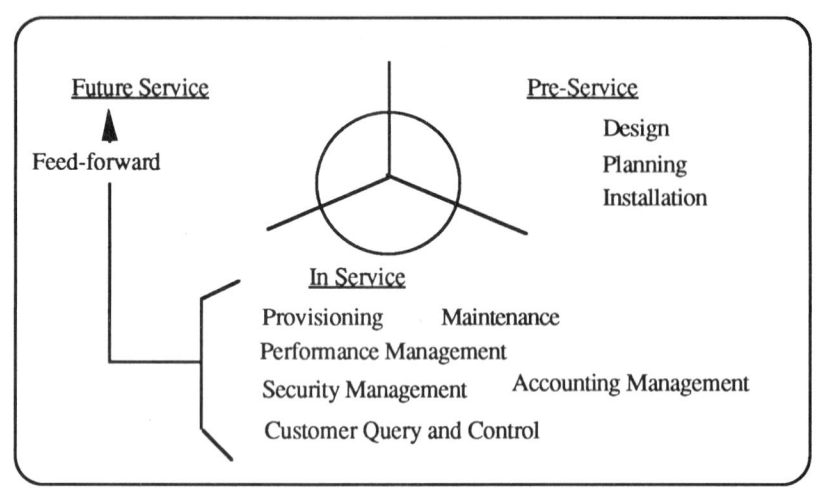

Figure 1 - TMN Life-cycle Model

In order to maintain uniformity in the work, a generic template for the functional specifications has been developed. This incorporates a nine-point specification plan which is iterative for the purposes of further decomposition of sub-functions. The steps are:

- Name of Functional Area
- Description of Objectives and Services of the Functional Area
- Scope statement
- Preamble
- Relationship to other Functional Areas
- Objects
- Static description
- Dynamic description
- Glossary of terms and references.

It is worth mentioning here that a considerable amount of validation work is carried out within the project by means of case studies. These studies apply the principles of the NETMAN TMN model to existing networks. A synthesis of these approaches has facilitated the decomposition of the nine TMFAs to a more detailed level as in figure 2. As examples of further decomposition of TMFAs, Accounting Management has been further refined into Charging and Billing, Cost Accounting and Interadministration Accounting; Provisioning into Customer Facing, Resource Assignment, Contractual Phase, Service Activation and Service Cessation or Interrupt.

```
Design          Provisioning       Security
    Planning        Maintenance        Accounting
        Installation    Performance         Customer
                                            Query &
                                            Control
            Customer Facing
            Resource Assignment
            Contractual              Cost Accounting
            Service Activation       Billing and Charging
            Cessation/Interrupt      Interadministration Accounting
```

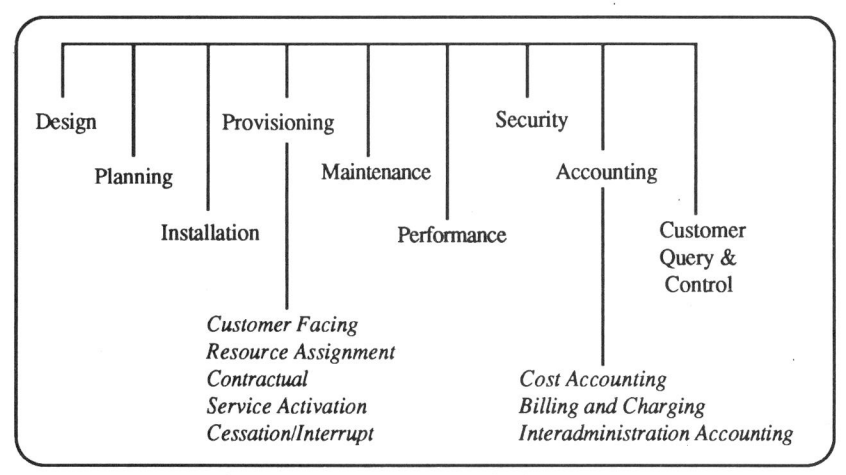

Figure 2 - Nine TMFAs with some further decomposition

3. THE NINE TELECOMMUNICATION MANAGEMENT FUNCTIONAL AREAS

A brief description of the nine TMFAs follows.

3.1 Design

Design encompasses the engineering and documentation of network elements, customer equipment and service providers equipment based on specified requirements. This functional area can be decomposed into service design, network design and element design. The object-oriented paradigm (discussed later) and CCITT Recommendations I.130 [4], I.310 [5] and Q.65, [6], have been considered in producing this specification. (H401) [7].

3.2 Planning

Planning is the organisation and control of the implementation of Enterprise or Business layer decisions for the introduction and management of communications systems. Its role is to bring about the introduction of services in a cost effective manner while being cognisant of the requirements of the customer and operator (H402) [8].

3.3 Installation

The placement of network elements and the physical and electrical (or optical as appropriate) interconnection of the network and supporting elements is the function of Installation. (H403) [9].

3.4 Provisioning

Provisioning is responsible for activating network resources which will supply the customer with his requirements. It is the first of the in-service functions of the life-cycle and as such is responsible for the activation of all other functional areas in that domain. The relationship between customer and provider is established at this point of contact and set out in contractual terms. Activities dealt with in provisioning are Customer request for service, Resource assignment, Contractual negotiation and Service Activation. Modification to and deletion of Services are also specified. This functional area will also provide an interface between the customer and other functional areas such as maintenance and accounting for the purposes of dealing with customer difficulties (H404) [10].

3.5 Maintenance

The commonly accepted interpretation of maintenance is usually on-line corrective Fault Management. This concept has been expanded, with particular consideration being given to Preventative and Surveillance-based maintenance as described in CCITT Recommendation M.20 [11]. Scheduling of maintenance tasks and the logistics of workforce management have also been addressed. The scope of this functional area is quite considerable and it has been decomposed into a number of sub-functions for ease of study (H405) [12]. A brief description of these follows.

Maintenance Management

A set of functions responsible for the planning, the operation, and improvement of the overall Maintenance of the IBC system. Most of the identified sub-functions appear at the higher levels of the responsibility model described in [3] and include:
- Maintenance strategies definition
- Maintenance Planning. This function is responsible mainly for the execution and update of a maintenance plan which includes not only the maintenance requirements imposed by the operator but also the maintenance requirements resulting from the operation of the IBCN itself
- Maintenance costs evaluation and optimisation
- Maintenance staff/resource management (recruiting, allocation, training programs development)
- Statistical evaluations
- To create, store and maintain historical records of faults, associated test and monitoring methods used to detect them and the repair procedures used for their correction
- Co-ordination with other telecommunication management functions (e.g. with provisioning in aspects related to stock inventory administration or in the selection of equipment and equipment vendors).

Preventative maintenance procedures

Preventative maintenance is that which is carried out at predetermined intervals or according to prescribed criteria and intended to reduce the probability of failure or the degradation of the functioning of an item. These procedures relate more to routine activities (e.g. cleaning or replacing old components) that are mainly applied to mechanical components (e.g. disk drives or printers) according to criteria specified by the equipment vendors.

Test

These functions include the operation of different mechanisms provided by the system enabling continuous monitoring, in-service and out-of-service routine tests, per-action tests, alarm reporting and performance management reports. Surveillance-based maintenance relies mainly on the continuous operation of monitoring functions in order to identify potential faults before affecting the network, e.g. threshold levels exceeded. A common usage of these test and monitoring functions by maintenance and performance management is envisaged although this issue needs further discussion.

Trouble Detection

Troubles within a system must be detected before corrective actions can be initiated. This function is performed either through the acknowledgement of a systems' user complaint or through the acknowledgement of event/alarm reports provided by monitoring and test functions. Detection can be categorised into primary and secondary. When a trouble is easily and promptly detected it is considered to be a primary detection function. Complex troubles requiring more sophisticated techniques are handled by secondary detection functions.

System Protection

This function encompasses the activation of a set of processes that prevent the propagation of a detected trouble through the entire system thereby restricting the impact of the underlying problem.

System Recovery

The capability of the system to recover as fully as possible to its nominal state of operation after the detection of a trouble (this capability can be provided for instance using redundant design, spare and back-up facilities).

Trouble Notification/Reporting

This aspect of the specification provides the capability of alerting and recording the existence and severity of a trouble **in a prioritised manner**. This function includes the activation of audible and visible indicators and the generation of both external reports (e.g. reports issued to the operator or maintenance staff) as well as system internal reports (messages).

Trouble Verification

The capability to determine if a previously detected trouble condition still exists (usually this function is performed before the activation of other functions - e.g. initiation of a repair or trouble reporting).

Trouble Diagnosis and Localisation

The capability of isolating the trouble (pin-point the trouble cause) to its most elementary source, usually corresponding to a logical system resource (ideally with a physical correspondence to a substitutable system resource unit - e.g. a slide-in-unit, software module). This function includes the selection, evaluation and running of tests, correlation analysis of different reported troubles and the capability to perform the mapping of logical system resources into their spatial addresses (e.g. shelf number, rack, etc.) using system configuration data.

Put System Resource Out-of-Service

Put identified system resource OOS (Out-of-Service) or TOOS (temporarily Out-of-Service). This is a generic function that can be triggered as a result of the execution of different functions (e.g. as part of a system protection or repair process).

Repair Scheduling

Capability of the system to schedule the necessary repair actions for different troubles according to their severity and assign the appropriate and most available skilled staff.

Repair/Replacement

After having verified if the trouble still exists (trouble verification), repairing/replacing the faulty substitutable system resource by a spare one. This function consists of four sub-functions, Therapy, Maintenance Supervision, Maintenance Verification, and Return-To-Service.

Therapy

Therapy will locate and repair/replace those substitutional entities (SE) necessary to clear alarm conditions and enable a return to service.

Maintenance Supervision

Capability of the system to provide guidance to the maintenance staff and to control the overall repair procedure including the provisioning of spare parts.

Repair Verification

Testing to ensure the operation of the substitutable system resource before returning to service.

Return to Service

Generic function triggered by other functions (e.g. as a result of a repair procedure). It basically puts the substituted system resource into service.

3.6 Performance Management

Performance Management is employed to regulate the IBCN in order to maintain the quality of service levels agreed with customers. The cost effective optimisation of resources is also an objective for this functional area. The activities undertaken in Performance Management are Monitoring, Analysis and problem alerting, Diagnosis, Optimisation and Control (H406) [13].

3.7 Security Management

Security management is defined as being the provision of security services for customers [14]. These services are defined as Authentication services, Non-Repudiation, Integrity and Confidentiality, services. Authentication services involves the transmission of given criteria which can be identified as valid by co-operating entities. Password recognition is one such service. Non-repudiation services are described as a capability to the user of the service being provided with proof of the origin or delivery of data or its contents. Integrity allows the originator of the data to provide the recipient confidence that the data has not been modified. Confidentiality is a security service which provides confidence that data transmitted in a particular instance of communication has not been modified (H407) [15].

3.8 Accounting Management

The usage of IBCN resources and services will be measured by Accounting Management functions for each customer and the customer will be billed accordingly. Usage records and Tariffs will be used in the calculation of charges to customers who will be presented with invoices at predetermined intervals. Billing intervals and payment methods will be agreed at the contractual phase of the service provisioning activity. Cost Accounting will be carried out to control costs and revenues and tariffs. Inter-administration accounting will also be part of this functional area (H408) [16].

3.9 Customer Query and Control

The facility to allow customers access to particular areas of TMN functionality is expected in the IBCN environment. Such facilities envisaged are reconfiguration of the customer access (by the customer), queries to accounting and other access similar to those available to the Operator. The security aspects related to this facility were considered of sufficiently high importance as to warrant a separate functional area (H409) [17].

Having decomposed the nine functional areas to a greater level of detail the interactions between areas of functionality have become clearer. Ultimately this will lead to a series of functional specifications which will define all the necessary functionality for a complete TMN.

4. THE OBJECT ORIENTED APPROACH

The use of object oriented analysis (addressed by Coad and Yourdon [18]) was applied to the requirements capture results. From this analysis strategies for a high level system design could be postulated. This contributed to the definition of the nine functional areas proposed by the project. Additionally, the object oriented approach to specification has been used by standards bodies for some time. The project also had to keep in mind the way in which objects were being used for the purposes of interface specification, particularly between Operations Systems and Network Elements. It should be made clear at this point that the specifications being produced are *informal*. Knowledge elicitation techniques which can be applied to requirements capture and analysis are well known but these differ from formal techniques in that there is no verification process for the analysis and authentication of results. Formal specification languages such as SDL [19] are supported by verification techniques but were not suitable for the level of abstraction which was appropriate to the NETMAN work. An object template has been developed for consistency purposes. The template contains the following sections.

- Object Class   (Mandatory)
- Class Description   (Optional
- Superclasses   (Mandatory)
- Subclasses   (Optional)
- Part-of   (Mandatory
- Has-Parts   (Optional)
- Associations   (Optional)
- Attributes   (Mandatory)
- Operations   (Mandatory
- Notifications(Events) (Optional)
- Comments   (Optional)

This object identification/specification work has led to the introduction of a tool which enabled the CFSs to be defined with greater precision. This tool is applied to specific instances of network management functionality for the validation of the (informal) functional specifications.

5. NETMAN HYPERTEXT MODELLING TOOL

As the CFSs were prepared the associated documentation expanded and the level of detail increased considerably. This resulted in the specifications becoming more useful but also increasingly difficult to maintain in a consistent manner.

Periodic quality reviews of specifications are undertaken to identify inconsistencies, gaps and overlaps. This can be difficult enough within a single specification but, when attempting to specify nine functional areas which interrelate with one another to some degree, the problems magnify considerably.

This problem has been eased significantly by the use of Hypermedia tools. These tools use objects which represent graphically the static part of the specifications. The objects are located on the appropriate layer of the CNA-M model described in [3]. The tools allow individual objects to be interrogated so that attributes and their value sets can be read. An example of the Service Layer model for Service Provisioning is shown in figure 3.

Scripts can be associated with the objects to illustrate the behavioural aspects of a specification. Using these tools particular instances of network management functionality can be modelled and the overall objective functionality can be represented. This promotes a clearer understanding of the specification and assists in the reduction of inconsistencies.

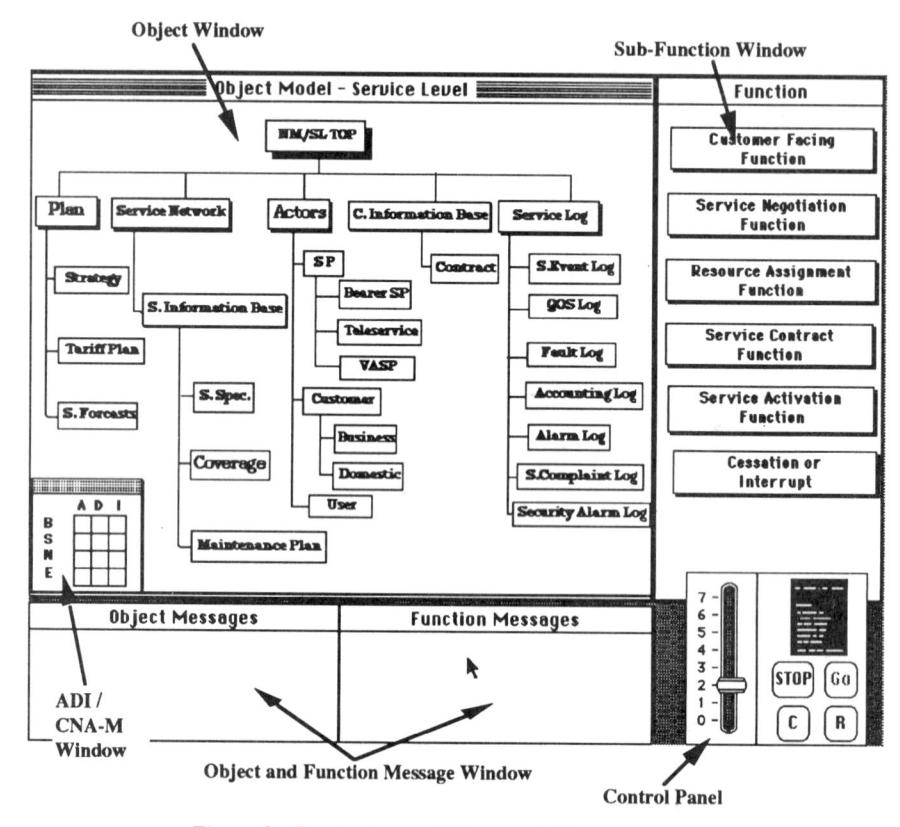

Figure 3 - Service Layer Object model for Provisioning

A process model showing the location of animation with respect to specification and design is presented in figure 4. The development of object-oriented techniques above an SDL platform are being considered as a method for future functional specification work. This would facilitate the use of formal techniques for the verification of specifications which currently does not exist.

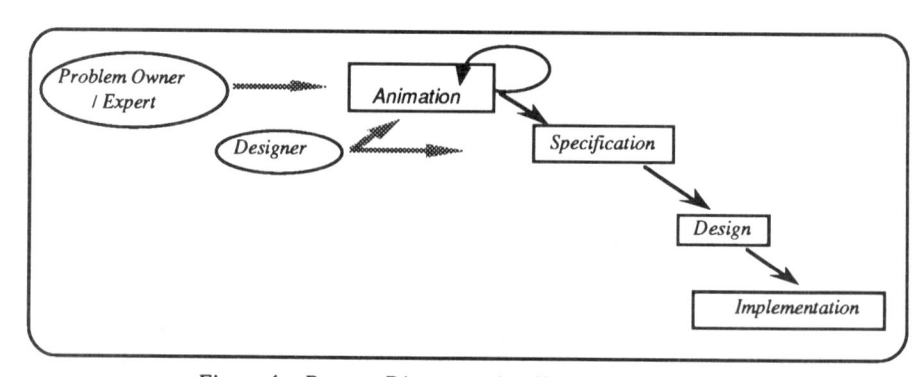

Figure 4 - Process Diagram using Hypertext Modelling

6. RACE CONSENSUS PROCESS AND STANDARDS CONTRIBUTIONS

A major aspect of the work has been the preparation of material for contributions to the European Telecommunications Standards Institute, ETSI. Contributions are made to ETSI after proposals have progressed through the RACE consensus process. This process allows the potential users of the specifications to make constructive criticism. Material from the specifications is then selected for input to ETSI if it is considered appropriate. Consideration is also being given to the inclusion of those functional areas which have been identified in addition to the FCAPS areas described earlier.

7. CONCLUSIONS

The material presented in this paper is the result of studies on the functional specification of an integrated Network Management system (TMN) for the proposed pan-European broadband multi-service network of the near future. Traditional islands of network management expertise have been acknowledged and conceptually integrated with original research to formulate a complete end to end approach. New proposals to encompass the required completeness of TMN functionality have been promulgated to the RACE TMN community and international standards bodies. The NETMAN approach to this functional specification work represents one of the few which takes account of the public domain requirements for TMN. Templates for developing functional specifications and objects have been produced to ensure consistency throughout the work.

The Hypertext modelling tool currently simulates the Installation, Maintenance, Provisioning, Performance Management, Accounting Management and Security Management functional areas. This work has been enhanced and it is now possible to demonstrate several instances of interworking between a number of the functional areas.

8 ACKNOWLEDGEMENTS

The author wishes to acknowledge the work of all colleagues in the RACE project NETMAN, particularly Stephen Plagemann. I further wish to thank the GUIDELINE Conference Organisers and Editorial Committee for their guidance and support.

9 REFERENCES

[1]   Commission of the European Communities : "Research and technology development in advanced communications technologies in Europe - RACE 1992", Brussels, March 1992.

[2]   RACE CFS H400 - Telecommunications Management Functional Specification Conceptual Models, Scopes and Templates

[3]   Plagemann, S. : "Telecommunications Management Conceptual Models", Sixth RACE TMN Conference, Madeira, September 1992.

[4]   CCITT - Recommendation I.130 "Method for the Characterisation of ISDN Services".

[5]   CCITT - Recommendation I.310 "ISDN Network Functional Principles"

[6   CCITT - Recommendation Q.65 "2nd.Method for the Characterisation of ISDN Services"

[7]   RACE Common Functional Specification H401 - TMN Design Services

[8]   RACE Common Functional Specification H402 - Planning Services in TMN

[9]   RACE Common Functional Specification H403 - Installation Services in TMN

[10]   RACE Common Functional Specification H404 - Provisioning Services in TMN

[11] CCITT - Recommendation M.20 "Maintenance Philosophy for Telecommunications Networks.
[12] RACE Common Functional Specification H405 - Maintenance Services in TMN
[13] RACE Common Functional Specification H406 - Performance Management Services in TMN
[14] RACE Common Functional Specification H411 - Security Management
[15] RACE Common Functional Specification H407 - Security Aspects of Information Exchanges in TMN
[16] RACE Common Functional Specification H408 - Accounting Management Services in TMN
[17] RACE Common Functional Specification H409 - Customer Query and Control
[18] Coad, P, Yourdon, E. : "Object-Oriented Analysis" - Prentice-Hall Inc. 1990
[19] CCITT - Recommendation Z.100 et.al. "Specification and Description Language"

ns Networks
R. Smith, E. H. Mamdani, J. G. Callaghan (Editors)
© Ellis Horwood 1992

TMN Architecture

James G. Callaghan, George I. Williamson (BT Laboratories, UK),
Ben Hurley (Broadcom, Ireland), Kevin Riley (UNIPRO, UK)
Gottfried Schapeler (Alcatel SEL - AG, Germany),
Derek Harkness (Roke Manor Research, UK),
Alex Galis (Dowty, UK)

ABSTRACT

This paper describes an implementation architecture for a Telecommunications Management Network (TMN) suitable for the management of the proposed Integrated Broadband Communications network. The architecture described is intended to support the future implementation of TMN systems within Europe and elsewhere. Having outlined some of the motivations for the development of a TMN architecture, the various functions associated with a TMN are introduced. An overview is given of the functional and physical views of the TMN. Also presented is an elaboration of the internal structure of an Operations System. One of the major components of any network management system is the Management Information Base (MIB). The MIB is described in some detail because of its central role in the architecture. Finally a list of some of the open issues still remaining concludes the paper.

1. BACKGROUND

This paper is based on the work of RACE project GUIDELINE. The main objective of this project is to fulfil a co-ordination role for the TMN technology projects in part II of RACE I [1]. Common issues among these projects are TMN architecture, TMN specification and standards, conceptual integration of different TMN applications, and evaluation and guidance on AIP for TMN. Co-ordination of work on these common issues is facilitated by GUIDELINE. This paper presents the results of the work performed on TMN architecture.

2. INTRODUCTION

Until comparatively recent times the management of telecommunications networks was approached on a rather ad hoc basis. Network management systems were designed specifically for each particular telecommunications equipment set and service. This resulted in a plethora of network management systems which were limited in their applicability, presented a number of interworking problems, and which limited the scope for increased automation of network management functionality. With the growth of complexity in modern telecommunications networks and the drive to reduce operational costs by increased automation in a more systematic and integrated fashion, the need for a more structured or architectural approach is evident. The TMN (Telecommunications Management Network) concept provides a framework under which this approach to telecommunications management can be investigated. A TMN is a logically separate network that interfaces a telecommunications network at several different points to send/receive information to/from it and to control its operations. A TMN may use parts of the telecommunications network to provide its communications. For example, TMN information may be supported by the Embedded Control Channel (ECC) in SDH networks.

However, before describing the TMN in detail it is necessary to outline some of the reasons why an architectural approach to TMN is more advantageous. This is presented in the following section.

3. ARCHITECTURAL MOTIVATIONS AND REQUIREMENTS

In general terms an architecture provides definitions of terms and concepts for explaining systems and models for reasoning about systems [2]. It provides the base specifications for the various building blocks of the system. The architecture should also identify the various functions of the system and provide guidelines for the decomposition of this functionality into modules. These modules, together with the definition of the messages and protocols for information transfer between the modules provide the framework for assembling the system.

The benefits of an architecture for TMN are as follows:
- provides an enabling conceptual and design framework
- supports problem decomposition / separation of concerns. For example, it supports the separation of conceptual levels of specification from the realisation aspects
- provides well defined reference points and interfaces
- supports flexibility, portability, and evolution
- supports reusability of generic piece parts by the definition of common building blocks.

Within the general TMN architecture there are three basic aspects of the architecture which can be considered separately when planning and designing a TMN. These three aspects are the TMN Functional Architecture, the TMN Physical Architecture and the TMN Information Architecture. These are addressed in the following sections.

4. FUNCTIONS OF THE TMN

A TMN is intended to support a wide variety of management functions which cover the planning, operation and management of a telecommunications network. It is clearly vital to all involved in the production of a TMN to understand the uses to which it will be put. At the most obvious level this translates into the need for a definitive functional specification. Many groups are making a contribution to such a specification, using a variety of methods, both formal and informal. General principles for the specification of the functions of the TMN are set out in CFS H100 [3]. The functions of the TMN are specified for RACE in the CFS H400 series specifications [4]. A description of these functions can also be found in [5].

4.1 Classification of Functions

One feature which is common to many of the specification activities is the use of a classification of the defined functions into groups which have aspects in common. Typical of such a classification is the use of five categories in the OSI specifications: Fault management, Configuration management, Accounting management, Performance management and Security management (FCAPS), which has been adopted by CCITT in Recommendation M.3010 [6]. Such classifications have a number of valuable attributes. They subdivide the whole of the "problem space", which is large and complex, so that attention can be focused on a sub-set of the whole. They serve to draw attention to similarities between functions, for example grouping those functions which address the needs of a particular group of users. They gather together functions utilised at the same stage in the life of the managed system or which have similar requirements on some attribute such as performance [7]. It is, unfortunately, inevitably true that no one classification can achieve all of these goals at the same time. The resulting value is sometimes not so great, particularly if the grouping is less intuitive and natural. Despite these drawbacks, one or more classification schemes should be adopted because, at the very least, they provide a partial framework for ensuring the consistency and completeness of the functions described.

4.2 Classes of Functions

One of the uses to which the designer of a TMN will need to put the functional specification is to identify which functional blocks are of general utility and which are specific to particular user oriented functions. To this end, a division into **user-specific** functions and **common** functions is recommended, these **common functions** are further subdivided into **infrastructure** functions and **user-generic** functions [8]. Some more information on this is presented later in this paper.

4.3 Non Functional Requirements

In addition to the need to specify the functional requirements on the TMN it is necessary to guide the implementers in a number of other areas. These non-functional requirements cover external and observable characteristics of the functional implementations, as well as purely internal or local ones. The former include such aspects as throughput and performance, which will affect, for example, the communications requirements and hence interface specifications. The latter embrace things such as the physical environment in which the host computer is to be located, which will affect the size and complexity of the sub-system and the measures taken to ensure its resilient operation. For a more complete list see [9]. Non functional requirements are also used in establishing reference configuration functional allocation strategies [10].

5. STRUCTURE OF THE TMN

One essential of any architecture is to describe the overall structure of the system being defined. One major use of such a structure is to permit the placement of the functionality required of the system in a reasonable number of blocks and the description of those blocks and of the interfaces between them.

A functional decomposition is first required, which deals only in functions to be performed and their relationships. The functional view describes the appropriate distribution of functionality within the TMN. It allows for the creation of function blocks from which a TMN can be implemented. To support this a range of physical organisations are possible. These must follow the functional structure but need to allow for a wide range of additional requirements, mainly non-functional. These add important aspects of distribution, cost, performance and so on, which have to be considered by the system implementer. The physical architecture describes realisable interfaces and examples of physical components that make up the TMN. One vital guiding principle to be followed is to make the commitment to a physical arrangement as late as possible. This does not prevent full exploration and elaboration of the functions required, which must be independent of placement, but ensures maximum flexibility to meet the additional requirements and provides maximum opportunity for their full discussion.

5.1 Functional Architecture

The overall TMN functionality is broken down into three main components, the operations system function blocks (OSFs), the workstation function blocks (WSF) and the mediation function blocks (MFs). These are related as shown in figure 1. The figure also shows the relationship of the TMN with the network element functional blocks (NEFs) and the Q Adapter functional blocks (QAF). Intercommunication between the function blocks is supported by the data communications function (DCF). The function blocks are interconnected by a number of standardised reference points: q, x, and f. The g and m reference points are not a subject for standardisation and are also shown. These are briefly outlined below, but have been described in detail [8]. Please note that the boundary of the TMN intersects the WSF, QAF and NEF. This is consistent with M.3010. In RACE TMN there has been considerable debate about the value of this choice, and many would prefer a cleaner boundary at a well defined reference point. A model of the TMN workstation function which proposes a clearly identifiable reference point between the TMN and the User has been described in [8].

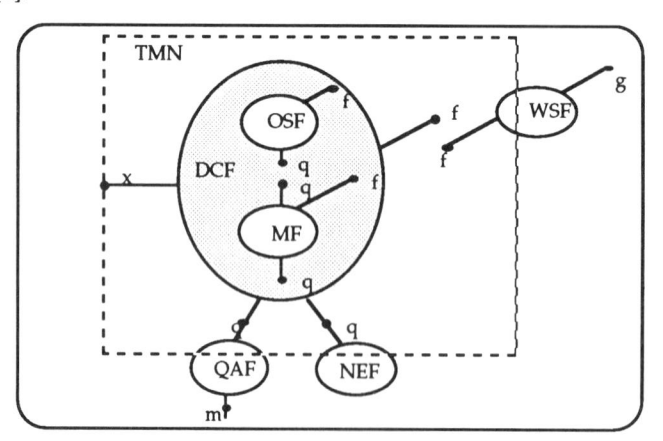

Figure 1 The generalised functional architecture for a TMN

TMN Functional Components

As just seen the TMN functional architecture is composed of a number of function blocks. The function blocks provide the general functions that enable a TMN system to perform the TMN management functions. The function blocks are:

- the *Operation Systems Functions* (OSF) which enables the management of the Network Element Functions (NEF). The OSF performs the information processing relating to management
- the *WorkStation Functions* (WSF) enables the user to interact with the OSFs
- the *Data Communications Functions* (DCF) is used for the transfer of information between the functional blocks
- the *Mediation Functions* (MF) act on information from NEFs to adapt, filter and condense it as required by the OSFs
- the *Network Element Functions* (NEF) which allow the communications with the TMN for the purpose of being monitored and/or controlled
- the *Q Adapter Functions* (QAF) allow the connection to the TMN of those network elements which do not support standard TMN interfaces.

Each of these function blocks are themselves composed of elementary functional components [8]. For example the OSF can be decomposed into application functions, management information base functions, and support or infrastructure functions. A significant amount of work on the decomposition of OS management functions has been performed by GUIDELINE and this work has not yet been mirrored in M.3010. Additional added value of the project work over M.3010 lies in the enhancement of the TMN architecture to make it more logically applicable to the management of IBC as a multi-service network. The separation of service functions from network functions, which is described at length elsewhere [11], provides the basis for the layering of OSFs. Without such a layering the OSF remains a very large and complex set of functions with no simplifying structure. GUIDELINE

has applied this layering approach to the OSF and it has been found to be very useful in supporting the separation of concerns and problem decomposition. The application of the layering approach leads to the organisation of the OSFs into four layers. These layers are known as:
- Business Management
- Service Management
- Network Management
- Network Element Management.

The abbreviations OSFB, OSFS, OFSN and OSFE are used to denote these four layers. The use of the layers also has implications for the reference points as discussed in the next section.

Reference Points

The reference points define the conceptual point of information exchange between non overlapping function blocks. A reference point becomes an interface when the connected function blocks are embodied in separate pieces of equipment. Briefly the reference points are as follows:
- *The q class of reference points:*
    - qx: The qx reference point connects NEF to MF, MF to MF and QAF to MF
    - q3 : The q3 reference point connects NEF to OSF, MF to OSF, QAF to OSF, and OSF to OSF
- *The f reference point* : This connects WSF to OSF and MF to WSF
- *The g reference point:* The g reference points are between the WSF and the User
- *The x reference point:* This connects a TMN to other management type networks including other TMNs.

Figure 2, taken from [6] shows the reference points between the management function blocks. This figure essentially summarises the TMN functional reference model.

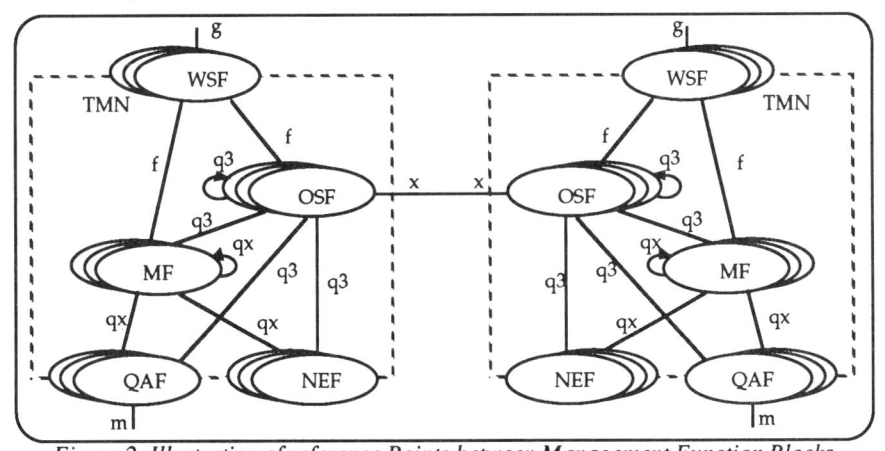

Figure 2 Illustration of reference Points between Management Function Blocks

As mentioned above, the application of the layering approach to the OSFs has implications for the reference points. Figure 3 illustrates the reference points for the higher layers of OSFs. GUIDELINE has also looked at reference points in a more abstract fashion, starting from the most abstract and general notions moving down to increased specialisation. The decomposition emphasises the differences in functionality at different positions. A taxonomy of reference points has been proposed [8] based on this approach.

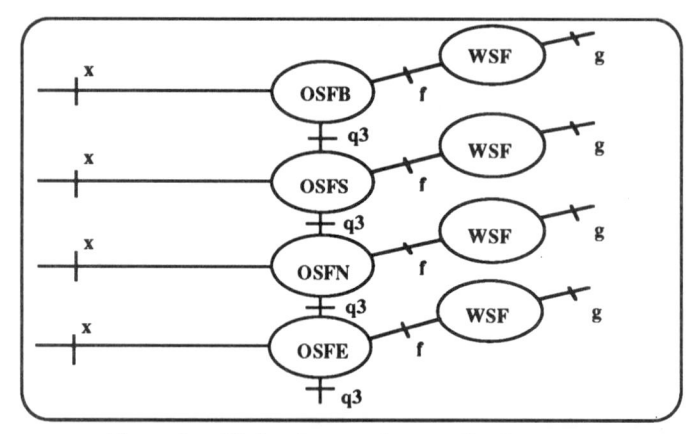

Figure 3 - Higher Layer reference Configuration

5.2 Physical Architecture

The physical or implementation structure of the TMN follows the basic shape of the functional structure. It, however, adds one or more placement options so as to assign functions to possible physical units. A wide range of physical arrangements are possible ranging from complete integration of function (everything in one box) to a highly fragmented and distributed approach, with small systems each supporting a minimal set of functions. It should be noted that the range of physical options based on the logical architecture is enormous. This allows a single logical structure to be reflected in many possible physical allocations which can be optimised to reflect local requirements. *Non-functional* requirements are particularly important here. What is required is a range of "sensible" placement scenarios which reflect today's thinking and current technology.

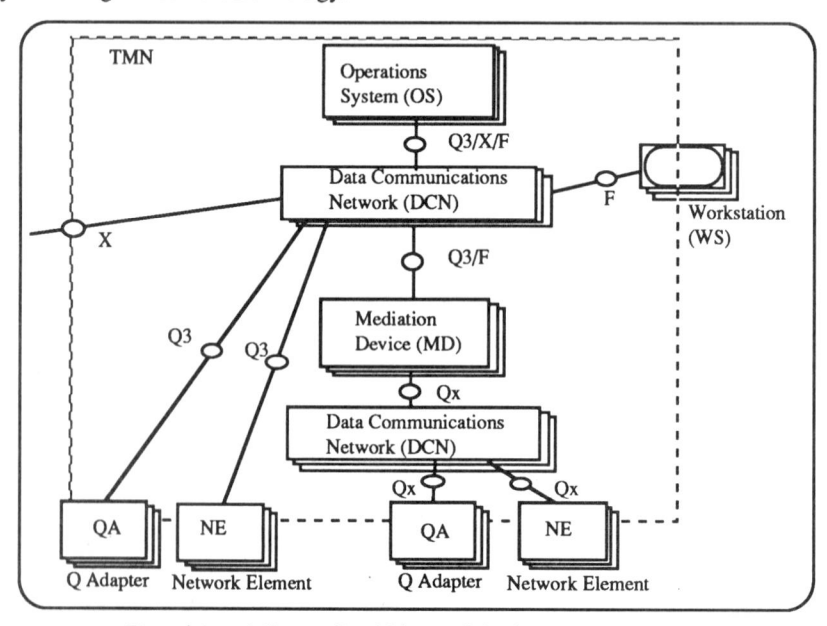

Figure 4 - A Generalised Physical Architecture (Example)

For further information on functional allocation the reader is referred to [10]. Another influence on placement strategy is the work on Open Distributed Processing which provides guidance for the design and implementation of distributed systems in general. Figure 4 shows a generalised physical architecture for the TMN. A TMN System provides the means to transport and process information related to the management of telecommunications networks and the physical architecture provides guidance on the implementation of such a system.

Each of the TMN Functional Blocks and Subcomponent Blocks map into one or more Implementation Blocks or Subcomponent Blocks. Thus the TMN System is seen as a set of interconnected Implementation Blocks - each Implementation Block representing a packaged set of implementation functions of the TMN System. An Implementation Block may itself be made up of other, lower-level blocks or components. Components are standard modules which can be used for many different applications. Figure 5 is a pictorial representation of an example of functional to implementation mapping for TMN components .

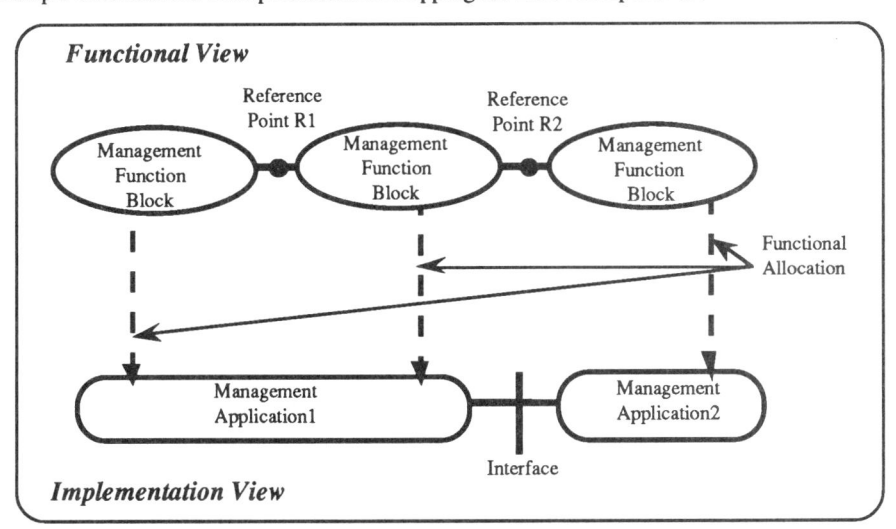

Figure 5 - Pictorial Representation of the Functional to Implementation Mapping

It is important to note that rules will be required to support the definition of the roles of OS, MD and NE in such a way that the vast range in scale and functionality possible may be supported.

Interface Definition

Considerable work on the definition of interfaces has been undertaken by ISO, T1M1.5, and OSI-NMF. The work has focused on the use of ASN.1 [12] definitions of managed objects and their attached operations (actions and events) to define input and output management messages. The management message sets are supported by defined Common Management Information Service (CMIS) primitives [13]. CMIS primitives are supported by Common Management Information Protocols (CMIP) [14]. CMIS services form the basis for the definition of message sets applicable to managed objects. CCITT TMN will define its particular interfaces to take account of available OSI management services and protocols.

Where possible the Q3, Qx, and X interfaces will be based on CMIS/P. These interfaces are also referred to as interoperable interfaces as they are associated with the interoperation of management systems. For a comprehensive analysis of the standardised approach to the definition of interoperable interfaces see [15].

5.3. Functional Structure of the OS

The purpose of this section is to elaborate the internal functional structure of the Operations Systems. It should be stressed that the OS internal structures definition is not intended to be fully prescriptive. An initial functional decomposition of an OS is given in figure 6.

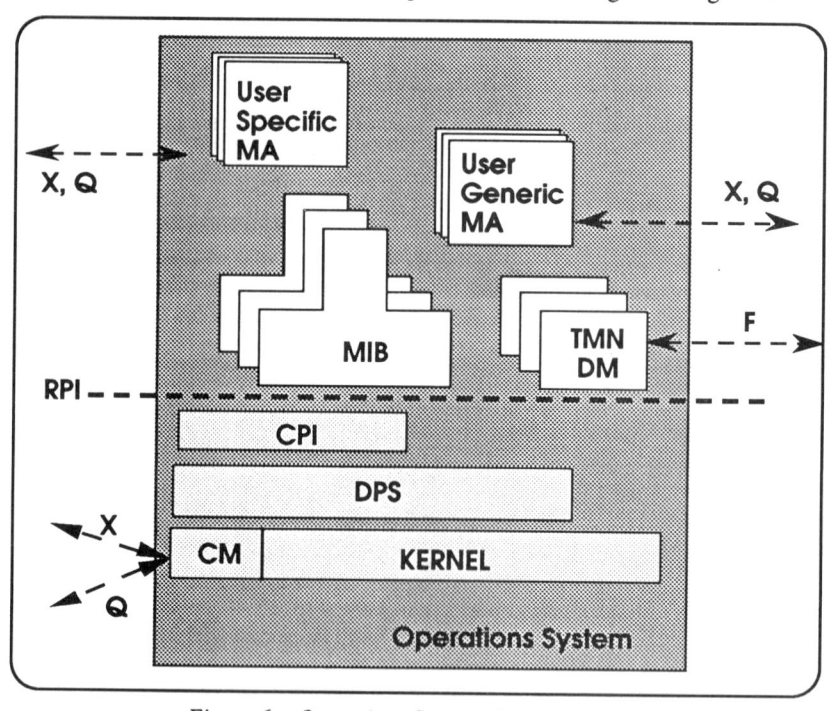

Figure 6 - Operations System Decomposition

The OS can first be decomposed into two major components, the Management Applications and the Information Base. The information base (IB) is in effect a realisation of a part of the information storage capability of the TMN. It should not be confused with the MIB which is the conceptual repository of all information on the TMN which could be realised as information bases located in OS, MD and NEs. A further decomposition of the MAs shows how the Management Application can be broken down into a User Generic Applications part which supports a number of User Specific Applications. It is mainly the latter which are "seen" by the (human) users, while the former take care of common TMN-related functions and, in particular, terminate the various intercommunications paths at the q, x and f reference points.

Examples of User-Generic Applications are Configuration Management and Event Management, and Infrastructure Applications, such as Database Management, Communications Management, User Interface Services, Security Services, Resilience Management, ADI (Awareness creation-Decision-Implementation) control loop support. The former group offer basic applications, which will usually be built upon or modified by User Specific Functions, while the latter group are more fundamental, lower level, building blocks. Both are candidates for some degree of standardisation, permitting multi-vendor systems to offer consistent services and enabling evolution and specialisation in a logical way. The rest of the OS functions are :

- A *Dialogue Manager* (TMN DM) to support user interface dialogue. The Dialogue Function (DF) interacts with dialogue functions of other TMN functional Blocks. It interprets the information from the Applications Functions, and from the dialogue

functions, from the other TMN Functional Blocks with regard to the dialogue state which includes:
- The boundary between the User Interface and the remainder of the application and system components
- The behaviour of the application from the view point of the user interface
- The types of objects that are supported by the application, the operations which can be performed on them, and the relationships between these types of objects

- A Management Information Base Function (MIBF).

The Management Information Base Function is the function managing the repository of management information. It is the data store that will capture the information model (i.e. Object related information). The set of managed objects in a system, together with their attributes, constitutes that systems MIB. Key examples of MIBFs are the View Manager Function, the Object Manipulation & Management Function and the TMN Directory Service Function. These functions are detailed in [8].

The application oriented functions would be upheld by a number of supporting functions grouped under the name of the TMN platform. Examples of TMN Platform Functions are Run Time Platform Interface (RPI), Computing Platform Interface (CPI), TMN Platform Kernel, Communications Management functions and Distributed Processing Support (DPS). These functions will support the distribution of TMN computation and will facilitate different types of transparencies. DPS is assumed to provide the following distribution transparencies:

- Access Transparency
- Concurrency Transparency
- Failure Transparency
- Replication Transparency
- Location Transparency
- Liveness Transparency
- Migration Transparency.

Further details of these may be found in [8].

6. MANAGEMENT INFORMATION BASE

The concept of the Management Information Base (MIB) in network management has been widely accepted by the standards making and similar bodies, viz. ISO with OSI Systems Management, T1M1.5 in ANSI, OSI Network Management Forum, various CCITT Study Groups and NA4 in ETSI. The MIB is a conceptual repository of all the information held in the TMN about the network and the TMN itself. An information base is required to store information on network and system configuration, customers and services, current and historic performance and trouble logs, security parameters and accounting information. An informal description of an information base implementation would characterise it as a collection of databases held in storage associated with the computers which make up the TMN. This collection could include large scale databases of customer records and their service and usage details. There will be a need for smaller scale databases containing regional configuration data and performance and maintenance information. It includes the tables, held in the memory of network elements such as switches, used to make routing decisions and to accumulate resource usage data. In principle then, the information base is the pool of data which any adequately privileged TMN user or TMN application may access.

Because of its crucial role the clear and unambiguous definition of the MIB is of vital importance in the progress towards a standardised TMN. Interworking at all levels will be simplified if a consistent model of the information available in the MIB is used by all TMN developers. The following sections describe an approach to modelling the management information base which has been developed by GUIDELINE based on extensive material available from standards bodies such as ANSI T1M1.5 and ISO's OSI Management work.

A common information architecture will provide one of the key integrating factors of the TMN system. International studies have shown that, in general, the longevity of the data is greater than the applications it will support. Therefore coherent and consistent approaches will be required. Some of the key information architecture requirements are outlined below.

- Techniques and methods are required for definition of a *conceptual schema* to support all aspects of service and network operations to the full range of TMN users. These approaches should support evolutionary enhancement of the network management systems.
- It is notoriously difficult to develop complex conceptual schemata, thus the information architecture should support *separation of concerns* in a manner which is meaningful and useful in the TMN context.
- Increasingly *knowledge rich* information bases will be required to support, for example, decision support applications – possibly using KBS (knowledge based systems) techniques. These could require definition of schemata with high orders of complexity.
- Techniques for *storage and retrieval of a large volume of information* are needed. MIBs should be realised in such a way that performance and access requirements may be met when the actual data is stored in databases which, in general, will be distributed and heterogeneous.

The TMN methodology makes use of the OSI systems management principles and is thus based on the object oriented paradigm. Hence the management systems exchange information modelled in terms of managed objects, where the managed objects are conceptual views of the managed resources.

6.1 Reference Model for a Distributed Database

In practice the TMN will be distributed and this obviously has implications for the management of the information in the TMN. It is thus important to have a general architecture for the management information base which supports distribution. Figure 7 shows a reference architecture for distributed databases [16] which has been used by GUIDELINE in the studying the MIB.

Global Conceptual Schema

The global conceptual schema describes the syntax and semantics of the objects that exist in the TMN. It should provide an authoritative integration of the information base objects required for an extensive range of TMN applications. It will also accommodate views of MIB required by its the various users. In this scheme the global conceptual schema will specify:

- Real-world objects of concern in the domain
- Relationships between the objects of concern to the applications
- Attributes and operations belonging to the objects
- Specific integrity constraints applicable to objects, relationships, attributes and operations.

The global conceptual schema should define the universe of discourse of the various parts of a TMN system and facilitate the use of the TMN system and its MIB by its various users and its integration with other systems such as network elements. It is to be used to define the concepts and terms which will allow TMN designers to interact effectively with each other. It will be used to define the interfaces between OSs, between OS and TMN users and between OS and NE. Further discussion on the Global Conceptual Schema can be found in [17].

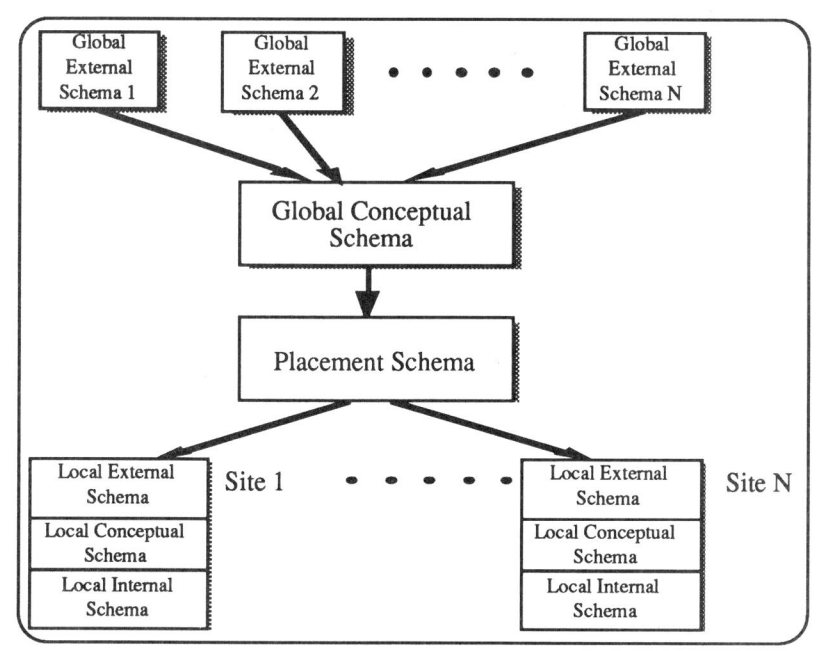

Figure 7 - Reference Architecture for Distributed Databases

All other schemata (including their data definition and manipulation languages) are derived from the global conceptual schema. Logical data independence is provided by the *Global External Schemata* each supporting different application classes or different classes of users. In general, their view is limited compared to the view of the global conceptual schema, and may of course involve a different representation of the managed objects. The global external schemata may have their own data definition and manipulation languages.

Schema Levels

The *Placement Schema* defines details of database fragmentation, location and replication and therefore supports transparency to these aspects above that layer. The schema levels within a database at one site are local and are based on those in the ANSI/SPARC architecture [18]:

- *Local External Schemata* : The local external schemata in this particular distributed environment supports the use of its corresponding local database within this distributed environment. The local external schema maps the fragments defined by the placement schema to the external objects of the local database. These provide the transparency required to support heterogeneous environments.
- *Local Conceptual Schemata* : The local conceptual schemata have similar properties to the global conceptual schema, but are restricted to their local databases.
- *Local Internal Schemata* : The local internal schemata define the local low level data representation of their particular local database.

All three local layered schemata of course have their specific data definition and manipulation languages. There are a number of implementation choices for the local databases: object oriented databases; the traditional types of databases: relational, hierarchical, network type; data tables; directly addressable data in memory or on disk; sequential files on disk.

6.2 The Federated Database Approach

The approach outlined above assumes a top down development process where the data requirements are identified before the databases are designed. In practical terms such a process may not be possible, for example where a number of pre-existing databases need to be integrated or where organisational constraints prevent the necessary top level agreements and definitions. In this case an approach based on *Federated Database* Architectures has been suggested [8]. It should be noted that since most of the MIB definition work currently being undertaken in standards fora is addressing small parts of the overall TMN problem area, these bodies are effectively defining external rather than conceptual schemata. In addition the integration of these schemata will pose considerable difficulties because each grouping is defining its views with a high degree of autonomy. Federated databases would not have a global conceptual schema rather a number of *federated schema* which would support the co-operation between autonomous databases. Effectively these databases would be integrated bottom up from a number of relatively autonomous developments. However in the communications management domain tight coupling between systems is often necessary and a high degree of agreement of common concepts is desirable. It is clear that some mix between the top-down and bottom up approaches to database definition identified above will be required. A preliminary analysis of a number of options for the scope of the MIB has been conducted [8] following the ODP viewpoints. The ODP/ANSA viewpoints have helped to separate some of the concerns in the MIB definition. The early results are encouraging. However, further development of the ideas is necessary before firm conclusions can be reached. Given that a post-hoc mapping of existing MIB architectural concepts to the ODP viewpoints is being attempted a clean and clear mapping to ODP may not always be possible.

7. TECHNOLOGY FOR TMN

An evaluation of the advanced information processing (AIP) technologies considered by the RACE TMN projects for the implementation of the TMN or its components has been produced by GUIDELINE [19]. AIP technologies for the TMN can be employed to address problems in three main areas:

- to support network management run-time applications
- to support TMN run-time platforms
- to support the development environment of the TMN.

For details of the AIP technologies please refer to [19] and papers in this volume.

8. SUMMARY AND OPEN ISSUES

The work performed by the architecture group of RACE project GUIDELINE has been presented. The results fall into three main categories as follows :

- functional architecture
- physical architecture
- information architecture or management information base.

Inevitably with work of the scope and magnitude of that undertaken, many issues have been identified which could not be addressed in the time available and with the level of resources available for this task. Some of the major open issues are as follows :

- *Methods* : Much work needs to be done before a satisfactory methodology for TMN architectural definition can be arrived at

- *Non functional requirements* : The relationship of the functional and non-functional requirements needs to be clarified and scaling issues for TMN implementations needs to be addressed
- *MIB Issues* : Improved understanding of the mapping from conceptual level models (specifications) to implementations of various kinds is required. It is well known that complex conceptual schemata are notoriously difficult to define. Definition of a single global conceptual schema may not be possible given the complexity of the domain and the number of actors involved in its definition. Federated schemata have been postulated as an approach to solving this problem. However, these could reduce the generality of the model. Further studies will be required in this area
- *OS as a distributed system* : The views of the OS as a distributed system need further elaboration. This involves thorough investigation of the various requirements that arise such as replication, performance and resilience.
- *Validation* : Some work needs to be done on methods for validating the architecture before proceeding to implementation of the system.

It is hoped that these issues will be addressed by the various other groups studying TMN including the ongoing work within the RACE II programme.

9. ACKNOWLEDGEMENTS

This paper is based on the results of the working group of project GUIDELINE concerned with TMN Architectures, Terminology and Interfaces. Previous members of this group who made a significant contribution to the development of this architecture include Paul Senior (BNR, UK) and David Brown (Schema, UK).

10. REFERENCES

[1] Smith, R. : "Broadband Communications Management : The RACE TMN Approach", IEE Conference on Integrated Broadband Services and Networks, London, October 1990.

[2] APM Ltd.: "ANSA Reference Manual - Release 01.00", April 1989.

[3] RACE Central Office : "Common Functional Specification H100 - An overview of RACE Telecommunication Management Common Functional Specifications", Brussels, March 1992.

[4] RACE Central Office : "Common Functional Specifications H400 Series", Brussels, March 1992.

[5] Walles, A. : "Functional Descriptions of Network Management", Sixth RACE TMN Conference, Madeira, September 1992.

[6] CCITT Draft Recommendation M.3010 - Version R5 : "Principles for a Telecommunications Management Network", Geneva, November 1991

[7] NETMAN Deliverable 2 : "Analysis of Network Management Functions", 24/BCM/RD2/DS/A/001/b1, RACE Project R1024 NETMAN, August 1989.

[8] GUIDELINE Deliverable ME8 : "TMN Implementation Architecture", 03/DOW/SAR/DS/B/012/b3, RACE Project R1003 GUIDELINE, March 1992.

[9] RACE Central Office : "Common Functional Specification H200 - An Architecture for the TMN", Brussels, March 1992.

[10] TERRACE Deliverable 7 : "TMN Functional Allocation : Key Parameters and Strategies", 53/IBM/WP1/DS/B/007/b1, RACE Project R1053 TERRACE, April 1990.

[11] BT : "CNA-M, Co-operative Networking Architecture - Management", 1990.

[12] Steedman D. : "Abstract Syntax Notation One - ASN.1 - The Tutorial & Reference", Technology Appraisals Ltd, 1990.

[13] ISO/IEC : "OSI Systems Management - Common Management Information Service", ISO/IEC 9595.

[14] ISO/IEC : "OSI Systems Management - Common Management Information Protocol", ISO/IEC 9596.

[15] Callaghan, J. G. et al : "TMN Interoperable Interfaces", Sixth RACE TMN Conference, Madeira, September 1992.

[16] Gardarin G. & Valduriez P. : "Relational Databases and Knowledge Bases", Addison Wesley, 1989.

[17] Turner, T., Callaghan, J. G. : "Towards Integrated TMNs - The Global Conceptual Schema", Sixth RACE TMN Conference, Madeira, September 1992.

[18] ANSI/X3/SPARC Study Group : "Data Management Systems, Framework Report on Database Management Systems", AFIPS Press, 1978.

[19] GUIDELINE Deliverable ME6 : "The Application and Integration of AIP techniques within the RACE TMN", 03/BTR/712/DS/B/009/b1, RACE Project R1003 GUIDELINE, April 1991.

The Management of Telecommunications Networks
R. Smith, E. H. Mamdani, J. G. Callaghan (Editors)
© Ellis Horwood 1992

TMN Interoperable Interfaces

James G. Callaghan (BT Laboratories, UK), Ben Hurley (Broadcom, Ireland), Kevin Riley (UNIPRO, UK), David Griffin (GPT, UK)

ABSTRACT

The material in this paper addresses a number of open issues relating to interoperable interfaces.

The first item addressed in this paper concerns the generic interoperable interface requirements. Issues covered in this section include TMN platforms and the API (Application Programming Interface), covering items such as Generic Agents and Generic Managers.

In order to have a common approach to the specification of interfaces it is necessary to identify a general approach or methodology to be followed. CCITT have tackled this problem by producing draft recommendation M.3020 [1] in which such an approach is outlined. This approach is introduced and the results of applying the approach to a real interface specification are presented. An example of an interface specification developed within the ADVANCE project is also presented and the retrospective application of the CCITT methodology, mentioned above, to this interface is examined. The paper concludes with a brief summary of the key results of the work.

1. INTRODUCTION

In order for two or more TMN elements to exchange management information, they must be connected by a communications path and each element must support the same interface onto that communications path [2]. This brings in the notion of the interoperable interface.

The interoperable interface defines the protocol suite and the message carried by the protocol. It is based on an object oriented view of the communication. Therefore all messages carried deal with object manipulations. The interoperable interface is the formally defined set of protocols, procedures, message formats and semantics used for the communication. The message component of the interoperable interface provides a generalised mechanism for managing the objects defined for the information model. As part of the definition of each object there is a list of the type of management operations that are valid for the object. In addition, these are generic messages that are used identically for many classes of managed objects. The scope of the management activity that the communication at the interface must support distinguishes one interface from another. This common understanding of the scope of operation is termed Shared Management Knowledge. Thus in order to interwork, communicating management systems must share a common view or understanding of at least the following information:

- Supported protocol capabilities
- Supported management functions
- Supported managed object class
- Available managed object instances
- Containment relationships between objects

The shared management knowledge ensures that each end of the interface understands the exact meaning of a message sent by the other end. Before going on to describe some of the issues associated with the interoperable interface it is useful to outline the benefits of open interfaces. This is presented in the following section.

2. INTEROPERABLE INTERFACE REQUIREMENTS

This section presents some of the interoperable interface requirements which have been identified. The generic requirements are presented in terms of TMN platforms and the API (Application Programming Interface). Further information can be found in [3] and [4].

2.1 TMN Platforms and the API

A platform is a set of facilities that provides a defined infrastructure upon which management applications can be built. The platform should exist so as to aid the management application designers by providing a well defined interface to the facilities of the platform and the underlying systems. From the communications interface perspective the API consists of the functions that provide the access to the communications services.

The DCN (data communications network) of the TMN will provide a communications infrastructure at a particular level (Q3) which will have an internationally standardised set of communications services which will almost certainly contain CMIS (Common Management Information Service [5]) for the exchange of management messages. From examining the requirements of management applications within the environment of a TMN it can be demonstrated that any application built onto the Q3 network will have common functions irrespective of the application. Two of the main functions are referred to as the Generic Agent and the Generic Manager and these are explained below. Some of the other common functions are also identified. The collection of all these functions could be called a TMN platform.

Generic Agent

The purpose of the CMIS agent is to present information to CMIS managers in the form of managed objects. A manager will direct a CMIS message to an object located in a particular agent and it is the job of the agent to decode the message to determine which information needs to be retrieved or modified and then to do the task. The agent should be able to represent a MIT (management information tree) and have functionality for performing event reports, scoping and filtering. The basic job of the agent is to provide a mapping (at run-time) between the internal form of information and the external representation in the guise of managed objects. From an implementation point of view there are two edges to an agent; the first is the CMIS edge which needs to be designed to interwork with the CMIP (Common Management Information Protocol [6]) protocol; the other edge will interface to the internal representation of information. The nature of the latter edge will depend on the chosen implementation technology and will be the definition of the API (see figure 1). Considering that the agent must present information in the form of objects it would be convenient to exploit an object oriented technology, e.g. the agent could communicate with C++ objects and hence define the API between the agent and the local management application to be C++.

All managers attached to the TMN are potentially able to communicate with any agent in the TMN to retrieve and modify the information stored at that agent. For any implementation of a TMN there needs to be a structure to the management interactions, e.g. some managers will never need to access certain objects or agents during normal operation. This structure will be defined during the design phase of the TMN and is dependent on the functions and capabilities of the applications and managed resources, the organisational structure and policies of the operator, the implementation of the global information schema (e.g. the level of distribution), etc. In addition there should be security mechanisms built into the agents to ensure that managers only gain access to the objects and attributes they are authorised to see or modify.

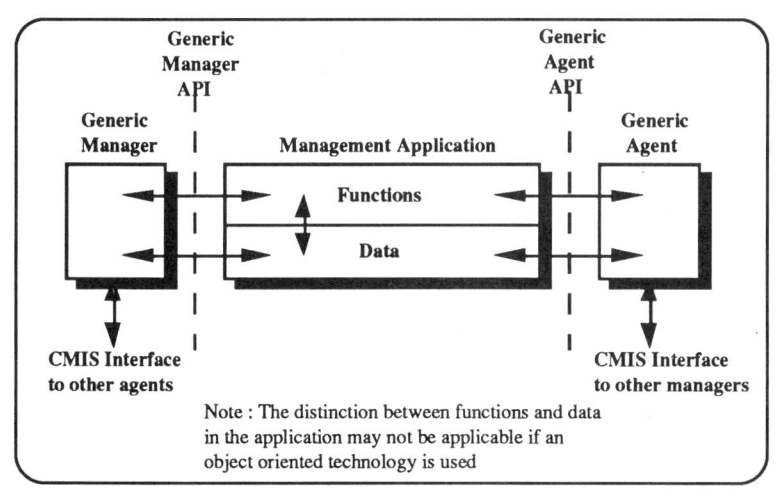

Figure 1 – Development View of the API

Authentication of access rights to an agent (for interactions within a TMN or between TMNs) is a topic that has not been fully addressed to date and needs further work. The mechanisms of authentication together with the structure of the management interactions will provide the ability to present different levels of abstraction to different managers accessing the same objects, e.g. managers with a high level of access rights will be able to see all attributes in an object, whereas others with a lower level of access will only be able to see a subset or summary. In order to simplify these polymorphic functions of the agent it is possible to conceive of more than one agent associated with a management application or managed resource. Each of these agents would represent local information in a different way to suit the requirements of different managers, e.g. performance and fault management views. This would mean that the directory services would need to exhibit polymorphism by providing addresses to different agents when given a request for the location of the same information by different managers.

Generic Manager

In the same way that a generic agent can be provided in a platform a generic manager can also be included. This would provide a mapping between CMIS managed objects and the internal data representation in the application. It should include functions to aid the management of remote objects, e.g. the handling of event reports, and possibly the instantiation of remote objects or whole Management Information Trees (MIT).

Higher Level Query Languages.

For some management operations it is not possible to articulate the complete operation in a single CMIS message and a sequence of related messages are required together with some processing of the responses. It may be possible to define a higher level query language that the management application uses for communication with remote agents. This component would perform the necessary translation into CMIS and vice versa. In this case an API would be defined consisting of an interface to the high level query language.

Other common functions of an application built on a Q3 interface include :
- Directory Services
- HCI Functions
- Database Support
- Distributed Processing Support.

Further details of these can be found in [3]. Facilities for the support of state management, transaction management and configuration management could also be part of a TMN platform and it would be desirable to attempt to create a complete list of management functions that are common to all or most management applications in the TMN. However, it will be the responsibility of the TMN designers to decide what functions can be separated from the applications themselves, and hence it will be their responsibility to define the API. Figure 2 illustrates the relationship between TMN platforms and the development time API.

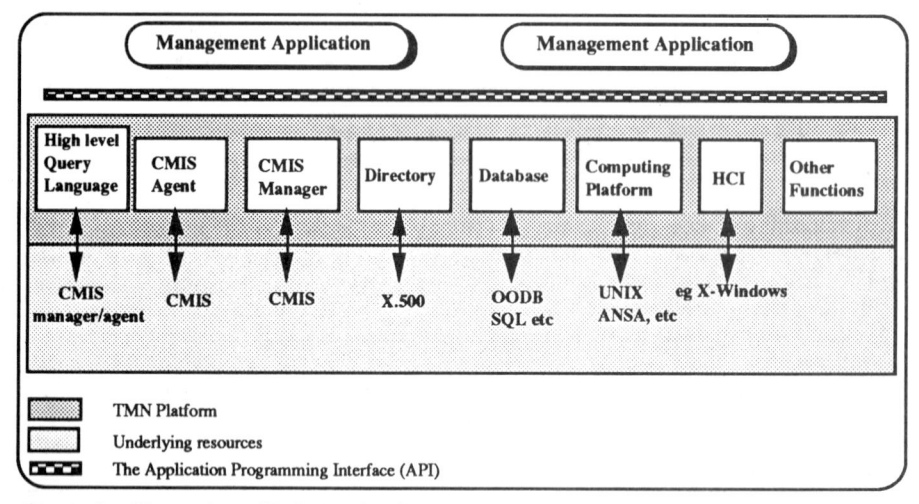

Figure 2 – Illustration of Relationship between TMN Platforms and development time API

In summary, although it is possible to define generically the collection of functions required in a TMN the detail of its implementation and the definition of the API will be the responsibility of the provider of the TMN.

3. GENERAL APPROACH TO INTERFACE SPECIFICATION

This section presents an introduction to a general approach to the specification of an interface. The specification methodology is based mainly on CCITT Draft Rec. M.3020 [1]. The design procedure outlined in M.3020 starts the design process by describing TMN management services which would be offered to an operator/user. This approach assumes that the objectives of the operator for managing the network have already been defined, so that TMN management services can be used by the operator to achieve those objectives. In studying the approach outlined by CCITT it was found that it is necessary to introduce additional design specification steps which describe the Telecommunications Management (TM) requirements, at a level which is independent of the human operator or organisational structure. This requires the provision of a set of top-down requirements and functions which can be used as a check list to assess the coverage achieved in the generation of TMN management services which are sometimes produced from a bottom up activity.

The CCITT methodology consists of 13 tasks and these are addressed in detail in [1]. However, in line with the above argument for steps which describe TM requirements, the CCITT methodology has been extended to include the necessary steps [7]. The additional steps are detailed in [3]. Rather than describing the various tasks, or steps, in detail it is more beneficial to summarise the tasks and their outcomes. This is contained in figure 3 (based on Table 1 of [8].

Task No	Task	Outcome of Task
0	Generate Generic Network Information Model	Generic Network Information Model
1	Describe TMN Management Services as perceived by the service user	TMN Management Services and Components List
2	Select TMN Management Functions	Function list
3	Define Objects and their attributes, operations, and notifications	Object Templates Object Relationship diagrams
4	Consolidation	Coherent and consistent model
5	Define Management Information Schema	Management Information Schema
6	Determine Communication Requirements	Requirements for Communications
7	Prepare documentation for protocol tasks	TMN Management Functional Profiles
8	Analyse message needs	Grouping of TMN management functional profiles.
9	Decide adequacy of existing protocols for each layer	Existing protocols and TMN management protocol suites
10	Define new protocol requirements	Requirements for new protocols
11	Define new layer services and protocols	New layer services and protocols
12	Select layer services	
13	Select layer protocols and form protocol suites	

Figure 3– Summary of CCITT Interface Specification Methodology

4. COMPONENTS OF AN INTERFACE SPECIFICATION

Following on from the methodology introduced above, the components of an interface specification are essentially the results of the various tasks of the methodology. These results are contained in the relevant Task Information Bases. Thus the components of an interface specification are as follows:

- List of TMN Management Services and Components used
- List of Management Functions
- Generic Information Model. This comprises Object Templates, Object Relationship diagrams and Management Information Schema
- TMN Functional Profiles (these embody the functional requirements)
- TMN Protocol Suites.

The two main facets of an interface specification are thus the Functional Recommendations and the Protocol Recommendations. Examples of descriptions of the above components can be found in [3]. In addition to investigating the CCITT methodology GUIDELINE has also extended the methodology so as to allow a consistent approach to the identification and definition of TMN Management Services which will support the overall Telecommunication Management requirements and not just the requirements which are supported by functionality which is automated.

5. EXAMPLE OF DETAILED INTERFACE SPECIFICATION

In order to gain experience from a deeper analysis of a particular reference point and interface, it was decided to pursue as a case study, the specification and implementation of an interface within a particular scenario. Thus, by restricting the scope in terms of breadth through the use of a scenario, and exploring the interface in terms of depth by bringing it to prototype level, a deeper knowledge of the requirements on an interface specification would ensue. The interface specification outlined here is concerned with the x reference point and X interface. It therefore explores the cooperation between management entities of two separate TMN domains. The scenario for the interface is drawn from the ADVANCE Complaint Handling Management Application [9]. It essentially attempts to resolve issues associated with providing a front desk approach, single point of contact for customer liaison, it being a vital component for automated interaction between provider and customer, or any two TMN domains.

The ADVANCE approach has been examined in connection with the AIM and NEMESYS requirements. Hence, it can be said that to some degree, the example is all-TMN encapsulating. It should be noted from the outset that the interface specification studied by GUIDELINE is an example. It is concerned with the X interface as viewed within a particular example scenario (the scenario has, however, been sufficiently generalised to span all three technology projects). The specification is thorough in that it includes implementation level details. The implementation aspects of the example verify the specification, but should not be viewed as prescriptive, i.e. there is no obligation to proceed from specification to implementation exactly as described in [3].

5.1 Case Study Background

The case study is based primarily on interface experimentation within the ADVANCE project. However the specifications and designs have been viewed from both an AIM and a NEMESYS perspective, in order to ensure compatibility with the specific requirements that these projects may have on the interface. Contributions from the latter two projects have enhanced the ADVANCE experimentation work to produce this case study. For detailed background information and a brief overview of the relevant ADVANCE experimentation work which leads to the description of the ADVANCE Customer Complaint Management Application, please refer to [3].

5.2 Interface Aspects

The interface aspects of the case study are considered under three headings as follows :
- The Interface Information Model
- The Communications Service
- Security Considerations.

Each of these aspects are explored in the following sections.

5.3 The Interface Information Model

GUIDELINE has adopted the Manager-Agent model of management and the associated concepts of Shared Management Knowledge as described in [3]. The ADVANCE Model of Other TMNs (MOT) also conforms to these concepts. In principle, the MOT is a representation of the basis for cooperation between two separate TMNs. It enables the transfer of management information through the use of an information service providing the appropriate operations.

The MOT as a model contains information object representations of management information which resides in other TMNs, see figure 4. Thus if TMN A wishes to cooperate with TMN B, then TMN A must establish within itself an MOT with information object classes and instances which represent the information within TMN B, which forms the basis for the cooperation. TMN B must provide TMN A with such representations, and the ability/functionality for such representations to access and modify the real information in TMN B under access and authority restrictions which are specified in contract and applied both in the MOT of TMN A, and within the TMN B.

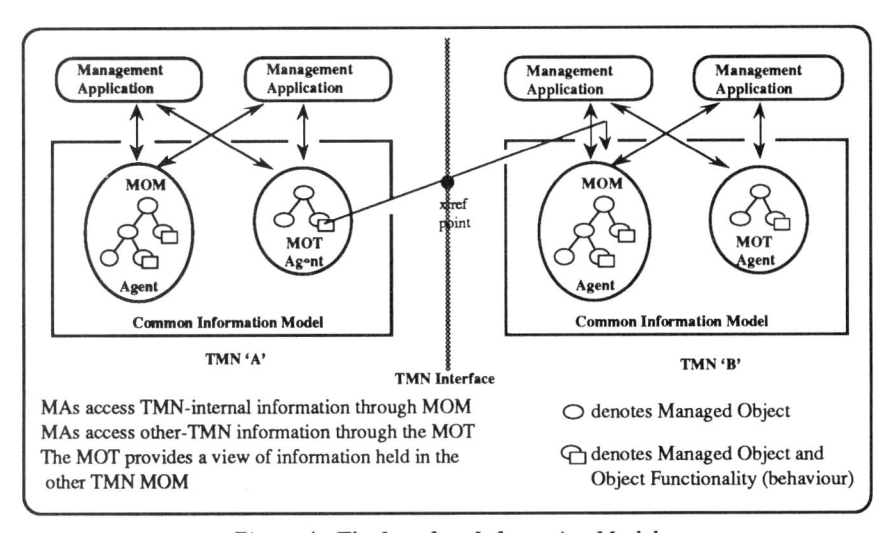

Figure 4 - The Interface Information Model

The inter-TMN management cooperation essentially boils down to (in the example) the provider exporting a representation of an object class which exists within his NCAS (Network and Customer Administration System). In the example, the complaint log is an off-line, exported representation of the investigated complaint log. The complaint log therefore becomes part of the Customer's MOT, representing as it does information which forms the basis of cooperation between the Provider TMN and the Customer TMN. The customer (or other TMN) is free to manipulate the attributes of the complaint objects within the confines of his MOT. When satisfied with the description of any particular complaint, the customer actions the complaint log to register a particular complaint instance within the provider system. This is executed by the complaint log establishing a management information service link with the provider system, and creating an investigated complaint log entry with the attributes associated with the appropriate instance from the complaint log.

Advantages of the ADVANCE Approach and Comparison to Manager-Agent Concepts

The MOT approach provides a standard interface to non-standard systems. Thus, the view a provider exports in the form of the MOT need not necessarily reflect the internal structuring of the providers system (this is also true of the direct manager-agent approach). The MOT, by the definition of the objects therein and their associated behaviour, specifies a minimum functionality which must exist across an interface for all TMNs. The MOT also allows partial views of a remote MIB.

The general view of the use of manager-agent concepts is shown in figure 5 with the inter-TMN cooperation aspect included. In this, a management application as manager is seen communicating directly with an agent. Such an approach to inter-TMN cooperation would require each management application to establish associations, and execute operations directly over the relevant association. However, the MOT replaces the requirement for strong coupling between specific applications of one TMN and agents of another. Instead, management applications are presented with local management objects which represent the basis for cooperation with another TMN. These objects, as part of their behaviour, contain the ability to interact directly with the other TMN.

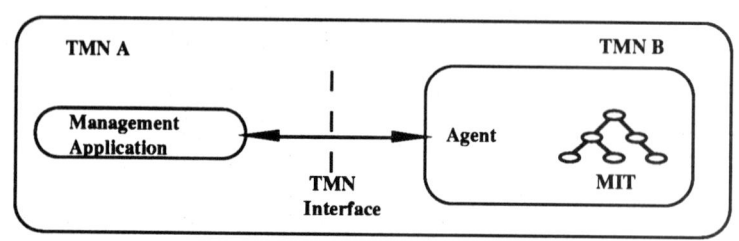

Figure 5 - General View of Manager-Agent Concepts

Thus the inter-TMN manager role is concentrated within one entity, e.g. the MOT, rather than being distributed over an arbitrary number of management applications. The TMN providing access provides the TMN requiring access with the appropriate MOT elements to do so. This has the twofold advantage of enabling the provider to develop the extent and method of access as he desires, while letting the user develop management applications locally based on the MOT interface. Furthermore, the precise nature of the communication between MOT and other TMN can change without adversely affecting management applications.

5.4 The Communications Service

The communications service which provides the ability to unambiguously transfer information and operations between the two TMNs is the Common Management Information Service (CMIS). The complaint log object in the MOT, which defines the information aspects of the interface for the purposes of this application, adopts a manager role and communicates with an agent in the provider system. This agent allows those attributes of the investigated complaint log object which also appear in a complaint log object to be updated.

Interface Messages

As a result of adopting CMIS, the messages which flow across the interface (unless otherwise stated, messages originate with the customer) are the following (note that not all the parameters for each message/operation are listed, as some are of relevance only to this actual example, and will only obscure the important aspects):

M_INITIALISE (Source, Destination, Context, Access);
This establishes the association with the other TMN, negotiating the context as the customer adopting a manager role, and the provider an agent role. The primitive also establishes access control for the operations to be carried out over the association.

M_CREATE (Complaint Object, Attributes);
This is interpreted as the creation of a new instance of the Investigated Complaint Log in the provider management system. Those attributes which Investigated Complaint

Log inherits from the Complaint Log object are modifiable in the create action. The create has the effect of transferring a copy of the complaint data held in the customer system to the provider system.

M_EVENT_REPORT (Customer, Information);
The provider system can issue several event reporting messages to the customer. The first acknowledges receipt of the complaint and indicates that it is now part of the complaint handling application of the provider. Subsequent event reports provide updates on the state of complaint resolution, and the estimated MTTR (mean time to repair).

M_TERMINATE (Source, Destination);
This terminates the association between the two TMNs, hence terminating communication and cooperation for the current session.

Other Communications Considerations

In the example outlined, the customer has the ability to establish communication with the provider at will, e.g every time he wishes to make a complaint. However, an alternative exists for the frequency in which the customer may use the cooperation links, namely batch processing. Batch processing may be more cost effective, in that the customer generates and stores information, and periodically associates with the provider TMN for transfer of information. However, in a time sensitive area such as complaint processing, immediate access may be required. As the inter TMN cooperation will be provided as a service, and no doubt charged for on the basis of availability and utilisation, enabling the customer to access the service at will results in potentially increased income on behalf of the provider, and greater control over service usage by the customer. Either batch processing or direct on-line access can be built into the MOT approach.

5.5 Security Considerations

The interface specification and design takes one of two approaches possible, in that the MOT object, being a representation of an object in another TMN, has the ability to access and modify that external TMN object. Access control and authorisation checks, two major aspects of security are carried out at several levels. Firstly, only authorised applications within a TMN may actually access certain objects in its MOT, and furthermore, their level of access may vary (i.e. read only, create only, etc.). Thus one level of security control is exerted even prior to any attempt at cooperation with another TMN is made. A second level of security control comes into effect when an object in an MOT attempts to establish a management link with the other TMN. At this point the responding agent of the other TMN applies access control and authorisation checks also, ensuring that a valid management cooperation entity is requesting access.

An alternative approach would be to have the MOT object establish a link with an application in the external TMN, and in a submissive role, report the modifications which are being made in the MOT concerning the external TMN. The application can then determine whether or not to modify its information as a result of the report. This approach may have the effect of enforcing greater authority on behalf of the external TMN. However it is the conclusion of this experiment that the two approaches are similar, and that effective security can be applied in either situation. The security considerations for this management application have been handled in detail within the ADVANCE project [10].

The scenario outlined above has been enhanced by the addition of the X interface requirements for both the AIM and the NEMESYS projects. The specific requirements that these two projects have on the X interface are highlighted in [3].

6. RETROSPECTIVE APPLICATION OF THE CCITT INTERFACE SPECIFICATION METHODOLOGY

Section 3 above introduced the 13 tasks of the CCITT interface specification methodology. GUIDELINE have explored the application of this methodology to the example of section 5 and the results are summarised in figure 6. It can be seen from figure 6 that the application of the methodology even though applied retrospectively, still mapped quite closely to the approach taken in the example outlined above. This prompts the conclusion that the methodology is sufficient for interface specification purposes. Also the CCITT methodology is essentially a top-down approach to interface specification. This is in contrast to the approach used in the case study which, as it is based on the implementation of the interface, may be termed a bottom-up approach.

Task No	Task	Outcome of Task
0	Generate Generic Network Information Model	Generic CIM (Common Information Model)
1	Describe TMN Management Services as perceived by the service user	Complaint Handling Services and Components list
2	Select TMN Management Functions	Complaint Awareness Complaint Preparation Complaint Transmission Automated Correlation
3	Define Objects and their attributes, operations, and notifications	Complaint Objects as depicted in [3] and described in ADVANCE IMA.3 [9].
4	Consolidation	Management Application objects of CIM found to be coherent and consistent
5	Define Management Information Schema	Views defined for both "Provider" and "Customer" TMNs
6	Determine Communication Requirements	Object oriented information service required
7	Prepare documentation for protocol tasks	Not required
8	Analyse message needs	Information transfer messages defined in terms of CMIS primitives
9	Decide adequacy of existing protocols for each layer	CMIP identified as being adequate hence lower layers also adequate
10	Define new protocol requirements	None identified
11	Define new layer services and protocols	None required
12	Select layer services	CMIS and its supporting services (e,g, ACSE, ROSE at application layer)
13	Select layer protocols and form protocol suites	Default selection due to choice of CMIS

Figure 6 - Outcome of Application of CCITT Interface Specification Methodology

However, it must be remembered that the example of Complaint Handling is just one management service and thus the example has a smaller scale set of requirements on an interface and hence on an interface specification methodology. However, this example has served to validate the CCITT approach and to illustrate that it does not conflict with the specification that would arise if a bottom-up or implementation focused approach was taken.

6.1 Case Study Results and Conclusions

The predominant result is that an example X interface specification has been successfully achieved and implemented, albeit in prototype form. However, the added requirement of having to actually produce an implemented MA meant that the Interface had to be specified. Hence the results are not just of a paper study, but from a real workable/working application. The retrospective application of the CCITT interface specification methodology proved successful in that the experience showed two things:
- The methodology is useful and complete
- The example interface is complete in so far as it satisfies the requirements of the application.

One of the main goals of the case study was to examine in depth as opposed to breadth the specification and implementation of an X interface. This it achieved satisfactorily, but what conclusions can be drawn? For example, did the specification help implementation? The specification certainly clarified and identified common requirements between the three technology projects involved (AIM, ADVANCE and NEMESYS). Furthermore, by using accepted and well understood models of management (manager-agent) and management communication (CMIS), in conjunction with the object oriented information modelling approach of ADVANCE (MOT), a detailed specification and implementation of an X interface was possible. The resulting interface is well structured and adaptable, and can be migrated to other systems through utilisation of the MOT concepts.

Restricted resources dictated that an achievable subject matter was chosen as the example interface specification task. Therefore, though the example is complete and reasonably comprehensive, it does not touch on all aspects of a full X interface specification. It is therefore important to highlight those aspects of the interface which require further attention. In particular, many of the security aspects described above, though catered for in the design, were not included in implementation stages.

7. SUMMARY

The main focus of the work presented in this paper was on the interoperable interface and the detailed requirements on that interface. These requirements were presented from the generic perspective. A methodology for specifying interfaces was studied and it was applied to a specific scenario from the ADVANCE project. This scenario, Customer Complaint Handling, was examined in detail and a retrospective mapping of the interface specification methodology was performed. The reason for not using the methodology initially was that the scenario being examined had already been progressed significantly within the ADVANCE project. The results of the retrospective mapping confirmed that the methodology is indeed useful and that it helps to achieve a coherent and consistent approach to the specification of an interface.

8. ACKNOWLEDGEMENTS

This paper is based on the results of the working group of project GUIDELINE concerned with TMN Interfaces. This group also participated in the study of TMN architectures as described in [11].

9. REFERENCES

[1] CCITT Draft Recommendation M.3020 - "TMN Interface Specification Methodology", Geneva, November 1991.

[2] CCITT Draft Recommendation M.3010 - Version R5 : "Principles for a Telecommunications Management Network", Geneva, November 1991.

[3] GUIDELINE Deliverable ME8 : "TMN Implementation Architecture", 03/DOW/SAR/DS/B/012/b3, RACE Project R1003 GUIDELINE, March 1992.

[4] GUIDELINE Deliverable SE2 : "Initial Report on IBCN TMN Interface Requirements", 03/BTR/712/DS/B/007/b1, RACE Project R1003 GUIDELINE, June 1990

[5] ISO/IEC : "OSI Systems Management - Common Management Information Service", ISO/IEC 9595.

[6] ISO/IEC : "OSI Systems Management - Common Management Information Protocol", ISO/IEC 9596.

[7] Geiger, G., Plagemann, S. : Contribution to ETSI NA4 on Interface Specification Methodology, February 1991.

[8] ECMA : "A Management Framework for Private Telecommunications Networks", ECMA Technical Report TR/54, December 1990.

[9] IMA.3 : "Management Applications Design", RACE Project ADVANCE, ADPL200, February 1992.

[10] O' Mahony, D. : "Security Considerations for the ADVANCE Complaint Handling Scenario", RACE Project ADVANCE, ADBC0199, November 1991.

[11] Callaghan, J.,G. et al : "RACE TMN Architecture", Sixth RACE TMN Conference, Madeira, September 1992.

II - The Evolving TMN

The Management of Telecommunications Networks
R. Smith, E. H. Mamdani, J. G. Callaghan (Editors)
© Ellis Horwood 1992

TMN Evolution

Enrico Bagnasco (CSELT, Italy)

ABSTRACT
This paper presents an overview of the key aspects and driving forces that will influence the evolution of telecommunications management, from today's systems to the target TMN for the IBC network environment. The TMN is described in terms of its structural framework and functional, information, communication and usage models. The concepts of gradual automation of management services and of cooperative management are also introduced. The paper ends with the presentation of a set of conclusions and open issues which require consensus on an international basis.

1. TMN REFERENCE CONFIGURATION DESIGN

A RC (Reference Configuration) describes the building blocks and their inter-relationships which are necessary to fulfil a specific task. The task of telecommunications management requires the design of a TMN (Telecommunications Management Network) RC. For a complete description of the TMN RC, a set of six component models (RCKC - Reference Configuration Key Constituents) has been defined [1]. These RCKC structure the conceptual viewpoint on TMN in anticipation of potential TMN implementations [2]. Concise definitions of the six TMN RCKC are given hereafter:

- RCKC-1 (TMN Structural Framework) defines the framework of TMN building blocks and their inter-relations. These blocks are dubbed "conceptual" (e.g. OSC - Operation System Conceptual block), because they encompass both the functional (RCKC-2) and the information (RCKC-3) aspects.

- RCKC-2 (TMN Functional Model) defines the TMN functional model and is based on the concept of TMN FE (Function Element).

- RCKC-3 (TMN Information Model) defines the object-oriented TMN information model in accordance with related international standardisation proposals. The key issues for the TMN information model are the identification, classification and description of MOs (Managed Objects).

- RCKC-4 (TMN Communication Model) defines the TMN communication model considering all layers of the OSI model and exploring message components as well as protocol components for TMN interfaces. On the lower OSI layers a set of networks for conveying management information are examined.

- RCKC-5 (TMN Usage and Management Service Model) defines the TMN Management Service model introducing the viewpoints and demands of TMN user groups. The key concepts for this model are MS (Management Services) and their re-usable MSC (MS Components).

- RCKC-6 (TMN Supplementary and Non-Technical Aspects) presents the TMN non-technical aspects related to e.g. legal and regulatory issues, considered significant for TMN design and implementation.

2. EXISTING TMN SCENARIO IN EUROPE

In order to identify the starting point of the TMN evolution path, a survey of existing management systems was conducted [3]. This survey covered 15 PTOs (Public Telecommunications Operators) operations of 67 Public Networks in 12 European Countries plus 19 privately operated networks.

The primary result of the survey is that the network management scenario in Europe is composed of a large set of varied support systems and that the status of management systems is strongly dependent on the status of each managed network: older and larger networks (e.g. PSTN) have a very complex variety of management systems.

Basically, two broad classes of management systems are in use today in the public environment: the manufacturer equipments and the "home made" systems. The first type of management systems tend to be installed closer to the NE (Network Elements), and are tied to the manufacturer's technology. They usually provide a large set of functionalities within the areas of Performance Management, Fault Management and general data collection for further processing. The scope of these systems tends to be very focused on a defined part of the whole network. The second type of systems tend to be installed in the higher layers of the hierarchy of the management systems, usually providing solutions tied to the PTO's organization. They are dedicated to a restricted set of functions like Billing, Planning and overall Performance and Quality of Service monitoring. The scope of these systems tends to be wide, covering the whole network.

The survey shows that open technical solutions are currently limited either by the manufacturers technology or by the PTO's organizational structure, and that the lack of management standards (only recently becoming available) identifies very few of the examined systems as "open systems". In particular, the management systems closely connected with the NE, are dependent on its technology, while the systems positioned in the higher layers of the management hierarchy are dependent on the PTO's organisation and needs.

With respect to the level of integration, there is a clear predominance of specialised systems developed for and tailored to specific needs, which results in a poor level of integration among these systems and in inconsistencies and overlapping among the management data stored in the databases.

Another general aspect to be outlined is the extensive involvement of human operators in many management aspects, so that the level of automation in some areas is quite low. This item is considered to be very important because a high level of automation will be crucial to cost-effective network management of the complex IBC (Integrated Broadband Communications) environment.

With regard to the level of interworking between the different management systems, the survey shows that the situation of today appears to be less integrated in those countries (e.g. Denmark, Italy, Portugal) where there are different PTOs, either operating on a particular part of the country or on a particular segment of the network (e.g. domestic, international, intercontinental).

The trend, up to now, seems to be the development of proprietary support systems, at each layer of the management hierarchy, the only few exceptions being systems within the Accounting Management area (e.g Billing Centres). In general, the level of interworking among the management systems of the different countries appears to be quite low, being restricted only to some procedures within the Accounting Management area and to the management of international circuits. Finally, it's worth pointing out that the communication process among different systems is often affected by physical exchange of paper files, therefore preventing the easy further processing of the management data.

3. TARGET TMN REFERENCE CONFIGURATION

This section describes the TMN RC for the target IBC environment. After a brief description of the network environment itself (access, switching and transport portions), an outline of he target TMN RC (the six RCKC) [3] is given.

3.1 IBC Network Environment: Subscriber Access Portion

The key aspect of the target environment in the subscriber access area relates to the transmission media: metallic subscriber access lines will be replaced by optical fibres. The very high capacity potential of these optical fibre access links favours their exploitation for several subscribers on the transmission path level in configurations such as MAN (Metropolitan Area Networks), multiplexed FTTH (Fibre-to-the-Home), and SDH (Synchronous Digital Hierarchy) add/drop arrangements. For industrial and administration business users, the MAN access option plays a key role in the target environment: MANs will offer connectionless mode services, connection-oriented data services and isochronous services. The concepts of FTTH and FTTC (Fibre-to-the-Curb) are proposed to offer broadband network access particularly to residential customers. Also SDH is a valid option for the transmission in the subscriber access area, in particular the relatively easy and economic add/drop options it allows. In addition, mobile communication networks form an area in which substantial growth rates are expected in the years to come.

3.2 IBC Network Environment: Switching Portion

IBC will be based on a standardised ATM (Asynchronous Transfer Mode) information transfer mechanism. All levels of network nodes in the IBC will apply ATM according to the same international standard. From the viewpoints of traffic handling and management, it is desirable that the transition between the ATM-based IBC and the conventional (pre-ATM) networks concentrates on few points of inter-working. Signalling capabilities are to be provided in the IBC environment to control the establishment, maintenance, alteration and release of ATM virtual channel connections and ATM virtual path connections for information transfer. The traffic characteristics of a connection are negotiated on establishment and may be re-negotiated at any time during the connection. Connection types include point-to-point, multipoint and broadcast. Symmetric and asymmetric simple calls, multi-party calls and multi-connection calls are to be supported.

3.3 IBC Network Environment: Transport Portion

Given that ATM is the choice for the transfer mechanism within IBC, two alternatives have to be distinguished in principle for transmission on the physical layer: the ATM transport is supported by an SDH transmission network (ATM-over-SDH), or the ATM transport network relies on a cell-based transmission mechanism (pure ATM). Along with these two alternatives, two principle kinds of cross-connect devices are to be distinguished: the SDH-based cross-connect, allocated to the physical layer of the ATM transport network hierarchy, and the ATM-based cross-connect, located on the virtual path level within the ATM layer of the transport network hierarchy.

The layered structure of the transport network portion applied to the target IBC environment leads to the identification of the following three classes of transport network layers: Circuit Layer Networks, providing bearer services applying leased lines, circuit switched connection, packet switched connections or ATM based connections, Path Layer Networks, utilised to support different types of circuit layer networks and Transmission Layer Networks, dealing with the transfer of information between two nodes and on the actual transmission medium. As far as the TMN is concerned, there is a need to deal with the management requirements related to each layer, as well as to the layer-to-layer associations.

3.4 IBC TMN: Structural Framework (RCKC-1)

Since the target IBC network environment is conceived as a comprehensive infrastructure for information transfer and telecommunication service offering, the set of available TMN building blocks must encompass a broad scope. The following types of NEC (Network Element Conceptual blocks) are proposed:

- NEC-LN covers a low-complexity, non-standard NEC. For supervision and control, the corresponding devices are usually assembled in groups. For the adaptation to the standard TMN, common QAC (Q-Adaptor Conceptual block) are foreseen. Examples of NEC-LN models are conventional, plesiochronous muldex units, line terminations, regenerators
- NEC-HN covers a high-complexity, non-standard NEC. For adapting to a standard TMN it relies on Q-Adaptor conceptual blocks. Examples of NEC-HN models are existing exchanges and cross-connect devices
- NEC-LS covers a low-complexity, standard NEC. For supervision and control, the corresponding devices are usually grouped together and connected jointly. For example this model covers SDH muldex units and line terminations, as well as MAN muldex and customer access units
- NEC-HS covers a high-complexity, standard NEC, directly linked to an OSC (Operations System Conceptual blocks). For example it models advanced cross-connect devices, ATM exchanges, and MAN switching and control units.

The OSC (Operations System Conceptual blocks) are introduced as the conceptual units key for the full implementation of the distributed information processing, storage and retrieval tasks related to TMN. The following types are proposed:

- OSC-E responsible for the network element level management and directly involved in the supervision and control of NECs: thus they constitute the lowest level of OSCs.
- OSC-N responsible for the comprehensive network level management and recursively for the management of the sub-networks constituting the overall target IBC
- OSC-S responsible for the coherent management of the telecommunication services offered in the target environment
- OSC-B introduced to model business activities of the enterprises at the highest level of strategic management within an organisation. Decisions on this level will influence the introduction and configuration of the target IBC environment.

The QAC (Q-Adaptor Conceptual block) constitutes a lower level of adaptation between those NECs representing NEs that do not apply standard TMN interfaces and standard OS. They cope with the inclusion into the standard management context of conventional NEs from the pre-TMN era, as well as low-complexity NEs not justifying the expense for Q-interface implementation.

3.5 IBC TMN: Functional Model (RCKC-2)

For the definition of the functional aspects of the target TMN RC, a list of FEs was defined [1], [4]. The main objectives of this model are to assemble TMN FEs, to elaborate function allocation strategies and migration paths. Possible migration paths towards the target TMN for IBC are explored, considering aspects of automation, centralization and technology evolution. Migration for TMN can in principle be seen as a re-allocation of entities (i.e. MS, MSC, MO) to other TMN building blocks. Having adopted the object-oriented approach, the displacement of functions (migration) appears as a displacement of MOs with their re-allocation to another TMN building block. The evolution for reaching the target TMN will be marked by the two processes of the creation of new MO with the definition of the pertinent FE and of the displacement of such MO. Principally, this re-allocation can lead to a higher or a lower degree of centralisation.

However, an analysis of the path towards the target TMN shows that decentralisation, i.e. the path from the centralized manager to the decentralized manager and also to the network element, will prevail.

3.6 IBC TMN: Information Model (RCKC-3)

For the definition of the information aspects of the target TMN RC, a list of MOs was defined [1], [5]. TMN Information Modelling for the highly automated target IBC environment will be based on the object-oriented paradigm. Implementers of management equipment want to make use of the standards as much as possible, but face the constant problem of unstable standards. This problem is very severe for someone who intends to design a system in a specific management domain. A PTO that has responsibilities in a large set of management domains is even more affected by the instability of significant details in the emerging information model standards. Usually, he has already invested into a number of equipments that must coexist with the new equipment that will be purchased for the broadband network. These are some of the problems that a PTO incurs when they want to perform evolution planning for their overall management solution. The sources that now are available for the information modelling work represent very different views of the network and service resources. This means that it is difficult to combine the different "models" into a consistent and homogeneous modelling framework. The PTO must use a framework for modelling that provides sufficient stability for the TMN evolution planning process. He cannot rely on a number of standardised information models to be adequate for his needs. Instead he must exploit the various models as valuable input when creating his management solution based on the sharing of management knowledge between different OSs and NEs. In order to make optimum use of the emerging standards, a PTO must work actively on the definition of the information that is common to the different management areas under his responsibility. He must create and maintain a repository of class definitions that suits his particular management requirements, extensively utilising available standards. He himself must create an organization that is responsible for keeping the model up-to-date and suited to its specific needs.

3.7 IBC TMN: Communication Model (RCKC-4)

The following main communication relations for the exchange of management information within the target TMN can be defined: communication between NEs, communication between NEs and OSs, and communication between OSs. Different options for the transfer of information are to be considered in these three cases. It must be noted, however, that different management tasks, e.g. spontaneous alarms versus transfer of billing data, may require different transport media to satisfy capacity as well as real time requirements. The communication between NEs will take place, for example, between a Host Switching Unit and Remote Switching Units when customer or routing data have to be downloaded, and between a large cross-connect and associated multiplex units when alarm data are filtered. If all the NEs have access to a transport medium of the sub-network with sufficient capacity, e.g. an ISDN B-Channel, a CCITT SS#7 signalling link or the Data Communication Channel in the SDH network, then this option should be chosen. If this is not the case then a LAN may be used provided distances allow this. NEs equipped with an implementation of the interface required by the DCN will be directly connected to the OS, otherwise this connection will be through a Q-Adaptor. OSs will communicate over the DCN. There are, however, cases where other means can be more suitable. An example is a cluster of co-located OSs for different management tasks sharing a common database. In this case, a LAN would be the choice; also a SNA-based solution is conceivable. In principle, for the transfer of network management data inside a sub network the available transport capabilities and interfaces should be used, including those of the sub-network itself. The number of interfaces that have to be provided by an OS should be very small. A consequence of these considerations is that access from an OS to the NE in a sub-network can be via gateways. These will have mediation functions and protocol conversion functions built in and may have some management capabilities at the NE level. These devices interface to the OS via an OSI protocol stack according to the Q3 standards, and can either be stand-alone units or integrated in a large NE with sufficient processing power.

3.8 IBC TMN: Usage and Management Service Model (RCKC-5)

Management Services (MS) are implemented in management systems in order to fulfil specific management needs of the Public Network Operators (PNO). MS are defined as assemblies of reusable functional elements named TMN Management Service Components (MSC). Definitions of the entities involved are presented in [6]. According to this definition then, the set of MS relevant to a specific environment are dependent on a number of factors such as the the medium (e.g. copper, optical fibre, radio), the technology (e.g. SDH, ATM), the network architecture (e.g. star, ring) the telecommunications services offered (e.g. POTS, video, CATV) and the different profiles of end-users served.

Several key issues in the network and service evolution will lead to a considerably higher number of TMN users, compared with the current usage situation: the increasingly competitive situation in the telecommunications business, the management cooperation on a supra-national (in particular on a pan-European) level, the emerging provider-consumer hierarchy for advanced services, potential joint implementations of TMN and IN (Intelligent Network), and the demands for direct customer access to the TMN.

A TMN user is defined as a person or process applying TMN Management Services for the purpose of fulfilling management objectives. The classical users of management services have been the O&M staff of the PTO responsible for managing the network, but this situation is likely to change in the future due to the factors mentioned above. Routine operations may be carried out more efficiently and rapidly by the customer himself, provided sufficient security safeguards and appropriate support systems exist. There is a potential for additional revenues obtainable by PTOs, by making MSs commercially available to their customers. Alternatively the existing public network operators may be legally compelled to provide inter-operability with their competitors (as in the UK) which implies that management information will have to be exchanged in a standard format. Furthermore, some groups of special users may in the future have access to TMN for particular limited purposes, e.g. Trusted Third Parties in the area of security (providing certification services) and regulators (involved in the verification of equal access).

3.9 IBC TMN: Supplementary and Non-Technical Aspects

The evolving scenario must take into account the CEC proposals for the deregulation of telecommunications presented in the "Green Paper", and in particular the concept of ONP [3] [7]. The requirements that ONP will put on the TMN RC are presented here as an evolution of the idea of the hierarchy of OSCs. The proposal is the definition of an open, standard interface to OS involved in the management of reserved services, where ONP conditions would apply. The decomposition into Reserved and Liberalized Services, see figure 1, allows for the introduction of the regulatory and organisational implications of ONP. In particular, OSC-RS provide the management of services realised as exclusive offerings by PTOs, while OSC-LS cater for the management of liberalised services offered in competition under ONP.

4. THE CONCEPT OF GAMS

One of the most significant differences between today's management situation and the TMN in the target IBC environment will be the degree of automation in handling MSs. The current fairly low automation level will gradually be transformed to highly automated processes according to the object-oriented, information-driven paradigm. TMN automation is crucial in attaining the key objectives of future network and service providers with respect to their management systems to provide cost-optimum MS solutions, to thrive in a competitive environment, to cope with the growing network and service complexity, to conceive common approaches to TMN and IN (Intelligent Network), to adapt flexibly to rapidly changing customer demands, to overcome a lack of skilled O&M personnel, to introduce standards, and to promote pan-European inter-operability.

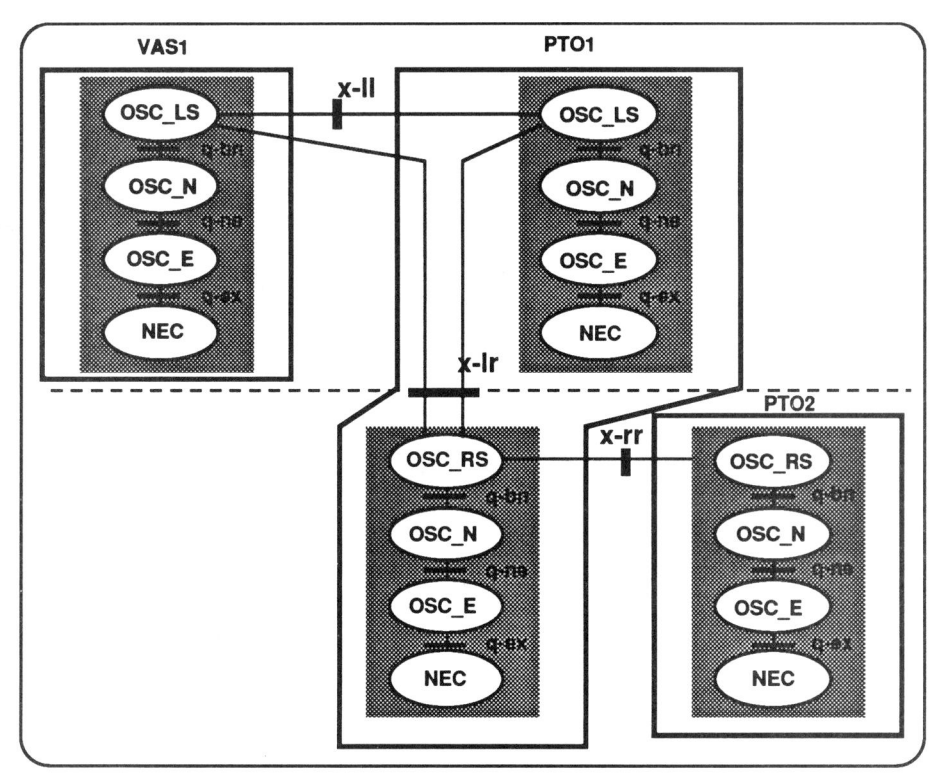

Figure 1 - TMN scenario under ONP conditions

TMN evolution will most likely focus on a phased transition to automated and standardized solutions. In order to provide a framework for the fulfilment of these objectives, the concept of GAMS (Gradual Automation of Management Services) was developed [8]. TMN automation benefits includes, for example, the reduction of demand for skilled personnel and the avoiding of delays in TMN evolution because of lack of skilled personnel. The current state of network and service management throughout Europe outlines that the considerable automation potential in the area of network and service management has only been exploited to a moderate degree up to now. Looking at the ADI (Awareness creation, Decision making, decision Implementation) cycle [4] of management processes, it can be stated that:
- Awareness Creation has currently reached the relatively highest degree of automation, including automatic supervision, alarm forwarding, performance data logging
- Decision Making still shows a moderate automation stage, but, considering the developments of Advanced Information Processing technology, offers the most extensive potential for automation
- the transition to automatic Decision Implementation largely depends on the development of networks and NEs to be managed. For some of the decision implementation tasks (e.g. in the areas of installation and repair) there are natural limits to automation.

5. ASPECTS OF COOPERATIVE MANAGEMENT

In the coming years, a number of factors will influence the telecommunications scenario, e.g :
- the implementation of Pan-European networks, e.g. GSM, ISDN and METRAN (Managed European TRAnsmission Network)
- the deregulation of the telecommunications market, leading to an increase in the number of network operators and service providers
- the adoption of concepts like One-Stop-Shopping
- the increased mobility of the subscribers.

These new types of scenarios will create a number of constraints on the management systems, create new demands for the accounting services, lead to an increase of information exchanged between PTOs and Service Providers, increase the demand for network management systems that can operate across more than one management domain: cooperative management will then become key [6]. See figure 2.

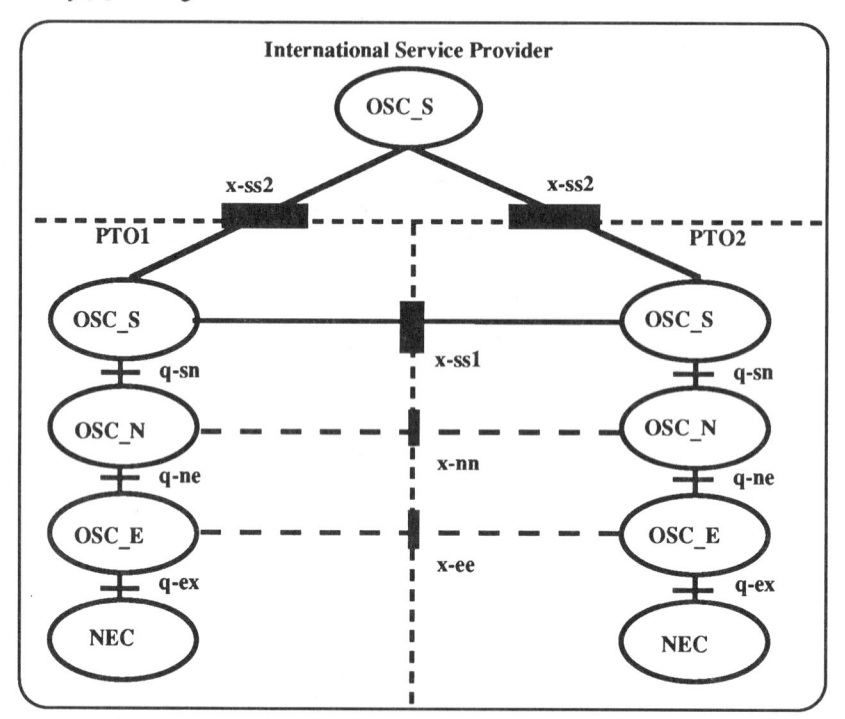

Figure 2 - TMN scenario for international service management

The management information needed to be exchanged between the different actors on the Pan-European telecommunication market will, very soon, assume a new dimension. It will probably be necessary for European PTOs to allow other PTOs and Service Providers to access a defined set of their own management facilities. In terms of management, the future network environment will then be composed of a set of network/management domains with peer-to-peer interworking but, most likely, with little joint management. Within each domain different options can and will be adopted, depending on the size and nature of the domain and services to be supported. In this way, fully centralized solutions, fully distributed solutions or hybrid solutions can be adopted. The most probable scenario, in the short term, will rely on the cooperation between management systems of different domains, while in the medium/long term some supra-national (Pan-European) service providers will appear. Among the major key points, the problem of shared management of networks and services in the deregulated Pan-European environment assumes pivotal importance, due to the need for clear demarcation of TMN domain ownership, management responsibility, security and inter-operability. Also the problem of full access to network data and managed objects is of key importance.

The X-interface definition (being the main vehicle for management information transfer) is of key importance, see figure 2, not only from the message set point of view (what functionality to achieve) but also from the protocol stack point of view (what protocols to use to cope with the desired functionality). The requirements of joint management can be addressed mainly by the definition of common management services addressing the main management domains.

6. CONCLUSIONS

Only a sophisticated and flexible structure of cooperating TMNs with shared responsibilities can cope with the challenge of IBC management.

The conditions set by regulations and laws demand deregulation, decree liberalisation and require equal access to services and to the communications infrastructure. A growing number of parties will be involved directly or indirectly in network and service management as operators, providers and customers. Only open, inter-operable solutions in conformance with international standards will fit into the target scenario. Customers, particularly business customers and enhanced service providers as customers of the basic service provider, will require direct access to TMN.

In the target IBC scenario, a common software platform will offer all necessary features and facilities for advanced applications in the areas of TMN and IN. The application interfaces are expected to be standardized to the degree that enables a multi-vendor environment, i.e. the installation and operation of applications from different vendors on the same platform. The object-oriented paradigm (applied to conception and implementation) forms the basis for standardised, open TMN solutions. A full range of MSs will be assembled from MSC as required, and offered to the customer.

TMN operators will offer their customers direct access to the TMN services, restrained by the necessary security rules. A consistent security policy will lead to the enforcement of corresponding security rules, and to the application of appropriate security mechanisms.

Pan-European cooperative management will guarantee the inter-operability between national domains. In the medium to long-term, a supra-national level of management centres may support the automated cooperation of national TMN domains.

7. OPEN ISSUES

Before the future TMN on the national as well as on the pan-European scale can be designed and implemented, some pending issues which will require a broad international consensus, have to be settled.
- International TMN standards have to be completed, accepted, enforced and applied universally. Open, interoperable solutions will then enable the cooperation on all levels via standard interfaces.
- The object-oriented design and implementation methods still require development efforts to cope with the management challenge in the long-term, especially in the areas of object-oriented information modelling and in implementing these models to distributed solutions.
- Existing, non-TMN management applications and databases will remain in operation for many years. This is required because of large investments in these systems and also in

the organisations that interact with them. Hence there is a strong demand to integrate those conventional management systems into the new TMN environment, rather than to migrate from them. The challenge in this respect is to develop comprehensive super-structures (encompassing databases and applications) that will allow new applications conforming to the standard TMN to cooperate with pre-TMN solutions.

- The shared management of networks and services in a deregulated supra-national environment will demand a clear demarcation of TMN domains from various viewpoints, e.g. ownership, management responsibility, security, and the consideration of pan-European inter-operability, including one-stop-shopping and one-source-billing requirements. The definition of X-interfaces between different TMN domains will become increasingly important, as the Pan-European networks are being set-up.
- The definition of Management Services and of the pertinent MS Components should follow a consistent and coordinated methodology oriented at the respective user needs: re-usability has to be the crucial criterion for the MSC conception and implementation.
- The detailed implications of the evolving regulatory and legal framework for the realm of telecommunications have to be defined. For Europe, ONP and related regulations of the CEC are likely to guide the general path; however, different deregulation paths and separate national interpretations may lead to different patterns of management cooperations (e.g. between network operators, service providers, customers).

8. ACKNOWLEDGEMENTS

This paper is based on original work developed by Project TERRACE under a RACE Contract awarded by the CEC. A special thanks to all the TERRACE Partners who contributed to this work.

9. REFERENCES

[1] RACE Common Functional Specification M201 - "TMN Reference Configuration: Abstract Framework Description."

[2] RACE Common Functional Specification H200 - "An Architecture for the TMN"

[3] RACE Common Functional Specification M200 - "TMN Evolution"

[4] RACE Common Functional Specification H400 - "TMN Functional Specification Conceptual Models, Scopes and Template."

[5] RACE Common Functional Specification H550 - "Telecommunications Management Objects."

[6] Jacobsson, M. et al : "TMN Reference Configuration Case Study Results", Sixth RACE TMN Conference, Madeira, September 1992.

[7] Study on the Application of Open Network Provision to Network Management, NERA/MITA study for CEC - DGXIII.

[8] RACE Common Functional Specification M210 - "Methods for TMN Reference Configuration Design and Evolution."

The Management of Telecommunications Networks
R. Smith, E. H. Mamdani, J. G. Callaghan (Editors)
© Ellis Horwood 1992

TMN Reference Configuration Case Study Results

Måns Jacobsson (Televerket, Sweden), Luis Alveirinho (TLP, Portugal), Andrew Kelleher (Broadcom, Ireland)

ABSTRACT

This paper presents how the concept of Management Services can be applied to case studies of real networks. The structure of the Management Services is first presented briefly. The methodology used in these case studies is then described. A presentation of the case studies and some results which can be drawn from these studies is also included. Finally, conclusions which can be drawn from the case studies and our experiences from this work are presented.

1. INTRODUCTION

RACE project TERRACE has studied methods for the design and implementation of TMN Reference Configurations. In particular, a set of case studies were performed in order to validate and consolidate the identified framework.

A Reference Configuration (RC) describes the building blocks and their interrelationships which are necessary to fulfil a specific task. The task of telecommunications management demands the design of a TMN (Telecommunications Management Network) RC. TMN reference configurations (TMN RCs) will constitute the conceptual framework for the design and implementation of the TMN [1], [2], [3], [4]. For the description of the TMN RC, six Reference Configuration Key Constituents (RCKCs) should be defined [4] in order to structure and depict the conceptual viewpoint on TMN in anticipation of potential TMN implementations.

The concept of Management Services, and the decomposition of a Management Service (MS) down to Functional Elements and Managed Objects.is included within the TMN RC design and is presented in this paper.

Within TERRACE, the TMN RCs have been defined at three levels (see figure 1):

- Level A   Time, network and country independent
- Level B   Time and network dependent, country independent
- Level C   Time, network and country dependent.

TERRACE has focused its work on level C for different timeframes, short term (until 1994), intermediate (1995-1998) and long term (after 1998). At a later stage these timeframes will be related to stages of network development rather than dates.

The level C TMN Reference Configurations definition started with a survey of current network management implementation in 67 networks. The conclusions confirmed the urgent need for harmonisation, due to low level integration and automation of TMN functions, and the short term trends which represent the first stage of the evolution process [5]. Level C TMN Reference Configurations were produced as a series of case studies of which the following have been completed:

- Short term scenario (PSTN) in Ireland [6]
- Intermediate term scenario (SDH) in Sweden [7]
- Intermediate term scenario (MAN) in Ireland [7].

In addition to this a study on how broadband communication, in particular ATM, might be implemented in nine European countries has been carried out [8].

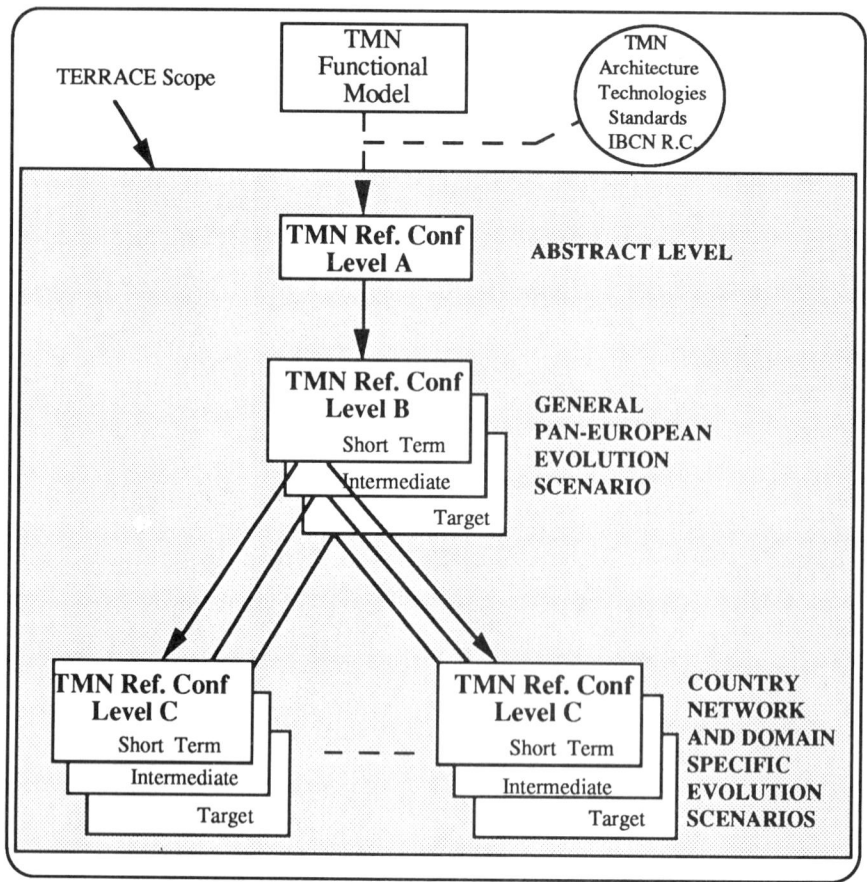

Figure 1 - Three Levels of TMN RC

2. DESCRIPTION OF MANAGEMENT SERVICES

This section presents the structure of the management services and the entities included in this structure.

2.1 The Structure of Management Services

In the scope of TERRACE work Management Services (MSs) consist of functionality which is grouped in order to be able to satisfy the management requirements of the TMN user. An MS can be decomposed into one or several Management Service Components (MSCs), which can themselves be decomposed into one or several Management Application Functions (MAFs). The MAFs consist of one or several Functional Elements (FEs), which provide access mechanisms for the Managed Objects (MOs). The complete structure is illustrated in figure 2.

As shown in the figure there is a many to many mapping between the levels. An MS is usually composed of several MSCs and MSCs are designed to support flexibly several MSs. The same applies to the MSC-MAF and to the MAF-FE relationship.

From a consolidated library of Functional Elements and a similar library of Managed Objects, a very large number of TMN user individual Management Services can be assembled. As the Management Services can vary widely, depending on the requirements from the individual TMN user, the degree of freedom for this assembly process is very high. It has to be mentioned here that this structure is the result of the work of the project and is not intended to replace any structure produced by any other organisation. It should rather be regarded as complementary to these structures and a possible input to future work.

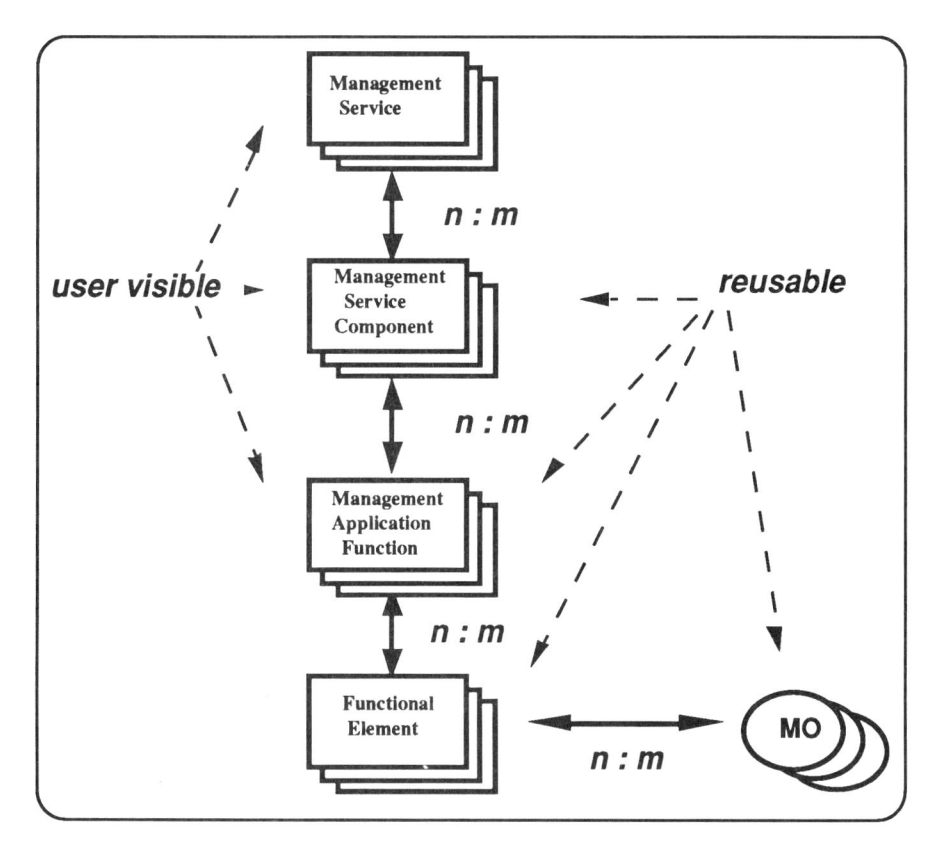

Figure 2: Decomposition of the Management Services.

2.2 Definitions of the Entities

The entities identified above are defined as follows :

Management Service

TMN Management Services (MSs) are offerings fulfilling a specific telecommunications management need of the TMN user. TMN users may be internal or external to the organisation of the respective TMN provider. MSs are predefined or customer designed assemblies of TMN Management Service Components. The significant activities of the TMN providers are supported by the TMN MSs, for example alarm handling, corrective maintenance and service provisioning.

Management Service Component

TMN Management Service Components (MSCs) are the reusable elements, constituting TMN MSs, and are visible to the TMN user. The user does not need to have any knowledge of the internal structure of the TMN MSCs in order to assemble or apply TMN MSs. Depending on user needs and user privileges, the same TMN MS can be decomposed into TMN MSCs to different degrees of detail, according to the required purpose. Possible ways of identification of MSCs include layering and partitioning of MSs.

Management Application Functions.

The TMN Management Application Function (MAF) is the lowest level MSC, with distinct characteristics that can access Managed Objects using a communication capability, for example, and FEs.

Functional Elements

TMN Functional Elements (FEs) provide access mechanisms for Managed Objects. They provide the capability of acting upon the object, referencing their operations and attributes. FEs are elementary TMN tasks required to supply the functionality of TMN MSCs. TMN FEs allow for the manipulation of MOs. Thus they map onto the operations to be performed on MOs.

Managed Objects

MOs represent any resource (abstract or concrete, simple or complex) visible to TMN and subject to management by the TMN. The universe of resources covers the entire telecommunications environment, including the TMN itself. One resource can be represented by several MOs under different management aspects, and one MO can model several resources under a common viewpoint.

This combination of Management Services, Management Service Components, Management Application Functions, Functional Elements and Managed Objects forms the necessary basis of the specification of a message set. A suitable working message set such as CMIP/CMIS [9], [10], but not necessarily that one, could be chosen, at implementation time. To confirm this a mapping of the functional elements onto the CMIP/CMIS has been carried out.

3. THE CASE STUDIES AND THE CONCEPTUAL TMN DESIGN APPROACH

In order to refine and verify the above method, it was decided that it should be applied to a set of case studies of existing networks, or networks which are about to be implemented. These case studies and the main conclusions are presented here.

For the TMN RC case study activities, the approach taken is described below. As this is not a step by step process but an iterative one the points listed below have not been numbered. Of course, some phases have to be completed, before proceeding to the next phase. It would for example not be fruitful to identify the managed objects and the functional elements before the network to be studied had been chosen.

Identified time frame. At this stage of the approach the time frame for which the case study will be relevant is identified. In the first attempt of applying the conceptual TMN RC design process the time frame chosen was the short term one. As this attempt turned out to be successful it was decided to continue with the approach and to apply it to the intermediate time period.

Identify the type of network. As the time period now being studied, was the intermediate, it was decided to focus on the network types that would be able to represent this period of time and the evolution stage towards the target scenario. As TERRACE had earlier performed case studies on SDH and MAN, it was found useful to use those results as the basis for the studies.

Identify country. In order to carry out the case studies in an effective way it was necessary to chose some countries where the network types already chosen are used or are about to be used. An important factor for these decisions was also the possibility of getting access to information. This fact limited the possible choices to the countries represented in TERRACE. The countries chosen for the MAN and SDH case studies were Ireland and Sweden.

Identify network. This stage of the procedure involved the selection of the particular networks to be studied. From a practical point of view, this was already decided when the country was chosen. The networks chosen for the SDH and MAN case studies were the forthcoming SDH network in Sweden and the MAN Telecom Éireann is planning to install in Dublin.

Identify TMN user needs (Management services). This stage of the procedure considers the TMN user and his specific requirements. These requirements could be represented by a set of management services. In order to handle these in a rational way the management services can be decomposed into one or several management service components to different degrees of detail, according to the respective purpose. These management service components represents reusable entities and can therefore be found in more than one management service It should be remembered that the TMN user may be external or internal to the respective TMN provider.

Identify management functionality (MAF and FE). This stage of the process identifies the functionality which will be required to support the management services identified above. As mentioned, these management services represent the TMN user needs. The functional elements are access mechanisms for managed objects described below.

Identify managed resources (MO). Having identified the managed functionality which will be required, the next step is to find the managed objects on which these act. One resource can be represented by several managed objects under different management aspects, and managed objects can model several resources under a common viewpoint.

4. PSTN CASE STUDY

The network chosen for the first case study was the PSTN network of Telecom Éireann. The choice of the PSTN scenario was made deliberately as it is a well understood and mature network and thus would greatly assist in the validation of the method [4].

This case study was carried out in two steps during 1990. First a decomposition of the 1990 PSTN scenario in Ireland was carried out. The impact of the current trends was then used in order to map the evolution in the short term. This resulted in a view of how the PSTN is likely to be at the end of 1994. It was then possible to do another decomposition for the 1994 scenario.

Eleven Management Services (or Application Services as they were called at that time for this case study) were identified. These eleven Management Services were;
- Network Service Provisioning
- Complaint Handling
- Alarm Handling

- Service Restoration
- Performance Information Handling
- Tariff and Charging Administration
- Management of Security Procedures
- Service Planning
- Network Planning
- Customer Service Provisioning
- Corrective Maintenance.

Of the eleven Management Services listed above, seven were described in detail and examined both for the case of the 1990 scenario and the envisaged scenario for 1994. The remaining four decompositions presented a more generic representation of the Management Service.

5. CONCLUSIONS FROM THE IRISH PSTN CASE STUDY

By comparing the decomposition for 1994 with that for 1990 in terms of the functionality and Managed Objects identified, it was possible to make some deductions about the increase of functionality that is likely to take place during this period of time. For example, if a group of functions appears in the decomposition for 1990 but not in that for 1994, then it may be assumed that the functionality represented by this group will not be required beyond the short term period. An example of where this would apply would be in the case of analogue transmission systems and their associated management functionality.

The case study made it possible to evaluate the structure of Management Services on a real network for the first time. The scope of each entity as well as the relationship between the entities were explored in detail and clarified. As a direct result of the case study, it was found that it was necessary to make some minor changes to the method. These changes included definitions, relationships and scopes. Of special interest were the interrelationships in the bottom of the structure.

This first evaluation also gave valuable experience for the forthcoming case studies as they were planned to be performed in a similar way. This first authentic case study lead to the refinement and consolidation of this concept. Therefore the SDH and MAN case studies focused on the case studies themselves and not on more general aspects.

The results from the PSTN case study were used as input into a study of the allocation of management systems functionality within Europe. The practical experiences could thereafter be compared with previous theoretical results achieved by TERRACE. It was then possible to verify and confirm these theoretical conclusions and to illustrate the path from the current functional allocation situation towards the target situation [11], [1], [7]. Functionality which has not yet been automated was of special interest and some identified trends indicated the kind of functions which are about to be automated.

Having validated the method and design approach during the PSTN case study, it was decided to apply the method to other networks and management systems in Europe.

6. SDH AND MAN CASE STUDIES

As this was found to be a powerful methodology TERRACE continued with this approach and applied it to different case studies. It has so far been applied to a MAN network in Ireland and the forthcoming SDH network in Sweden [7].

Since the PSTN case study was carried out, the structure of the Application Services had been subject to minor changes and replaced by the structure of the Management Service presented above. The reason for these changes was to bring TERRACE more into line with other organisations.

The following Management Services have been found relevant for the case studies of SDH in Sweden and MAN in Ireland.

Management Services related to SDH
- Embedded Control Channel Management
- Alarm Surveillance
- Performance Information Handling
- Management of Security Procedures
- Network Service Provisioning
- Customer Service Provisioning
- Service Restoration
- Network Planning
- Corrective Maintenance
- Complaint Handling.

Management Services related to MAN
- Alarm Surveillance
- Performance Information Handling
- Management of Security Procedures
- Tariff and Charging Administration
- Billing and Accounting Administration
- Network Service Provisioning
- Customer Service Provisioning
- Service Restoration
- Network Planning
- Corrective Maintenance
- Complaint Handling.

As for the case study of the PSTN in Ireland the work commenced with generic decompositions of the Management Services. Taking into account the general circumstances for MAN and SDH, each Management Service was decomposed down to the Functional Element and Managed Object level.

The results from these generic decompositions were then applied to the real networks. Detailed information about the networks investigated were provided to the participants. These documents were studied and compared with the generic decomposition already performed. The management services could then be given a more network specific approach, as it now was possible to consider the structure, purpose, use, etc of these two networks. Existing networks and management policies were also taken into account as these are likely to have a large impact upon the implementation of these two new networks. Additional MSCs, MAFs, FEs and MOs were identified as a direct result when the network specific situation was considered.

7. CONCLUSIONS FROM THE SDH AND MAN CASE STUDIES

The SDH and MAN case studies were not carried out in a two step approach as the PSTN case study was. The main reason why only one decomposition was carried out was that MAN and SDH are new types of network technologies compared to PSTN.

When these case studies started in the beginning of 1991, no real SDH or MAN networks existed in the countries represented in the study group. Thus it was not possible to use the two step approach which had been successfully applied to the PSTN case study.

This method showed itself to be very useful as it was applied to these two new types of networks. It was also possible to outline and structure new types of functionality, not used in existing networks.

Management Services, Management Service Components, Management Application Functions, Functional Elements and Managed Objects which are not currently being supported could be identified and described as a direct result of the work. Any entity that is not currently being supported represents functionality that it will be necessary to implement during an introduction of SDH or MAN.

These case studies also highlighted the extent to which the various MSCs, MAFs, FEs and MOs are being used at the moment. Functionality or groups of functionality already implemented in PSTN and PDH networks could therefore be identified. This functionality might be reused in SDH or MAN management systems. The importance of reusability of both functions and objects was therefore stressed even further. It was found to be very important that functionality is structured in a proper way during the specification of new application. If reusable modules are used, a lot of effort and resources can saved in the future.

As mentioned above the concept had been subject to some changes since the PSTN case study. It was now possible to apply this revised structure to real networks which were about to be implemented. The case studies verified that the changes had improved the concept and the methodology.

The two case studies provided the participants with useful experience of structuring and analysing the functionality required by SDH transmission networks and MANs, as early implementations of these technologies now exist. Most network operators in Western Europe have started to install such networks.

These studies were also an opportunity to investigate and clarify the differences and the relationships between the different levels in the structure. The MAF-FE-MO relationship was of special interest to the group. The scope and definitions of the entities could also now be investigated and confirmed.

The results from the case studies were also compared with NETMANs Functional Areas. This comparison is interesting since NETMAN is a top down decomposition of the Functional Areas and TERRACE is an aggregation process from basic network management entities [8], [12].

8. SUMMARY.

When applying the concept, of Management Services to real networks, useful results were achieved. It has been possible to verify theoretical results previously achieved by TERRACE. A number of new MSCs, MAFs, FEs and MOs have been identified as a spin off effect of these studies.

The method has been further refined and detailed during the work. It has also given the partners of the TERRACE consortium valuable experience not only in the field of reference configurations but also in the area of management of PSTN, PDH, MAN and SDH networks.

It has been possible to analyse and forecast various TMN evolution scenarios in Europe.

9. ACKNOWLEDGEMENT

This paper has been prepared under the auspices of the Project TERRACE under a RACE Contract awarded by the CEC. A special thanks to all the TERRACE Partners who contributed to this work.

10. REFERENCES

[1] TERRACE Deliverable 8 : " Intermediate TMN RC Design", 53/SEL/WP1/DS/A/008/b1, December 1990.

[2] TERRACE Deliverable 10 : "Target TMN Reference Configuration Design: Reference Configuration for the IBC TMN.", 53/SEL/WP1/DS/A/010/b1, July 1991.

[3] TERRACE Deliverable 12 : "Target TMN Reference Configuration Design: Reference Configuration for the IBC TMN.", 53/SEL/WP1/DS/B/012/b1, March 1992.

[4] Bagnasco, E. : "TMN Evolution", Sixth RACE TMN Conference, Madeira, September 1992.

[5] TERRACE Deliverable.6 : " Final Survey on TMN in Europe: Current Situation and Trends in Public and Private Networks", 53/XLT/WP2/DS/A/006/b1, May 1990.

[6] TERRACE Deliverable 9 : "Pan-European Evolution Scenario; Short term national plans", 53/BCOM/WP2/DS/A/009/b1, December 1990.

[7] TERRACE Deliverable : "Pan-European TMN Evolution", 53/TVT/WP2/DS/B/011/b1, July 1991.

[8] TERRACE Deliverable 13 : "Country-specific Target TMN Reference Configuration", 53/TVT/WP2/DS/B/013/b1, March 1992.

[9] ISO/IEC : "OSI Systems Management - Common Management Information Protocol", ISO/IEC 9596.

[10] ISO/IEC : "OSI Systems Management - Common Management Information Service", ISO/IEC 9595.

[11] TERRACE Deliverable 7 : "TMN function allocation: Key parameters and strategies.", 53/IBM/WP1/DS/B/007/ b1, July 1990.

[12] NETMAN Deliverable 10 : "Telecommunications Management Specifications", 24/BCM/RD2/DS/A/010/b1, September 1991.

The Management of Telecommunications Networks
R. Smith, E. H. Mamdani, J. G. Callaghan (Editors)
© Ellis Horwood 1992

Towards Integrated TMNs - The Global Conceptual Schema.

Terry Turner (Broadcom, Ireland), James G. Callaghan (BT Laboratories, UK)

ABSTRACT

The integration of applications covering a wide range of Management Application areas in TMNs (Telecommunications Management Networks) is judged to be very important for the future efficiency of advanced communications from the network operators' point of view. RACE TMN projects established a means of jointly examining integration concepts and a case study was performed and key issues identified. This paper describes the background to the work in question and elaborates upon the concept of a global conceptual schema for a TMN. The contents of such a schema are suggested as well as the approach that would be appropriate to developing it. It was concluded that there are additional requirements upon the design of such a schema in a TMN design context compared to the design of databases for typical commercial and administrative applications.

1. INTRODUCTION

Current Network Management Systems are not on the whole well integrated. However, integration of network management functionality is required by users and hence integration is one of the key issues facing designers and implementers of the TMN. The Integration Task Force (ITF) was established within the RACE TMN programme to develop a consensus on approaches for the development of a methodology for the realisation of integrated TMN systems. The main premise of the ITF was that *"the realisation of an integrated TMN system is dependent on having a clear strategy for the modularity and integration of its sub-functions"*. Essentially, the ITF's approach consisted of the identification of TMN functions and the realisation of those functions on the basis of an adopted integration methodology. The task was to find the ingredients of such a methodology.

The ITF co-ordinated the integration related work jointly undertaken by the GUIDELINE, ADVANCE, AIM, and NEMESYS projects. This work was of the form of a joint case study which was used to achieve common understanding of those key concepts needed to ensure that the RACE TMN projects adopted approaches which were consistent with the aims of building integrated TMN systems. This activity has to be seen as an endeavour which was complementary to the main architectural development work of GUIDELINE [1]. Its terms of reference emphasised giving feedback to the team defining the TMN implementation architecture. This paper provides information on the initial reasoning which the joint case study promoted about TMN integration, thereafter it presents a brief description of the focus of the integration issues and finally presents a description of one key mechanism which will enable TMN system integration, the Global Conceptual Schema for a TMN. This was described in terms of its definition, its scope, its contents and the use to be made of it.

2. INITIAL UNDERSTANDING OF INTEGRATION APPROACHES

The key objective adopted by the ITF was the identification of *functionality, information and working methods* which ought to be concentrated upon in order to make integrated TMN systems covering a wide span of functional areas a realisable goal.

The "Initial Report of the ITF Common Case Study" [2] addressed functions and information for Provisioning, Maintenance and Performance Management in the context of Broadband Networks. It clarified how concepts, such as managed object and management information base, were being used and which were most in need of joint understanding among the participants in the case study. In order to address the concept of TMN integration it was necessary to investigate both a top-down decomposition approach and a bottom-up re-aggregation approach. This process must be performed iteratively in order to achieve a fully integrated system. Figure 1 illustrates the top down approach. It was considered that following the decomposition, the important tasks are the identification of the common functions and the common information (data).

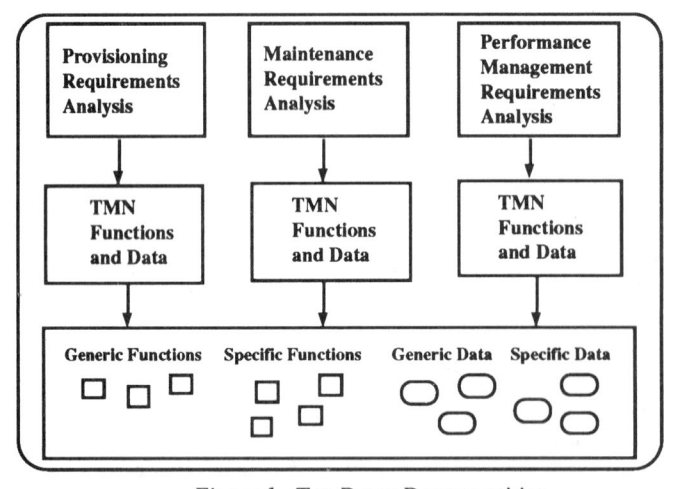

Figure 1 - Top Down Decomposition

The identification of the common elements (i.e. common functions and managed objects) motivates the bottom up re-aggregation approach as shown in figure 2.

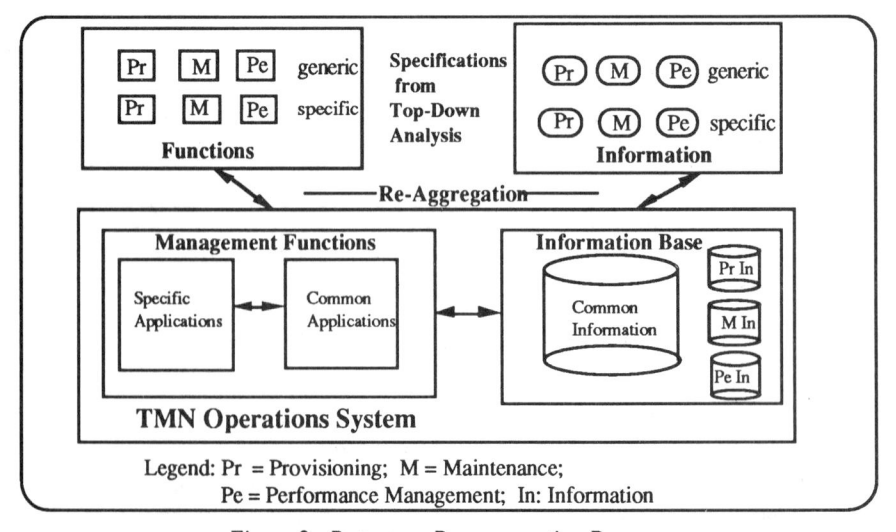

Figure 2 - Bottom up Re-aggregation Process

Figure 2 shows that the views of functions and information created by the top-down approach are collected together and assembled by a design methodology to form sets of management functions and an information base. The global conceptual schema described in this paper would be an important mechanism guiding the re-aggregation process.

Explorations of the constituent parts of an integration methodology for TMNs concluded that concentrated efforts by TMN designers on a knowledge base pertaining to the managed network to be shared by a range of applications would be needed. It was also suggested that applications ought to be supported to interwork with each other by the creation of a model representing the communications behaviour of the applications functions. The information needed for the building of this "integration" model was considered to be best derived from a top-down analysis of the application domain. The network knowledge base and the integration model were judged to be common components of critical importance to a TMN integration methodology.

3. FOCUSING THE INTEGRATION ISSUES

As part of the focusing process for exploration of integration procedures an examination was made of the key open issues listed in the TMN implementation architecture document [3]. The following open issues were judged to be relevant to one of the main concerns for TMN integration: that of management information structuring :

- improved understanding of the mapping from conceptual level models (specifications) to implementations of various kinds
- The TMN Architecture Group approach to Management Information Base (MIB) definition and the practicality of developing both database and interface definition from a single global conceptual schema must be tested.

Three key aspects of the TMN architecture germane to integrated TMNs are:

- The layering of TMN applications according to the so called *responsibility* layers labelled, Business Layer, Service Management Layer, Network Management Layer and Element Management Layer
- The approach to Management Information Base Definition
- TMN Interface Specification Methods.

The mission of the ITF was not confined to examination of the TMN architecture. It has been previously described that ITF addresses "conceptual integration" which means the common adoption of concepts whose meaning and application are fully understood and shared in the RACE TMN community. It does not concern itself with the more physical levels of integration of TMNs . The specific concepts of concern in this context were:

- Layered Management Applications
- Global Conceptual Schema (and Sub-schemata)
- Managed Object Class
- Managed Object Instance
- Intra TMN Interfaces (OS-OS and OS-NE).

The first of these concepts maps to the first of the key application areas of the TMN architecture mentioned above; the second, third and fourth map to the Management Information Base related application area; and the fifth concept maps to TMN Interface Specification area. Considerable input is made by standards bodies to the definition of some of these concepts. The ITF was building on such work but taking a more implementation oriented view-point than is typical of standards bodies. In so doing, it was intended to add clarification to assist TMN implementers. The five key concepts selected were considered to be the most fundamental concepts needed for TMN development. The achievement of consensus on, and the gaining of experience from the application of these fundamental concepts were the means chosen to perform the ITF work. This paper is devoted primarily to the presentation of the consensus on one of those key concepts, the Global Conceptual Schema. This concept was judged to be particularly important as it provides a container for information on another key concept, the set of managed object classes. Furthermore it provided a basis for the development of the network knowledge base and TMN integration model described as an essential precursor to integration in [2].

4. GLOBAL CONCEPTUAL SCHEMA FOR A TMN

4.1 Background

The TMN implementation architecture [3] indicated that databases within TMN Operations Systems are to be made explicit even at the conceptualisation stage of TMN design. It is somewhat unique in this regard. It is important to understand that the term "database" is to mean a database system including a database management system. The application of the TMN architecture requires that the use of database technology is made explicit in order that issues concerned with distributed information modelling be tackled. Figure 3 shows the fundamental conceptual view of TMN Operations Systems (OS) decomposition according to [3]. It was observed that management functions are seen to access both databases and network elements. It was argued that this placed new requirements on system design that need to be fully recognised. This places a requirement that the models used by the applications (management functions) cover information, both in the databases and in the network elements, as well as representing the resources in the network elements which are to be manipulated for management purposes.

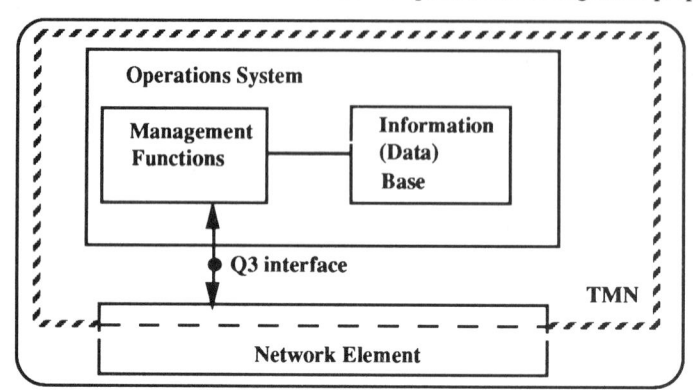

Figure 3 - GUIDELINE View of TMN OS Decomposition

4.2 Joint Case Study Management Information

The first step taken in the case study towards the design of an ITF-TMN (the name give to the TMN specifically designed in the case study) global conceptual schema was to examine the list of managed objects made in a previous case study [2]. The examination had aims of better structuring the objects, having a consistent approach to definitions and the elimination of overlap. This examination determined that four groups of information types existed in the list:

Group 1: information describing fundamental managed objects which are representing and have facilities for accessing the resources of the telecommunications infrastructure. These are fundamental in the sense that they are the primary means of interfacing to the managed resources.

Group 2: information describing data that would typically be held in an administrative data base (such as customer records).

Group 3:	information describing abstract classes of objects that could be used to generate classes of fundamental managed object. These classes are such that instances are not expected to ever be created from them. They might be more accurately described as templates for class definitions.

Group 4:	information describing (managed) objects which perform "application-like" activity. Examples of this class are *repair log* and *CamFault Hist* (representation of the history of faulty call configurations) They were not defined explicitly in the case study as application-oriented but they can be deduced to be different to group 1 above. They can be envisioned to be designed to co-ordinate the operations on the fundamental managed objects.

The structure implied by these groups helped clarify the use that would be made of the information, all of which had been described as managed object definitions. This analysis indicated that the term managed object was been used in a less that discriminating manner by the projects concerned. It is suggested that the term *managed object* be reserved for entities referenced in uses of management information services. This would be fully in keeping with the prescription of the OSI management documentation such as ISO/IEC IS 10040. Furthermore, it clarified that a Global Conceptual Schema would reference other entities besides implemented managed objects classes.

4.3   What is The Global Conceptual Schema ?

As it is a *schema*, it is a specification of information that would be pertinent to several possible implementations. As it is *conceptual*, it is expected that considerable abstraction from the full set of information of the TMN is achieved and that it is biased towards being a representation that would match the TMN end-users view of the overall management task. On the basis that it is *global* in nature, each TMN has one and only one such schema even if a distributed systems approach is used for realising the TMN functions.

Thus it is the set of information that covers at an abstract level what needs to be known for the TMN to perform the management tasks for which the TMN is intended.

It is to be seen in the context of a set of schemata normally referred to in database design methods. These are *External Schema* (the information collection that is specific to a particular user or application program of the database system); *Conceptual Schema*; *Internal Schema* (the physical structure of the data set in the computer) and, if it is a distributed system, a *Placement Schema* is needed which defines how the data has been partitioned, allocated to individual nodes and replicated. Distribution also brings in notions such as *global* and *local* to distinguish between the overall database seen as one entity (global) and the database specific to one site (local) in the distributed system. All such schemata are needed for a TMN of any reasonable scale. The global conceptual schema should be designed in such a way that its durability (stability over time) is assured.

4.4   What is the scope of a TMN Global Conceptual Schema ?

The global conceptual schema covers information concerning the managed domain meaning those entities which are the responsibility of a TMN to manage. It covers all (albeit in an abstract form) the information needed to perform management of the entities. It also covers the management activities which reflect the solution of the overall management task. Such activities may be human executed, machine executed or split between both humans and machines. It is not yet clear if the precise split of management activities into human-based and machine-based is needed in the schema. It would be better if it was not required as this split is bound to change as policies and technologies evolve.

When addressing the scope of this schema it may be beneficial to state what should not be expected of it. It was not seen to cover:
- the physical arrangement of the information in individual databases. This is covered by the internal schema
- A full listing of the managed entities at the instance level. This is contained in the Management Information Tree which is an important data structure which is complementary in nature to the global conceptual schema
- information that is needed for the very specific manipulation of managed entities which may be dictated by particular equipment vendor solutions
- the full statement of requirements upon the TMN concerned. For this other documentation besides the global schema is needed as well.

It was open to further discussion whether or not the management of entities that are specific to the TMN itself should be within the global schema or if such a task merits a global conceptual schema in its own right. It is likely that there will be common usage of the same entities by both the TMN services and other telecommunication services and so by managing these latter services, part of the TMN is also managed. Other matters relating to the management of TMNs was left for further study.

It is worth reflecting on the relationship of this schema to the CCITT recommendation M.3100 for a Generic Network Model (gnm). As gnm is not intended for any specific TMN, it cannot be a substitute or candidate for the global schema of a specific TMN. Nevertheless, the information contained in gnm is highly relevant to the designers of a global schema for a TMN solution. Other models which have a similar relationship to the global schema are the SNMP-MIB (Simple Network Management Protocol-Management Information Base), ETSI NA4 Draft Network Model, ISO-OSI System Management Model of Management Information and OSI NM Forum List of Managed Objects. It was recognised, of course, that none of these models claim to be TMN global conceptual schemata.

4.5 What does the Global Conceptual Schema contain?

According to the joint case study work it was concluded that the information in the global conceptual schema includes the following:
- Managed Object Class definitions
    - (n-) layer-oriented
    - system management-oriented
    - relationship-oriented
- Database Conceptual Data Models
- Application Oriented Entities
- Consistency Control Information.

Figure 4 summarises what the global conceptual schema for a TMN contains. The various components are described below.

Managed Object Classes

The majority of these object definitions will be externally sourced. ISO System Management Framework distinguishes between the objects that can be allocated unambiguously to one of the seven layers of the OSI reference model and other objects. These (n-) layer-oriented objects can be considered to be direct representatives of managed open system resources. In telecommunications systems such resources are the network and service infrastructures. Other objects are termed by ISO-OSI work to be system management managed objects (ISO/IEC IS 10040). These are objects which directly assist management activity such as logs. For network and service management we also need to represent relationships (such as trails) between resources and to use managed objects to do so.

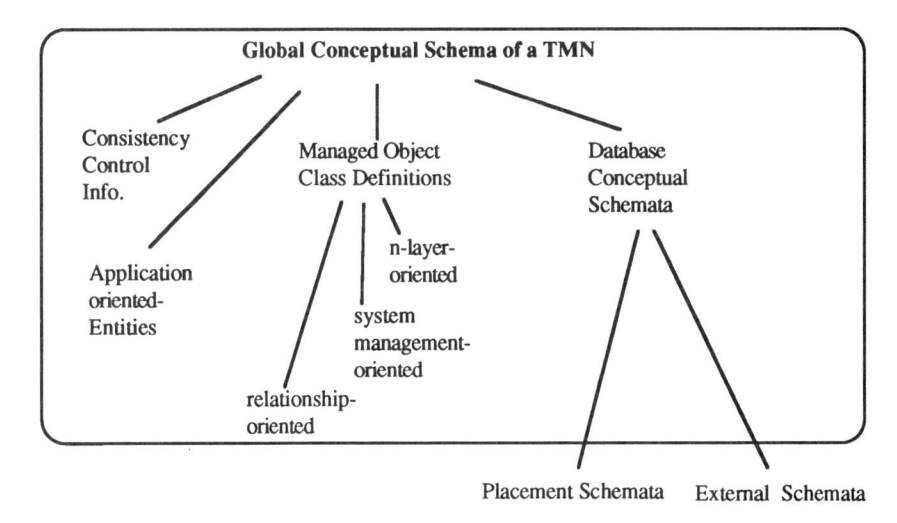

Figure 4 - Contents of a Global Conceptual Schema for a TMN

Database Conceptual Data Models
In those situations where databases are to be used as part of the TMN solution conceptual data models for such databases will be required. Creating such data models has become a normal part of database design methodologies, [6] and [8].

Application-Oriented Entities
Information on application-oriented entities such as Operation System Management Application Functions, will be needed to define how applications are accessed. It is inevitable that applications will be added and deleted as the responsibility of a TMN is adapted to changing circumstances. It is worthy of reflection that certain system management oriented managed objects can have similar roles to application-oriented entities. On an intuitive level, it was considered that it was likely that the kinds of information that would be needed included the roles (manager, agent, dual manager-agent) that the application satisfies, its authority to create/delete objects and its interface definition.

Consistency Control Information
It was clear from the foregoing that several bodies or groupings of information will be present in the global schema. There would be little point in regarding these to be part of one schema if there were not logical links between them. This is the case and information is needed which can be used to ensure consistency across the full information set. Undoubtedly, there would be internal consistency checking as needed in the various individual models but this is unlikely to be sufficient. Included here will be definitions of constraints to be applied when changes are to be made such as instance creation/deletion to the TMN information set. Some of this information resides within managed object definitions. The full understanding of the risks to consistency that will occur in a global schema such as this was yet to be achieved.

The foregoing description of the global conceptual schema suggests that the distributed database architecture included in the TMN Implementation Architecture can be used but would be augmented to reflect the fact that management data is present in network elements as well as in databases of the TMN. The placement schema is important in that it will not only reflect the distribution of data in databases but will also cover the presence of data in the network elements. Figure 5 shows a simple, but important modification for TMN design purposes of the normal distributed database architecture.

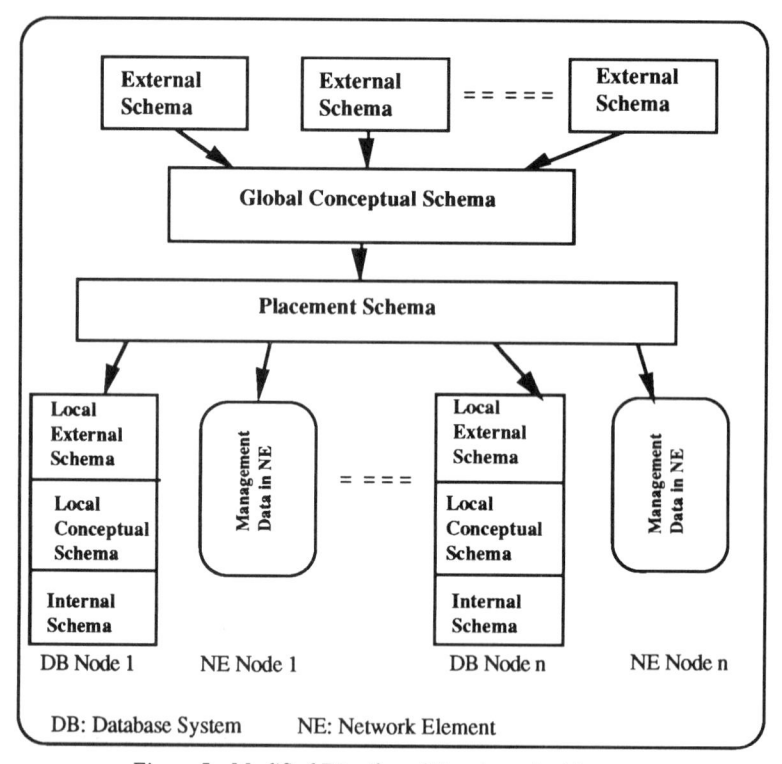

Figure 5 - Modified Distributed Database Architecture

4.6 How is such a schema created?

The design of TMN schemata would be a part of structured TMN design methodologies. However, no such methodologies have been published. Section 4.5 indicated an idea on how a global conceptual schema design could be started by referencing a number of generic or standardised management information models. The joint case study experience in schema design has been limited to a small number of management activities in the maintenance, configuration (provisioning) and performance management. None of these functional areas have been completely covered in this process. Nevertheless the approach taken in making a paper-based prototype solution for even a limited set of problems was considered to be a good starting point for schema design.

Rapid prototyping is known to be a non-trivial assistance to clarifying requirements [10]. It is argued that this is also true for the global schema, although it is inevitable that it would be pointed out that the global conceptual schema should be derived from requirements statements approved by the individuals responsible for acceptance of the TMN solution (the traditional approach). This is accepted but the RACE-TMN work has indicated that the conceptual schemata play a role in the formulation of such statements.

Experience in RACE projects has shown that a step-wise refinement approach would be the most likely route to a high quality schema. By such an approach is meant that progressively more complete prototypes of the schema are developed, reviewed and enhanced until an approved schema version is designed. An eclectic approach is recommended to choosing what background material, e.g. generic network model, is taken into account and when. The typical case will be that some management information items extracted from external sources such as ETSI will be included in the schema as mandatory content. Declaration of the parts that will be mandatory cannot be done at the outset however.

It has been suggested [4] that the starting point for the TMN design, and hence the TMN global schema design, is the specifications of the nine functional areas defined by RACE Project NETMAN [5]. These management areas cover *Design, Planning, Installation, Provisioning, Maintenance, Performance, Security, Accounting and Customer Control*. A complementary starting point is the definition of the services and network infrastructure to be managed. The ITF joint case study contains substantial material on such definitions [2]. As the design progresses specific requirements concerning the TMN to be realised become important. This specific information would come from the main stakeholders in the TMN. Figure 6 defines the process at an abstract level.

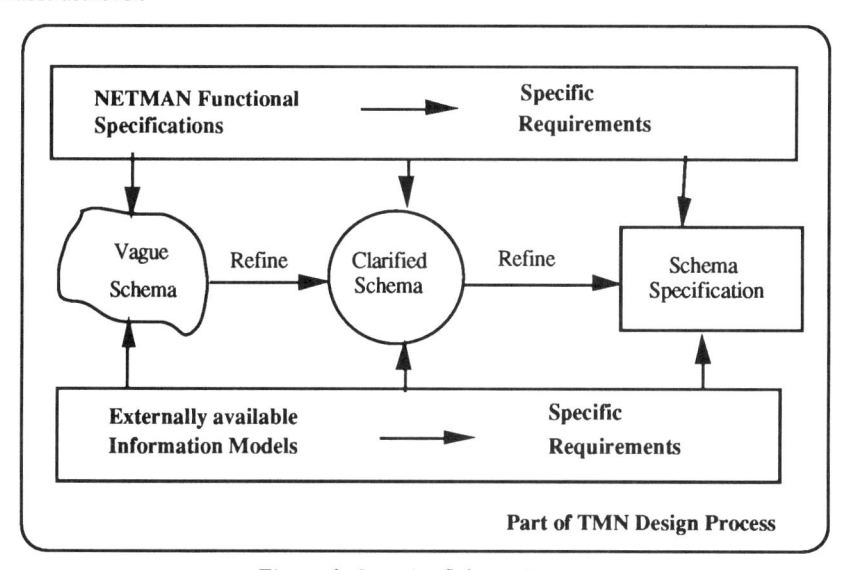

Figure 6 - Iterative Schema Design

4.7 Generation of Database Schemata

Due to the novelty in the TMN research area of the notion of database schemata design, this paper now describes how the generation of database schemata for a relational system can be conducted from the global conceptual schema for a TMN. In generating a global conceptual schema for distributed database systems there are essentially two approaches which can be taken.

The first of these approaches may be termed a bottom-up approach where a number of sub-schemata already exist and an attempt is made to unify them into one global schema. This approach relates to the development of federated and export schemata for federated databases.

The second approach which is more appropriate to the TMN design in RACE may be termed a top-down approach. This approach is to be used where an existing schema is not available and a schema has to be made from other available information such as an object model of the domain, the managed domain in the case of a TMN. This is illustrated in figure 7 which shows a mapping from object models including a global model such as the TMN global conceptual schema to the three schema architecture for distributed databases.

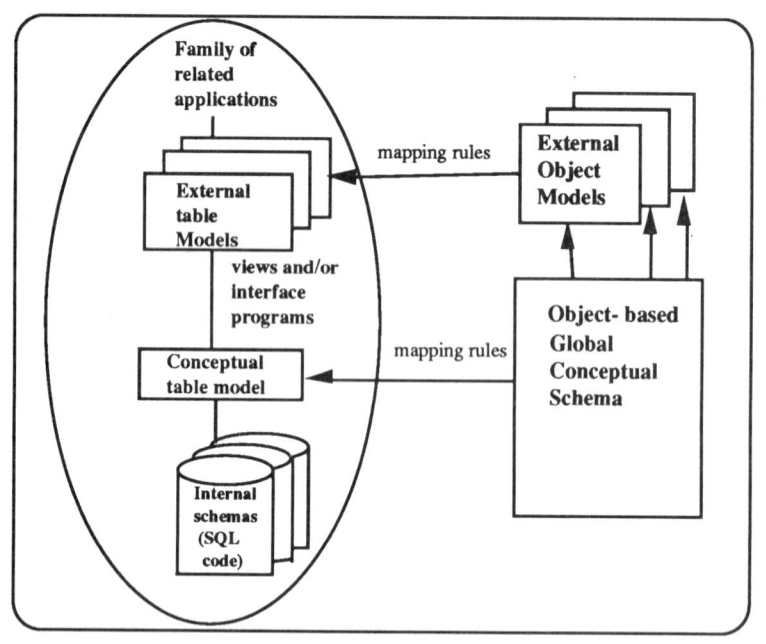

Figure 7 - Object Based GCS for TMN and the Three Schema Architecture for RDBMS

Within this top-down approach there are two options. Firstly, one can design the conceptual object model based on assumptions of what the external object models could be. In the case of the joint case study this involved the design of a conceptual object model based on what we assume the object models of provisioning, performance management and maintenance to be. The second option starts with the external object models and then infers a conceptual object model from these. This was the option that the joint case study has taken, i.e. attempting to build one object model based on the models for provisioning, performance management and maintenance. In both options the object models are then mapped onto the architecture for **RDBMS** (Relational Distributed DataBase Management Systems) resulting in the required schemata. For further details on how this mapping is done the reader may refer to [6].

4.8 How is the Global Conceptual Schema used?

According to the GUIDELINE architecture [3], two components of the TMN design are derived from the global schema, namely OS to OS and OS-NE interfaces specifications and local conceptual schemata for particular database types (after decisions on which solutions such as relational or object-oriented databases has been made). Local schemata are also referred to as data models [7]. The process of completing a relational database design from a conceptual schema has been described in the database literature [8]. CCITT and ETSI, among others, have defined methodologies for defining OS-NE interfaces. However, the authors have not found reference material which combines database design and system level interfaces design in one methodology. The GUIDELINE TMN Architecture can be seen as a first step towards the basis of such a methodology.

As the TMN design matures, the issue of having copies of the same data accessible both in OS resident databases and across OS-NE (and inter TMN) interfaces has to be addressed. The role of the global schema in recording such decisions is still undefined.

Finally the schema would be an essential part of the name-binding process which would be a natural part of any TMN design. The objective of name binding is to ensure that each entity of concern to the management task has a name which is clearly distinguished from other entities' names and that a user-oriented naming convention is followed.

5. CONCLUSION

This paper has presented very briefly the work performed jointly by four RACE projects in the area of TMN integration. The main content of the paper presented a consensus view of what we mean by a global conceptual schema for a TMN. This would be an almost essential ingredient in achieving highly integrated TMNs which cover a number of functional areas.

TMN implementation architecture [3] brought distributed data base architecture within its scope but left some key data management issues to be worked on further. This requirement coincided with the needs of TMN integration studies among RACE projects to be more focused on particular issues, hence the global conceptual schema was chosen for deeper analysis and consensus building among TMN projects. It has been clarified that considerable new requirements on such a schema exist due to the prospect that data will not only be stored in, and retrieved from, database systems in a TMN but will also be available in network elements. Large volumes of management data will also reside in the non-database function blocks of switching systems. This situation requires a revised view of the traditional database schemata architecture. This presented the need for a schema definition language which can support a variety of database types (relational, etc.) and the information structures promoted for use for management data definition [9].

It can also be concluded that the global conceptual schema design method will have to be a central part of formalised TMN design methodologies which can be expected to emerge over the coming decade.

6. ACKNOWLEDGEMENT:

This work reported in this paper has been partially funded by the CEC RACE programme projects GUIDELINE, AIM, ADVANCE and NEMESYS. The authors thank members of the ITF and colleagues in BT plc and Broadcom for their constructive comments and criticism.

7. REFERENCES:

[1] Callaghan, J. G. et al. : "TMN Architecture" Sixth RACE TMN Conference, Madeira, September 1992.

[2] RACE project GUIDELINE Deliverable ME 7 : "Initial Report on the ITF Common Case Study", 03/BCM/ITF/DS/C/011/b1, February 1992

[3] RACE project GUIDELINE Deliverable ME 8 : "TMN Implementation Architecture", 03/DOW/SAR/DS/B/011/b3, March 1992

[4] Azarmi, N. & Turner, T. "A Proposal for an Integration Methodology for a TMN, Proceedings of 5th RACE TMN Conference, London, 1991.

[5] RACE project NETMAN Deliverable 10, 24/BCM/RD2/A/DS/010/b1, September 1991.

[6] Rumbaugh, J. et al : "Object-Oriented Modelling and Design" Prentice Hall International, 1991

[7] Gardarin, G. & Valduriez, P. : "Relational Databases and Knowledge Bases", Addison Wesley, 1989.

[8] Ceri, S & Pelagatti, G. : "Distributed Databases Principles and Systems", Mc Graw-Hill International Editions, 1985.

[9] CCITT recommendation X.722, Guidelines for the Definition of Managed Objects, ISO/IEC JTC1 SC21 WG4

[10] Hekmatpour, S. & Ince, D. : "Software Prototyping, Formal Methods and VDM", Addison Wesley Publishing Company, 1988

III - Modelling Aspects of TMN

The Management of Telecommunications Networks
R. Smith, E. H. Mamdani, J. G. Callaghan (Editors)
© Ellis Horwood 1992

Object Oriented Modelling in RACE TMN

Manoochehr Azmoodeh (BT Laboratories, UK), Colin Enstone (GEC-Marconi UK)

ABSTRACT

A major area of concern in the RACE programme is information modelling to satisfy the requirements of complex communications management functions and to provide a basis for open interoperable management systems. These management systems need to integrate all functional areas across all phases of the life cycle from pre-service (planning, installation) to in-service (maintenance, configuration) to future service (forecasting). The standards bodies notions of managed objects and managed information bases are found to be too limiting in this context. Based on the RACE experience, this paper elucidates some of the requirements for information modelling in the TMN context and draws upon standards and RACE TMN work to shed light on future directions in the area.

1. INTRODUCTION

Modelling is used in many areas of Communication Management (CM) where 'descriptions' of systems have to be given in order to facilitate reasoning about these systems. Modelling is used to define abstract descriptions of communication systems for many purposes, amongst which are analysis, design, implementation, specification of software systems, descriptions of information structures and definition of standardised reference points/interfaces.

Application of object orientation in modelling activities is also investigated in diverse areas of software systems of CM. For example, in the area of modelling of data structures for large persistent data and knowledge bases, object orientation is used to capture structural semantics which cannot be captured adequately using the relational data model. Programming languages emphasise structuring and reuse of code through object orientation. Distributed processing systems require an API (application programming interface) independent of operating systems. Objects have been used in these areas to define API as well as interoperable interfaces for distributed systems. ISO ODP and also the Object Management Group (OMG) are endeavouring to define an open platform and an 'object API'.

The TMN has introduced the notion of managed objects and Management Information Base (MIB) as the repository of all management information. Standards bodies are active in the area of defining object modelling guidelines and libraries of managed object classes for defining abstractions of networks and services to be managed and also to some extent libraries of resources within a TMN such as event sieve and event log. These managed objects are defined primarily to specify interoperable interfaces and to cater for TMN functions such as event management and simple configuration management. These models do not meet the requirements of more complex TMN functions such as fault management, traffic and quality of service management.

The RACE TMN work has taken a much broader view of the use of the object paradigm than existing standards bodies. The RACE view spans, among others, technology, implementation, architectures, specification, design and analysis of CM systems. The standards notions of the MIB do not adequately portray the requirements of TMN functions in all stages of the lifecycle of a TMN. As there is currently no proven integrated framework where all these diverse modelling activities can be described, some of the requirements for object modelling in the context of TMN will be elucidated and based on experiences of initiatives such as ISO-ODP and the OMG, an attempt is made to shed light on the future directions in this area.

The work described in this paper was carried out under the Object Modelling Special Interest Group of the RACE GUIDELINE project [1].

2. OBJECT ORIENTED MODELLING IN TMN - OVERVIEW

In automated communication management systems, modelling has been used to tackle at least the areas in table 1.

Field of Study	Modelling Purpose
Specification	Describe standard definitions prior to implementations (implementation independent)
Analysis	Describe requirements for a problem domain
Design	Describe a solution to tackle a problem domain
Database modelling	Describe data structures for flexible access to persistent data (query languages; separation of external, conceptual and internal descriptions; deductive and constraints capability; concurrency; integrity; recovery)
KBS	Describe knowledge structures for intelligent applications (ontological and epistemological)
Programming Language	Describe structures and behaviour for implementing a system (defining computation)
Interoperable interfaces and API	Describe the interactions between systems. Define protocols for distributed systems interactions
Simulation	Behaviour modelling of TMN systems for diverse applications such as forecasting, provisioning, etc

TABLE 1 - Application areas of modelling in TMN

A description or abstract model in any of the above categories has three essential orthogonal components, namely structure, function and dynamics [2] (see figure 1) which are integrated into the notion of an object.

- Structure : This refers to a description of what there is, denoting existence of things of interest. The structure of a model describes aspects such as aggregation, association and generalisations of objects of interest in a given domain. Existing formalisms include relational model, entity-relationship model, semantic networks, semantic data models and data structures of programming languages.

- Function : Describing what is being done. For instance, VDM [3], Z [4] and programming languages are among this category.

- Dynamics : Describing when things happen. This is an expression of the responses of a system to events in terms of actions and states and the propagation of new events. Formalisms include state transition diagrams, finite state machines, statecharts [5], message sequencing charts, petri nets, temporal logic, modal logic [6], CSP [7].

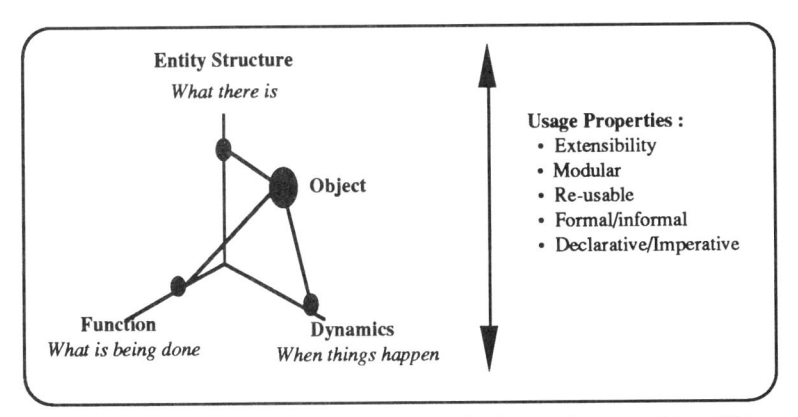

Figure 1 - The Three Components of a Model of a System Integrated into Objects

Application areas with specific focus and aims concentrate on one or more of the above orthogonal components to various degrees. However, more complex applications have led to increased expectations and hence the need for incorporating all three components expounded above and the resulting adoption of the object paradigm.

For example, in conventional database modelling, only simple structures are employed to organise the logical and physical aspects of large bodies of data. In many areas, such as CAD/CAM and network management, conventional relational models of data have proved inadequate [8] and the adoption of object orientation is evident through the advent of OODBMSs. Most conventional programming languages employ simple structuring mechanisms such as arrays, records, etc and concentrate heavily on function decomposition and description. Influences and benefits of object orientation in this area is evident from the plethora of object oriented programming languages now in use. These languages use objects as the unit of computation and reuse is made available via various inheritance algorithms. Object oriented design is being promoted [9] to capture all three components of modeling, albeit with little degree of formality. Object oriented analysis techniques are emerging [9], [10]. In the specification area the object oriented paradigm appears using objects to represent observable processes in system models. In the interoperable interfaces and the API area, the managed objects of TMN and OMG's API are evidence of application of the object oriented paradigm.

3. MODELLING REQUIREMENTS FOR MANAGEMENT APPLICATIONS

In this section, some of the major requirements of management applications (MA) with respect to object oriented modelling are described. Management applications are the automated functions of a TMN which need models of the managed network and services as well as models of other TMN functions in order to carry out their tasks. The TMN modelling requirements primarily fall into six categories :

Category 1 : Modelling the structure and behaviour of networks and services to be managed.

The models in this category capture structure and simple dynamics. Structural aspects cover associations and aggregations amongst objects [2]. Simple dynamic behaviour is defined in terms of legal sequences of events received and emitted by an object, and propagation of the effects of incoming events. The declarative descriptions of part of the dynamics of these objects are required in order to be used, for example, by the knowledge-based system components of diagnostic applications [11].The function provided by objects in these models are minimal and are mostly restricted to translation of the operations, alarms and other notifications from/to the real resources.

To enable adequate representation of many generic components (resources) of a network, an effective and flexible set of modelling concepts are required to accommodate the complex structures of networks' connectivities of different types (for instance, containment hierarchies, functional dependencies, powering dependencies, etc). The notions of types and type hierarchies are essential to capture the specifications of these generic components. The concept of constraints in first order logic is used to specify these structural types formally [11], [1] (see example in section 4). As demonstrated by the RACE AIM project, a simple formalism such as KBS rules can be used to specify the functions (operations) of these objects [12].

Category 2 : Modelling the structure, function and behaviour of components comprising a management system of a network.

In this category, the emphasis of modelling is more on the specification of functional aspects, interface definitions of objects and the dynamics of these interfaces rather than on structural aspects. These are required firstly to enable design and implementation of management modules, secondly, to define interoperable interfaces of each module and thirdly, formal descriptions of dynamics are needed for formal reasoning. The behavioural specification of these objects should be represented in a formal language such as temporal logic.

Category 3 : Programming environments (languages)

In the implementation phase of management applications, it is required to minimise the semantic gap between the specification formalisms and the implementation environment formalisms (languages). The programming language should support objects, classes, incremental inheritance and procedural attachment to allow relatively simple implementation of structure, function and dynamics of objects [13].

Category 4 : Access to models in categories 1 and 2

Managed objects form a basis for defining two types of interfaces. Objects are used to define Inter-Operable Interfaces among the open sub-systems constituting a TMN and for defining Application Programming Interfaces (API). These interfaces need to be defined embracing the object oriented paradigm independently of programming language used, operating systems, computers, etc.

A pragmatic issue in the use of objects is the ability of applications to access object structures. In particular reference to API within a TMN environment, there are strong requirements for applications to have advanced retrieval capabilities [1] for objects in the MIB. These applications dictate a move away from object-at-a-time mode of access and towards using a more powerful language where properties of many objects as well as invocation of many operations can be achieved through a single high level command. Other pragmatic issues such as operational requirements (speed, communication capacities, etc.) mandate that a query language be embedded into the API as well as perhaps the inter-operable interface protocols for TMN components.

As part of the RACE ADVANCE project, the language OBSIL [13] is being developed with the above requirements in mind. The proposed language is a simple, concise and powerful language which is specifically defined to meet the TMN requirements. OBSIL provides mechanisms for expressing complex qualifications on objects using the concept of 'path expressions' and quantifications such as 'for all', 'there exists' and 'at most n'.

Category 5 : Construction, Maintenance and Administration of MIB

The MIB provides the repository of objects in categories 1 and 2 above. In the life cycle of a management system, the MIB goes through a process starting with initial construction and followed by evolution stages. The RACE TMN projects have demonstrated that the use of object types and classes and hierarchies together with the use of a language of constraints [11] provides a suitable basis to define specialisations and variants of generic component types in a network. These type descriptions are then used to generate (construct) object models for new managed network and service components.

MIB evolution falls into two categories: firstly, object changes such as addition/deletion of objects and update of existing objects and secondly class hierarchy changes such as adding new classes and modifying properties of an existing class. The evolution of the first type can be controlled with the use of constraints. Upon insertion and deletion of objects, constraints are used to check for consistency of object structures.Other necessary requirements with respect to organising and managing an MIB need to be addressed including support for views, authorisation and authentication, support of transaction security, support for persistence, distribution and federation of object bases (to solve the 'legacy' problem [14]) and multi-user access.

Since an MIB is defined as a set of objects, to minimise the complex transformations between the object oriented paradigm and conventional implementation paradigms (such as relational technology, etc), persistent object implementations are necessary. There are a number of Object Oriented DBMS (OODBMS) emerging in the market place. However, no formal definition of OODBMS system exists and there are a lot of differences between systems described as such. An interim consensus [15] exists.

Category 6 : Analysis and design of management applications and methodologies

Descriptions of the problem domain and requirements (analysis) and descriptions of solutions (design) are needed. The analysis process makes informal or incomplete problem statements more precise and exposes ambiguities and inconsistencies. Context, assumptions and performance needs are elucidated in the analysis stage. The design phase describes solutions by emphasising decomposition of systems into subsystems, concurrency features inherent in the problem, allocation of subsystems to processors, management of data stores, implementation of subsystem control and trade-off priorities. Object oriented analysis and design methodologies are emerging to aid these two stages in modelling for CM. There are a few methodologies in use [9][2][16][10].

4. BASIC OBJECT MODELLING CONCEPTS

In this section, a basic description of object oriented modelling is given. This is based primarily on the ODP modelling and specificational concepts [17] and also on the language TROLL (Textual Representation Object Logic Language) [18]. The main reason for using this approach is that these modelling concepts are independent of particular applications, systems, implementations and programming systems and allow for specifying active/passive, static/dynamic, autonomous objects, etc. These ideas are used to specify the EventReportManager object of the TMN. Furthermore, the fundamental distinction between object types and classes are highlighted. In the following, the bold and italics terms refer to ODP and Troll terminology respectively.

4.1 BASIC MODELLING AND SPECIFICATIONAL CONCEPTS

An **object** is a model of an entity. An object is characterised by its behaviour (this refers to function and dynamics defined earlier). The behaviour of objects can be observed through their **attributes**. An object is encapsulated in that it protects its integrity and decides what interactions it takes part in and when. Objects perform functions (or offer services).The values of attributes are instances of a particular *abstract data type* (**data elements**).

The behaviour of an object is defined as the set of all sequences of **actions** which an object can perform (or *events* it can take part in) subject to dependencies that are defined to exist between actions. Specification of behaviour constrains the sequences of actions which may or may not occur. The **state** of an object is the condition of that object which determines its behaviour. A **static interface** is a specified subset of the observable action sorts of a given object. **Abstract interface** is the behaviour observable at a given static interface.

An object **type** is a predicate which evaluates to true for an object of that type. Thus, types define the 'intension' of an object class by defining all possible objects of that class. This usually refers to a formal specification of the common features of a set of objects. A type is defined using a template (textual description of the common features of objects) together with definition of identification of its objects. A type is **subtype** of another type if all objects which satisfy the first type also satisfy a second type.

A **class** is the set of all objects satisfying a type. The elements of the set are called its members. A class is a time variant entity. The **subclass/superclass** relation corresponds to set/subset relation. A **derived class** (**parent class**) template is derived by incrementally modifying an existing class template. The latter class is the parent class.

The distinction between types and classes is a fundamental issue. Types refer to the formal specification of a collection of objects whereas a class is a collection of objects in their physical representation.

Incremental inheritance (*syntactic inheritance*) is the derivation of a new class template by incrementally modifying an existing class template. If a template has more than one immediate parent template, the inheritance is called multiple. **Inclusion polymorphism** is the property that an instance of a class can behave as an instance of a second class.

A combination of two or more objects yields a composite new object. The characteristics of the new object are determined by the objects being combined as well as how they are combined. An example in the TMN area is of sub-components of a 'maintenance manager'. Figure 2 shows its decompositions into instances of types `event manager', 'fault localiser', 'test manager', 'repair supervision' and 'repair scheduler'.

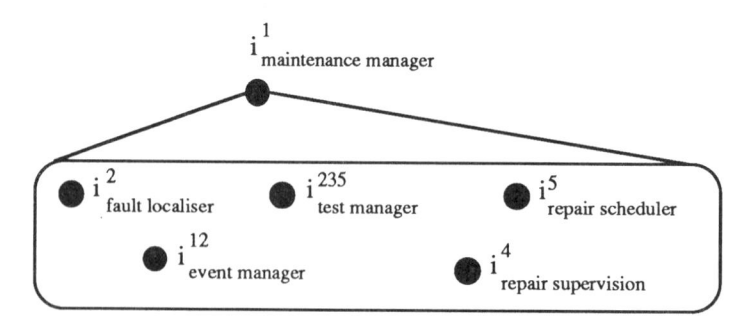

Figure 2 - The Decomposition of the Object `maintenance manager'

4.2 Case study : Object Oriented Specification of EventReportManager

In the following, the language TROLL [18] is used to produce a template which describes an EventReportManager object.. The definition given is not complete. A complete definition can be found in [1]. The template defines the signature (interface) and the structure and behaviour of the object.

A typical TROLL template contains :

    template [template name]
        data types   import of data type signatures
        attributes   attribute name and type declarations
        events   event names and parameters declarations
        constraints   static and dynamic constraints on attribute values
        derivation   rules for derivation of attributes and events
        valuation   effects of events on attributes
        behaviour
            permissions enabling conditions for event occurrences
            obligations completeness requirements for life cycles
            commitments state dependent short term goals
    end template [template name]

The event report manager is responsible for storing and managing event (alarm) reports to be used by other maintenance objects (fault localiser, test manager, repair supervision and repair scheduler). Three queues are maintained by the event report manager : events received but not yet sent to fault localiser, events sent to event localiser, and cleared events. When a clear is received for an event that has not been sent for fault localisation, the event is removed from the first queue and added to the cleared event queue. When a clear is received for an event that has been passed to the fault localisation object the event will be added to the cleared events queue and a notification issued. Periodically un-necessary event reports are removed. It is deemed that only historical event reports not relevant to the operation of maintenance objects will be erased.

In the following partial specification, it is assumed that class EVENT is defined with attributes such as date, state, resource, etc. Thus, the abstract data type (ADT) |EVENT| := **tuple**(timestamp: date, state: op_state, resource : |MO|). MO is the class of *managed objects* abstracting the resources in a network to be maintained. Also, the ADT date is used with usual operations (later(d1,d2), etc). The following notes refer to numbered sections of the specifications :

(1) The lists new_events, event_in_fault_loc and event_cleared are disjoint
(2) The clear_notification event is generated by the instances of this class
(3) All the new events are sent to fault localisation
(4) After an erase event, all events with a timestamp before d are deleted from lists.

Object class EventReportManager
    identification
        data types nat;
            ERM-No : nat;
template
    data types |EVENT|, date;
    attributes
        new_events : **list**(|EVENT|);
        event_in_fault_loc : **list**(|EVENT|);
        event_cleared : list(|EVENT|);
    constraints
        forall e:|EVENT|
            (e in new_events) **xor** (e in event_in_fault_loc) **xor** (e in event_cleared);   (1)
    events
        birth create;

           death kill;
           new_event(**in** e : |EVENT|);
           clear_event(**in** e : |EVENT|);
           send_event_to_fault_loc(**in** S : set(|EVENT|);
           active clear_notification(**in** e:|EVENT|);   (2)
           erase_events(d : date);
       valuation
           variables e:|EVENT|, d :date, S:set(|EVENT|);

           {e.state <> clear} [new_event(e)] new_events = append(e, new_events);
           {e.state = clear **and** (e in new_events)} [new_event(e)]
              new_events=remove(e, new_events), event_cleared = append(e,event_cleared);
           {e.state = clear **and** (e in event_in_fault_loc)} [new_event(e)]
              event_in_fault_loc = remove(e, event_in_fault_loc),
              event_cleared = append(e, event_cleared);
           {not empty(S) **and forall** e:|EVENT| (e in new_events) => (e in S)}
              [send_event_to_fault_loc(S)] new_events = clear(new_events),
              event_in_fault_loc = append_set_to_list(S, event_in_fault_loc);   (3)
           [erase_events(d)] **forall** e:|EVENT| later(d, e.time_stamp) =>   (4)
              new_events = remove(e, new_events) **and**
              event_in_fault_loc = remove(e, event_in_fault_loc) **and**
              event_cleared = remove(e, event_cleared);
       behaviour
           permissions
              variables e:|EVENT|, d :date;
              {e.state = clear} clear_notification(e);
           commitments
              {sometimes after(new_event(e)) and e.state = clear and (e in event_in_fault_loc)}
              =>
              clear_notification(e);
end class EventReportManager

5. REVIEW OF THE WORK OF STANDARDS AND NORMATIVE BODIES

There are many initiatives either in standards, or influencing standards in the area of modelling for TMN. These span ISO, ODP, CCITT, ETSI, NMF, OSF Distributed Management Environment (DME), TINA (Telecommunication Information Networking Architecture), Object Management Group (OMG) as well as standardisation activities for languages such as C++, CLOS and SQL to formal description languages such as CSP, CCS, VDM, Z, OOZ and LOTOS. In this section we shall give a brief account of major influences in this area and an evaluation of their recommendations.

5.1 Standards for the use of Objects in TMN Interface Definitions

OSI-CCITT

There are a number of joint ISO and CCITT SG VII standards or recommendations which refer to managed objects. The "Systems Management Overview" (ISO 10040, X.701) provides an introduction and overview of the family of systems management standards. The "Management Information Model" (ISO 10165-1, X.720) describes the model of management information in terms of managed objects, operations and notifications. Concepts of inheritance, allomorphism, containment, naming etc are defined. "Definition of management Information" (ISO 10165-2 X.721) defines generic managed object classes together with their notifications, operations etc.

The "Guidelines for the Definition of Managed Objects" (ISO 10165-4, X-722) provides guidance on how to produce managed object definitions and on how to describe protocol-specific aspects of managed objects. This recommendation gives a template for defining managed object classes and their properties, name bindings etc. ASN.1 is used to describe the syntax of the template. A major shortcoming is that the behaviour of managed objects is specified as text. there is no recommendation for the formal definition of the behaviour of managed objects.

CCITT SGIV have produced a number of draft recommendations in the M.3000 series. "Principles for a TMN" (M.3010) adopts OSI management concepts but the role of managed objects is at present almost entirely confined to the Operations System to network element interface. The network element contains management communication functions, agent processes and managed objects, in addition to the network resources. In the "TMN Interface Specification Methodology" (M.3020) management activities are divided into TMN management services, TMN management service components and finally TMN management functions. The managed resources are modelled by managed objects which are linked by an interface to the TMN management functions. The "Generic Information Model" is contained in M.3100.

NMF - Network Management Forum

NMF have defined a number of classes that describe the managed objects that can be accessed across an interoperable interface. The definitions concentrate on the syntax of an object rather then describing its behaviour. The classes are chosen to give some separation of concerns in so far as physical and functional resources appear in different classes. However the forum classes are primarily aimed at alarm surveillance and configuration management and in their current form are not adequate for the range of management functionality envisaged in the RACE TMN. In particular the classes do not provide support for defining behaviour, for equipment and function hierarchies and for causal relationships expressing functional dependency relationships.

ETSI

There are a number of ETSI NA4 European Technical Standards or Technical Reports which correspond to the CCITT SG IV recommendations referred to above. ETSI are producing object models for customer administration and for routing management which do not yet have CCITT equivalents. Work on Traffic Management is due to start shortly.

5.2 OMG - Object Management Group

The Object management Group (OMG) was formed in 1989 by a consortium of software and hardware manufacturers, database vendors and other interested parties. The goal of OMG is to promote the use of object technology through maximising portability and interoperability of software regardless of the computer type or operating systems in which objects are stored. One of its objectives is to define a 'language independent object model' to be used for OMG-compliant components. The Common Object Request Broker Architecture (CORBA) [19] is defined to specify objects and object communications. Products which are CORBA-compliant are already emerging. The definitions of the modelling concepts are not yet concrete and still include some ambiguities. No distinction between types and classes of objects are made. Currently OMG does not advocate formal descriptions for modelling.

5.3 ODP - Contribution to modelling

The basic reference model of Open Distributed Processing (ODP) defines descriptive models for describing distributed systems at different levels of concern (viewpoints of ODP). The descriptive model embodies the notions of viewpoints, aspects, modelling concepts and specificational concepts. The latter two areas are of particular relevance to our modelling activities. A brief resume of the fundamental concepts was described in section 4. The ODP prescriptive model (ISO-ODP, Part 3) defines a set of languages for writing specifications of ODP functions for each viewpoint. These languages typically use a subset of terms and concepts used in the descriptive model. Part 5 of the ODP basic reference model defines a set of Formal Description Techniques as possible languages for defining formal semantics of objects. Z, SDL, LOTOS and Estelle are considered in the latest draft.

5.4 Database standards

Some of the main standards in the database area are : ANSI-SPARC three layer DBMS architecture, SQL standards and Draft Remote Database Access (RDA) by ISO/ANSI for heterogeneous DB access.

RDA is based on a client-server model and uses the well defined OSI services. The SQL Access Group formed in 1989 expedite definition of RDA and SQL and validate standards through prototype implementations. The SQL Access Group have implemented a prototype and demonstrated interoperability across OSI stacks and are working on defining RDA services for TCP/IP transport. The use of RDA does not solve the problems outlined in section 3. In RDA, the client processes needs to know the exact type of the server database before making requests. Consequently, the burden of translating to many different database query languages is delegated to client processes. The OBSIL solution provides a single language for all database accesses regardless of database type, etc.

In the OODBMS arena, there are still no normative recommendations. Despite many products, the applications pull has not been strong enough to warrant standardisation.

6. SUMMARY AND CONCLUSIONS

The importance of modelling and justifications for object modelling where structure and behaviour are seamlessly integrated have been explored in the TMN context. The notion of modelling is used in a very broad sense in order to discover the commonalities in modelling concepts as well as identifying the criteria for their realisations. Some of the major TMN requirements for modelling are identified : declarative specification of TMN models (functions, information, etc) to provide clear, complete and unambiguous descriptions; need for advanced databases for storage, manipulations and processing (deductive capability, constraints management) of static objects of a model; the need for a query language to form part of the API as well as the interoperable interface; provision of a platform to support object interactions and distributions; and standardised network, service and TMN models.

It is likely that different modelling activities in the TMN will require different specification languages with various degree of formality :

> *Structure* : objects, types, classes, attributes (relationships), constraints (in first order logic), deductive capability
>
> *Functions and dynamics* : VDM, Z, state machines, temporal logic (Troll, etc) , Harel's statecharts, CCS, CSP, LOTOS, object oriented programming language.

There are now increasing applications of the object oriented paradigm in diverse fields (OODBMS, object oriented programming, KBS, object oriented-API, platforms, etc). Integrated systems such as the TMNs of the future will firstly require the unification and coexistence of such diverse variations of the object oriented paradigm. Secondly, methodologies and formal systems are needed to organise distribution of models and applications of a TMN. In the light of these issues, appropriate dialogues are required between OSI/CCITT and initiatives such as ODP, OMG, etc to reach an agreed level of understanding.

7. ACKNOWLEDGEMENT

The authors wish to thank members of the OM-SIG -Mark Newstead, Bernd Stahl, Paul Stern, Rob Davison and Nader Azarmi and other RACE TMN community colleagues for many valuable discussions.

8. REFERENCES

[1] Azmoodeh , M. : "Object Modelling Special Interest Group Report", GUIDELINE project, 1992.

[2] Hodgson, Ralph "Subject-Orientation - "A context for Object Oriented Design", in Exploiting Object Oriented Technologies, UNICOM Seminar, 26-27 June 1991.

[3] Jones, C. : "Software Development: A Rigourous Approach" Prentice Hall International 1980

[4] Spivey, J. : "Introducing Z: A Specification Language and its Formal Semantics" Cambridge University Press 1988

[5] Harel, David "On Visual Formalisms", CACM, May 1988, Vol 31, No 5

[6] Jeremaes, P, et al : " A modal (action) logic for requirement specifications", Proc of Software Engineering 86 Conference, pp278-294, IEE, 1986.

[7] Hoare, C.A.R. : "Communicating Sequential Processes", Prentice Hall International, 1985.

[8] Manola, F. : "An Evaluation of Object Oriented DBMS Developments", research report TR-0066-10-89-165, October 1989, GTE Laboratories.

[9] Rumbaugh, et al "Object Oriented Modelling and Design," Prentice Hall International, 1991.

[10] Coad, P., Yourdon, E. : "Object Oriented Analysis", Yourdon Press, 1990.

[11] Manning, K., Spencer, D. : "Model Based Management", 4th RACE TMN conference, Dublin, 1990.

[12] Azmoodeh, M. : "Generic Representation of Structure and Behaviour of Networks", TMN 4 Conference, Dublin, Nov 1990.

[13] Azmoodeh, M., Shomali, R. : "OBSIL : An object based system interaction language" in the RACE-TMN Object Oriented Modelling SIG proceedings, June 1991.

[14] M Brodie, Invited Presentation, TINA conference , Japan, 1992.

[15] Atkinson M, et al "The Object Oriented Database System manifesto", Technical report Altair 30-89, 21 August 1989. Also, in proceedings of DOODS, Dec 1989, Kyoto, Japan.

[16] Booch, G. : "Object Oriented Design with applications", Benjamin Cummings, 1991.

[17] Recommendation X.9yy : Basic Reference Model of Open Distributed Processing - Part 2: Descriptive Model, Editor Draft.

[18] Gunter Saake, Ralf Jungclaus and Hans-Dieter Ehrich "Object Oriented Specification and Stepwise Refinement", ODP Conference, Berlin, October 1991.

[19] The Common Object Request Broker: Architecture and Specification, draft 26, Aug. 1991, Ref. OMG Doc 91.8.1

The Management of Telecommunications Networks
R. Smith, E. H. Mamdani, J. G. Callaghan (Editors)
© Ellis Horwood 1992

Service and Network Model Implementation

Kevin Manning, Ed Sparks (Roke Manor Research, UK)

ABSTRACT

The current trend in telecommunications network management is to use object-oriented techniques to specify and implement integrated network management systems with open interoperable interfaces. A number of languages exist to describe such objects. ISO has defined the object description language GDMO (Guide-lines for the Definition of Managed Objects). However, GDMO currently only gives a formal account of data, behaviour is described by natural language descriptions that may be incomplete, inaccurate and ambiguous. This paper describes MODEL an object description language that is capable of specifying the structural and behavioural semantics of managed objects. It discusses the implementation of service and network models that supported a number of Management Applications in the NCAS (Network and Customer Administration Systems) domain. Experimentation has demonstrated the descriptive power of MODEL and that object-oriented deductive databases could play a central role in the provision of future open and integrated network management systems.

1. INTRODUCTION

RACE project ADVANCE is producing a series of recommendations on the applicability of Advanced Information Processing techniques for the specification and design of Network and Customer Administration Systems (NCAS) that forms an integral part of a TMN. Central to the recommendations is an NCAS Architecture [1], [2] that provides a framework for the design and implementation of an integrated management system. The present phase of work has focused on the investigation of this architecture through the prototyping of an example NCAS system [3]. The results of this work are reported in a number of papers on different aspects of this prototype.

This paper describes the implementation of the service and network models that support the prototype demonstrators together with the additional objects, defined by management application developers, to allow information to be passed between the applications. A number of languages exist to describe managed objects. ISO has defined the object description language GDMO (Guide-lines for the Definition of Managed Objects). However, GDMO currently only gives a formal account of data, behaviour is described by natural language descriptions that may be incomplete, inaccurate and ambiguous. This paper describes the logic based object modelling language (MODEL - Managed Object Definition Language) used to implement the models. This language extends the modelling language reported in [4]. It is capable of expressing the behavioural semantics in addition to the structural semantics of managed objects.

2. MODEL BASED MANAGEMENT

Management applications (MA) [5] cover a broad range of management activities. Planning, and Provisioning of both services and networks; customer query and complaint handling. Each of these heterogeneous MAs requires knowledge of the managed domain: networks, services, and customers. The view of the managed domain for MAs can be quite different from different MAs. The different views provided by each of the MAs must be mutually consistent. One way of ensuring this mutual consistency is to incorporate the different views into a common shared representation of the managed domain. The consistency of this shared knowledge can be ensured by incorporating mechanisms that enforce correctness of the representation. This consistency can be further ensured by providing a query interface which forces MAs to communicate explicitly with the common information through the use of a single common query language. Much of the complexity of the managed domain can be hidden from MAs by the use of appropriate abstractions.

In ADVANCE the shared knowledge is represented using an information model, a specification of the rules according to which data is structured and manipulated. The technique used to achieve this is object-orientation. This shared knowledge is known as the Common Information Model (CIM). The concept of model based management describes a scenario, in which a conceptual model supports MAs in performing their tasks. This support includes mechanisms for manipulating the real resources – by manipulating the CIM. Consistency and correctness of each operation are ensured by mechanisms in the model so that only correct, consistent operations are applied to the real resources. In addition alarms, or other events occurring in the real resources, update the model, which may in turn notify MAs of the change in circumstances. Thus the CIM is kept up to date and mimics the behaviour of managed resources. In this way operations are applied to a model which represents the real resources.

3. MANAGED OBJECT BEHAVIOUR

An object is a collection of data together with the operations that are used to manipulate the data. In a fully encapsulated scheme an operation invoked within an object may interrogate or change data within the same object or invoke operations either internally or on other objects. This ability to write operations which cause changes in the state of a number of objects is named behaviour. Behaviour refers to state transitions and dynamic properties (ie. operations and relationships between operations).

A second form of behaviour is displayed by the CIM while carrying out operations requested by the MAs. This behaviour only exists when compound operations are provided by the object classes. The basic operations of reading and writing data define the most primitive behaviour. Compound operations effectively take functionality out of the MAs and place it within the CIM.

A third form of behaviour exists whereby the model has built in rules that ensure that the changes made by the first two forms of behaviour are consistent with rules held within the model. Thus the model could determine when the state of the model no longer reflects the state of the managed resources, and when a task required by a MA is not permitted due to the state of the model.

The models, developed under ADVANCE, exhibit the second and third forms of behaviour, this ensures that management requests are permissible.

4. MANAGED OBJECT DEFINITION LANGUAGE (MODEL)

MODEL is a development of the object definition language reported in [4]. The main enhancement to the language is the inclusion of operations into the class definition that define the behaviour of managed objects. The network and services models are both implemented in MODEL. The language MODEL has been implemented as a meta-interpreter in Prolog and provides a prototype object-oriented deductive database [6], [7], [8], [9]. MODEL defines two kinds of objects: classes and instances. An object class is a template or type for a set of objects (instances) that have similar properties. Every object instance is related to a single object class. All instances of an object class have the same characteristics and conform to the same rules ie. they can store the same kind of information and can perform the same operations. In database terminology the class model is called the schema and the instance model the database.

Classification provides the mechanism to distinguish and define object classes. Classes do not exist in isolation, relationships between classes may be defined. Two special relationships exist, superclass and subclass which defines class generalisations and class specialisations of an object class. Relationships come in three basic types generalisation / specialisation, aggregation, and association. These relationships are used in combination to form a simplified description of something, termed an abstraction. Objects are abstractions of domain entities and concepts, emphasising information that is significant to the user.

4.1 Managed Object Definition

A managed object is the view of a resource or set of resources that is subject to management. A managed object is defined by the attributes visible at its boundary, its relationships with other managed objects, the management operations that can be applied to it, the notifications emitted by it, and the behaviour it exhibits in response to operations and notifications.

The next section presents how these various aspects in the definition of a managed object are represented using an object description language. A full definition and an explanation of the BNF syntax is in [10].

4.2 Class Definition

As stated earlier the model is defined by a class model which consists of a number of object class descriptions which have the following form:

```
Object-Class:      <name>
         superclass    <class sequence>
         [subclass     <class sequence>]
    Description:   "<string>"
    [Types:        <type descriptions>]
    Relations:     <relation descriptions>
    Attributes:    <attribute descriptions>
    Constraints:   <constraint descriptions>
    States:        <state descriptions>
    Operations:    <operation descriptions>
    Notifications: <notification descriptions>
End-Class
```

Structural properties of a class fall into three categories: relations, attributes and constraints. Three types of relation exist : class, primitive and derived. The class relations, superclass (generalisation classes) and subclass (specialisation classes), define inheritance. An object-class inherits properties from all its superclasses, and its own properties are inherited by each of its subclasses. These relations are fixed and identical for all instances of each class. Primitive relations reference other entities in the information model, the value is stored explicitly. The value is a structured term defined by a complex type descriptor. Derived relations depend on the values of other relations and attributes and their value is inferred. They are defined in logic as Horn clauses. Attributes are distinguished from primitive relations by being referred to atomic terms such as integers, reals, strings and booleans. Attributes use the same type definition language as primitive relations. Constraints are defined in first order logic. They define dependencies within and between relations and attributes.

States, operations, and notifications are the dynamic properties of a class. States are named combinations of attributes and/or relations which have specified values. Operations and notifications are types of functions which move an object from an initial state into another state, as such they are said to perform a state transition. Operations and notifications have the same form and are written as a series of logical conditions which must succeed if the operation /

notification is to complete. Types provide a means of defining a complex type for multiple use within an object, this avoids duplication.

4.3 Type Definition

Object classes define a type for object instances. Object instances have the same properties as the object classes except that primitive relations and attributes have an associated value that satisfies the type descriptor defined in the class. Type is not synonymous with class, types are not first class objects. The fact that these concepts are distinguished helps to maintain a clear conceptualisation of the model though this does mean that the language loses some representational power.

Each primitive relation and attribute has a value defined by a complex type. This type specifies completely the structural semantics of the instance of the property. The type definition language has a number of operators that convey the concepts of necessity, optionality, cardinality and naming.

aggregation:	a , b	Objects of type a and b must both be present
optionality:	a l b	Objects of type a or b but not both must be present
	a ! b	Objects of type a or b or both may be present
cardinality:	{<type>}# <cardinality>

Cardinality specifies the bounds on the number in a set of related classes. It can express uniqueness. Cardinality can be a fixed integer, a range specified as an upper and lower bound, a special symbol or a variable identifier. Special symbols are ++, denoting one or more, and * zero or more. Cardinality captures the primitive ideas of necessity and optionality. For example {wheel}#4 represents a set that will contain four wheels; whilst {friend}#* may be empty. An example from the network model would be composition of a network which may have a number of nodes and links (connections and trails are subclasses of link).

Object-Class:	network
Relations:	contains {node}#N , {connection}#C , {trail}#T

Each type defines a value. The values supported by the language are the structured values sequence and set, the simple values integer, real, char, string, boolean and complex structured values that incorporate other values. A BNF description for the definition of a value is given in [10]. The aggregation operator defines a sequence as does the optionality operator when both elements are present.

Sets are defined by the set designator { }. The type of the set is evaluated for each member of the set and not once for all members. For example the set type {male l female}#5 defines a set with five members that are either male or female. It can have any combination of males and females up to a limit of five. It does not limit the set to be a set of five males only or a set of five females only.

4.4 Queries and Logical Conditions

Derived relations, constraints, states and actions are expressed using object-oriented logical expressions. The expressions are constructed using the usual logical connectives *and*, *or* and *not*. The connectives have the usual precedence *or*, *and* and then *not*, the last is the most tightly bound. Parentheses () can be used to override the precedence. Variables communicate values and begin with an upper case letter.

The basic building block for logical expressions is the predicate. Relations, attributes and management states are treated as predicates. A number of built-in predicates are also available [10].

Predicates have the form:

<object predicate>	::=	<object>.<property identifier>[(<actual arguments>)]
<object>	::=	<object identifier> I <variable identifier> I self
<arguments>	::=	<actual argument>[,<actual arguments>]
<argument>	::=	<value> [:<type> I :<variable identifier>]

An argument of a predicate can be a value or a variable. A variable is used to retrieve the value of the property. The value returned is not the structured value but each individual simple value that is part of the structured value. In effect the predicate in its unmodified form represents a binary relationship between the object and the objects that are members of the composite relation. This provides for more natural and clear conditions. The keyword self refers to the object instance of the class in which the definition appears.

Assuming the previous definition of the Network *contains* relationship and an instance *network* composed of Nodes, $n1$ and $n2$, and Connection $c1$, the following query retrieves the immediate components of *network* in variable P.

   network.contains(P)

P is bound to the values $n1$, $n2$ and $c1$. The read built-in predicate is used to retrieve the unmodified structured value if needed. The query

   read(network.contains(P))

binds P to the single value $(\{n1;n2\},\{c1\},\{\})$ (a sequence of three sets). The class identifier modifier specifies the type of the objects that are to be retrieved. Again this reduces the number of conditions needed for many common queries.

   network.contains(P: node)

binds P to the values $n1$, $n2$. It is equivalent to the query

   network.contains(P) and   P.all_instance_of(node)

To test a relation or attribute for a specific value a value may be given as the argument.

   network.contains(n1)   succeeds   but   network.contains(n3)   fails.

The use of a variable to denote the modifier allows for more flexible querying in that the class identifier can be communicated from previous conditions. The value of the relationship (or attribute) may be assigned values by two ways. A write operation overwrites the value, or an append operation. adds extra values.

Using the previous example the query

   write(network.contains(({n4;n5;n6},{c2;c3},{t1})))

will replace the previous definition so that

   network.contains(n1)   fails   but   network.contains(n5)   succeeds

Using the query

   append(network.contains(({n4;n5;n6},{c2;c3},{t1})))

will add these values to those already defined so

   read(network.contains(P))

binds P to the single value $(\{n1;n2;n4;n5;n6\},\{c1;c2;c3\},\{t1\})$ (a sequence of three sets).

The model is queried using a logical expression.

4.5 Primitive Relations

Primitive relations are built-in references from a class instance to other class instances within the model. The related instances are stored within the instance definition and may be retrieved immediately (instances referred to by derived relations are located each time the relationship is invoked).

Default values may be specified for a primitive relation in the class definition. When created a new instance will take this default value if no value was defined explicitly during creation. If no value has been given either default or explicit then the transaction has finished (see section 4.10) the instance is held to be in an illegal state. The values that a primitive relation may take is determined by the definition of the relationship and is constrained by the rules set out in section 4.3. The value of a derived relation may be accessed and assigned by methods given in section 4.4

Inverse relations may be specified for primitive relations. These inverse relations are specified on the relation name and apply for all instances in the model. For example if an inverse relation was defined for the contains relation and called contained_in, then the following statement holds

network1.contains(n1) iff n1.contained_in(network1)   (iff is read as if and only if)

If this statement did not hold, the model would be deemed to be inconsistent.

4.6 Attributes

Attributes are similar to primitive relations, the value of the attribute being stored in the definition of each instance. Like primitive relations default values are permitted, and all attributes without a default value must be given a value during the creation of the instance. Access and updating attributes values occurs in the same manner as access and updating of primitive relations. Attributes differ from primitive relations in that the may have numerical, boolean, or strings as values. They hold the data that is associated with an instance. Inverses are not defined with attributes.

4.7 Derived Relations and Constraints

Derived relations take the form of Horn clauses where the consequent is a single atom or a n-tuple of atoms. Names given to components of a type are treated as variables. The scope of term name variables is the object class definition within which they appear (see section 4.3 Type Definition). Other variables used in the definition of derived relations and constraints have their scope limited to the clause in which they are used. Variables can be unified as in Prolog.

Object-Class: node
Relations:
    connected_to(Node:node)   if
        node.contains(P1:port) and
        P1.terminates(L:link) and
        Link.terminated_at(P2:port) and
        P2.contained_in(Node) and
        self \== Node

The derived relation *connected_to* represents the connectivity to an adjacent node. Its value is queried in the same way as primitive relations and attributes.

Constraints have the following forms:

    <constraint> ::=   <condition> iff <condition> |
                            <consequent> if <condition> |
                            always <condition> |
                            never <condition>

The first form is read as A if and only if B. Whenever A is true B must be true and whenever B is true A must be true. In the second form whenever the condition is true the consequent must be true. Unlike the consequent of a derived relation constraints have a consequent that can be a complex logical expression. The other two forms are intuitive. The first says the condition must always be true and the second that the condition should never be true. Following on from the example in section 4.3. A network has constraints to ensure that the network is non-trivial.

Object-Class: network
Relations: contains {node}#N , {connection}#C , {trail}#T

Constraints: non_trivial   always N >= 2

This constraint is equivalent to

        non_trivial
        always self.contains(Node1: node) and
                    self.contains(Node2: node) and
                    Node1 \== Node2

4.8 States

Before describing the states definition section of a managed object definition it is necessary to make clear the definition of a state. A managed object state is a particular combination of values of the various attributes and relations of the object. This should not be confused with the management states defined in the States section. Management states declare particular management views of an objects state. Different management aspects can be represented such as configuration, usage, operational and administration. They define a subset of valid states against which the objects behaviour is defined. It is through the definition of management states and operations and notifications (that move the managed object between management states) that a management policy is defined. Management states are defined as a logical condition on attributes, relations and other management states. There is no restriction on the definition of one management state being dependent (derived from) on the state (attributes, relations and management state) of another object. Management states can be parameterised to convey a similar management view on particular objects or values.

4.9 Operations and Notifications

Operations and notifications are defined in terms of state transitions which describe the effects on the managed object and related managed objects.

<operation/notification>
    [conditions <condition>]
    <current state>
        [actions <action>]
    <new state>

The supplementary conditions and actions are optional. Conditions and actions are defined by logical conditions as described in section 4.4 Queries and Logical Conditions.

The actions define the effects of the operation or notification by invoking basic operations in the model (write, append and remove for primitive property values and create and delete for objects) and actioning operations on or sending notifications to other objects. The actions form an atomic transaction. The constraints must all be valid at the completion of the action and it must achieve the required change of state. An action that evaluates to false indicates failure to carry out the required operation. To interpret the action definition an imperative reading can be given to the logical expression assuming a left to right execution of the condition (cf. Prolog). Backtracking on failure would undo any changes that were made by the basic operations. To illustrate consider the operational status from the X731 standard [11], see figure 1.

It is sometimes desirable to have notifications which do not require to be invoked. For instance to send a warning message whenever a certain state is entered. These notifications have *trigger* as their name, and so are named trigger notifications, or automatic notifications. All automatic notifications are tested after an operation is completed and before the constraints are checked.

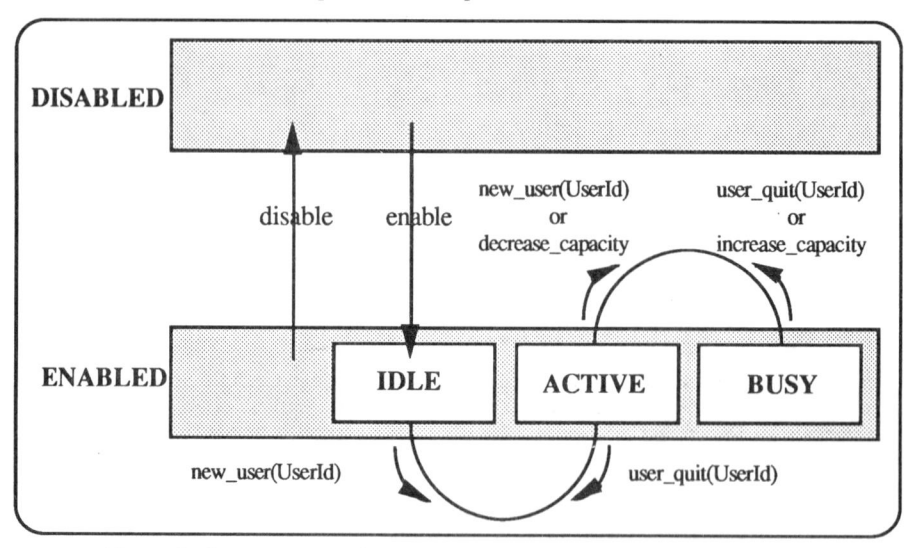

Figure 1 - State transition diagram for the X731 operational and usage states

4.10 Transactions

It is often the case that to perform a given task a number of operations must be invoked, and that while the whole task may leave the objects in a consistent state an individual operation within the task leaves an object or objects in a inconsistent state. An object is in an inconsistent state is when the values of the attributes and relations are such that one of the constraints is broken.

In this example setting Port1 termination function to "output" could temporarily put linkA into an inconsistent state. Links must have ports with termination functions that allows information to pass down the link (eg an input port and an output port). To prevent a temporary inconsistency from causing a task to fail the concept of a transaction is introduced. A transaction is a number of operations or queries that are performed in series, each depending upon the successful completion of the previous operations and queries. During the course of a transactions the constraints and automatic notifications are suppressed until the last operation has completed. Upon completion of the last operation all the automatic notifications in all the effected objects are tested. Finally all the constraints are checked.

5. COMMON INFORMATION MODEL

The Common Information Model (CIM) is the shared representation of the managed domain. This information which consists of both data and functions (termed common services) is represented using object-orientation. The CIM, as the name suggests, holds a shared representation, it does not preclude additional information or functionality on the managed objects being held elsewhere. Thus the MAs may store private data on CIM objects and contain functions that manipulate CIM objects. Three sub-models have been identified within the CIM, these are the Managed Object Model (MOM), Management System Model (MSM) and the Model of Other TMNs (MOT). The role of the CIM is more fully discussed in [1],[2]. In the current ADVANCE work, only the MOM has been implemented.

The MOM contains a representation of the managed system; network, services, and customers, each of which have their own schema within the overall framework of the MOM. These schema also reflect the responsibility layers of BT CNA Model [12]. In addition there exists the adaptation layer, the purpose of which is to isolate the service layer from the specific characteristics of a particular network.

5.1 Generic Network Model

The generic network model [13] covers both logical and physical networks. It is loosely based upon the H550 [14] generic network and likewise does not support the connectionless networking concepts. A telecommunications network consists of a number of network strata, which are related to each other by client-server relationships. The term strata comes from NETMAN and is used to replace layer and level concepts (to prevent over-use of these terms) used elsewhere. Examples of strata are the ATM virtual channel level, ATM virtual path level, see figure 2.

Higher Layers	
ATM Layer	ATM Channel Level
	ATM Path Level
Physical Layer	Transmission Path Level
	Digital Section Level
	Regenerator Section Level

Figure 2 - Hierarchy of the ATM Telecommunications Network

A network consists of nodes and links, constraints upon networks ensure that a network must be non-trivial (there must be at least 2 nodes) and that it is connected (there is a no part of the network without links to the remaining part of the network). Links transport information between termination points (object class port), these termination points ensure the integrity of the link and include the termination functions. All termination points are located at a node. A node may contain an arbitrarily large number of termination points, however, the connected constraint imposed on the network object means that each node must contain at least one termination point.

There are two basic types of link, a connection, a simple link directly joining nodes, and a trail, which is formed by an ordered sequence of connections. Both types of link are associated with a unique class of termination point named a connection termination point and a trail termination point respectively. A consequence of the definition of a trail is that a node must be capable of assigning relationships between its connection termination points and hence the connections so as to form a trail. A node may therefore be used to represent the H550 cross-connection object. Links transport information either from point-to-point, or from point-to-multipoint (broadcast). The direction of information flow is dictated by the termination points. A termination point may be assigned to have termination function of type input, output, or input/output. Input means that the information is arriving at the node, output means that the information is departing the node, input/output allows information flow in both directions. Valid combinations of input, output, or input/output termination functions are ensured by constraints in the link class. A sub-class of port called a network access point provides an entry point (and/or exit point) to the network.

A sub-network is a network that is contained within another, larger, network. Sub-networks do not appear directly within a network but are represented by a node. The network access points in the sub-network map to the ports in the node representing the sub-network. This abstraction allows a network to be a sub-network in several different networks and, additionally, the appearance of the sub-network may be different depending on which network is accessing it. This could allow MAs to have their own view of a network yet still ensure shared and consistent information.

Every network is contained within a single stratum of the telecommunications network together with all its sub-networks. Relationships exist between the objects in one stratum and objects within the stratum immediately above and below. The principle relationship is that a connection is supported by a trail in the lower level stratum. A connection is supported by a single trail, although a trail may support any number of connections. Termination points are related in a similar way.

ATM Network

Network layer MAs concentrate on ATM (Asynchronous Transfer Mode) technology. ATM is very likely to be chosen for implementing broadband ISDN (B-ISDN). It is independent of the means of transport at the physical layer. Thus the ATM levels are purely logical. B-ISDN, and hence ATM, supports switched, semi-permanent, and permanent point-to-point and point-to-multipoint connections, it provides on-demand, reserved and permanent services. To support the MAs the network model includes the ATM path level and ATM channel level. All ATM classes appear as specialisations (subclasses) of the generic network model objects. The relationships and constraints in the ATM classes are refinements upon those defined in the generic objects. This refinement is illustrated by the containment relationship shown in figure 3.

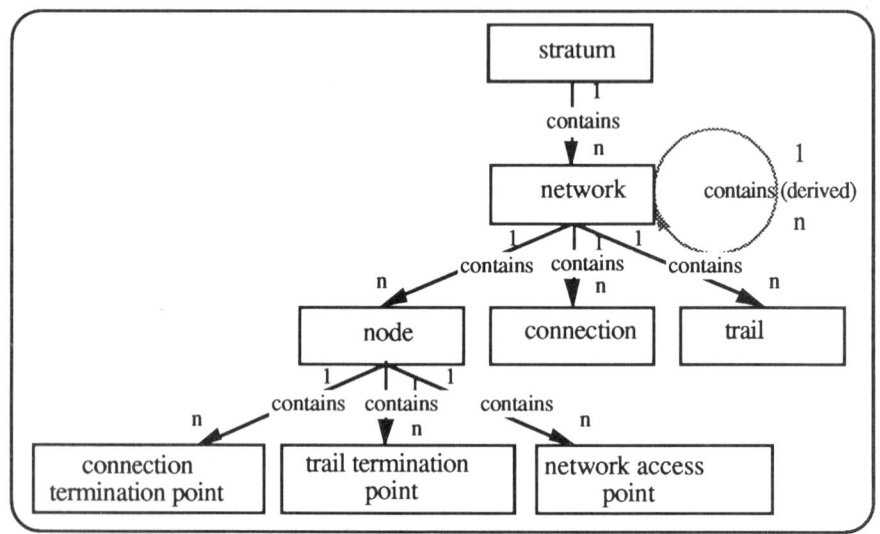

Figure 3 - Containment relationships of the generic network

5.2 Service Model

The service model supports a range of service management layer applications prototyped within ADVANCE. It has concentrated on providing a representation of IBC telecommunications services at a level of abstraction suitable and useful for those applications. The service model is based on [15] which takes many of its concepts from the CCITT Blue Book recommendation I.211.

Within the service model, service components represent the basic building blocks of IBC services and service control elements represent the primitives that invoke and manage them. Service components provide management with a description of the particular information types that can be transferred during service use. The description of a multi-media service will therefore specify a number of service components as being required. Service control elements present management with a view of an interaction that would take place between a service user and the network resources for the control of service components.

The taxonomy of IBC services for the service model is based upon the mode of information transfer between users during service operation. Five generic service classes have been identified, namely conversational, messaging, collection, retrieval and distribution. Service components and service control elements are related to specific IBC services which are specialisations of the generic service types.

5.3 Adaptation Layer

The adaptation layer is based upon CCITT draft recommendations I.362 [16] and I.363 [17] B-ISDN ATM Adaptation Layer. The adaptation layer supports relationships between a network layer and the service layer. Services are provided at service access points, and require various functionality to be provided at the network access point(s) that the service uses. The type of functionality required is dependent upon the service to be provided.

The network-service interface object has relationships with both service component objects within the service layer and network access point(s) contained within a particular node within the network layer. Whenever a new type of service is required which cannot be supported by the existing network-service interface objects a new instance must be created. The creation is successful only if the network elements represented by the node are capable of supporting the service. Examples of the functionality required are conversational connection control or messaging connection control.

6. CONCLUSIONS

This paper has presented an object modelling language that supports the implementation of the Model Based Management concept. It has successfully supported the information and interworking requirements of a demonstration NCAS system.

MODEL has been shown to be capable of representing the structural and behavioural semantics of the managed objects in the NCAS domain. Use of an executable formal language proved to be an effective way of specifying managed object models that are precise and unambiguous. Direct implementation has enabled the consistency and correctness of the shared information to be guaranteed.

Complexity is reduced because domain knowledge resides in the model rather than the application, and due to the declarative nature of the language, provides a better conceptual model of the managed domain.

The CIM supports the information requirements of a number of heterogeneous MAs by providing abstract views of common information and ensuring mutual consistency of these views. It facilitates integration and reusability. It provides for the interworking of MAs acting as an intermediary in communications between MAs.

This work has indicated that future commercial implementations of object-oriented deductive database technology could play a central role in the provision of future open and integrated network management systems.

7. REFERENCES

[1] Farley P., Strang C., Harkness D.,: "An Implementation Architecture for Network and Customer Administration Systems", 5th RACE TMN Conference, London, November 1991.

[2] RACE Project ADVANCE "Deliverable 6: Draft NCAS Architecture", February. 1992.

[3] RACE Project ADVANCE : "Prototype Version Two Specifications", 09/BCM/RD3/AR/B/027/B1, October 1991.

[4] Manning K., Spencer D., "Model Based Network Management", 4th RACE TMN Conference, Dublin 1990.

[5] Carvalho F., DaSilva L., Girvin J. and Wright P., "Implementation of Management Applications for Network & Customer Administration Systems", Sixth RACE TMN Conference, Madeira, Sept.ember 1992.

[6] Lloyd, J. W. :"An Introduction to Deductive Database Systems", Australian Computer Journal, Vol 15, No 2, May 1983.

[7] Gallaire, H., Minker, J., Nicholas, J. M. :"Logic and Databases: a Deductive Approach", Computing Surveys, Vol 16, No 2, June 1984

[8] Lloyd, J. W. , Topor, R. W. : "A Basis for Deductive Database Systems", The Journal of Logic Programming, 1985:2: 93-109

[9] Lloyd, J. W. , Topor, R. W. "A Basis for Deductive Database Systems", The Journal of Logic Programming, 1986:1: 55-67

[10] Sparks E., Manning K., Hardwicke J., Quist B., "Integrated Network and Service Model and Modelling Mechanisms", RACE ADVANCE (R1009), ADPL198, January 1992

[11] CCITT Recommendation X.731 : "State Management Function".

[12] British Telecom Plc, "Management Reference Model: Management Layers and Domains", Co-operative Network Architecture, Vol. 7, Infrastructure, PA 0036:Part 2:1.00, 1991

[13] CCITT M.3100, "Generic Network Information Model for TMN (M.gnm)", November 1991

[14] RACE Common Functional Specification H550 : "Telecommunication Management Objects".

[15] RACE Common Functional Specifications "General Aspects of IBC and IBC Services", Document 1, CFS B210-C304.

[16] CCITT Draft Recommendation I.362, "B-ISDN ATM Adaption Layer (AAL) Functional Description, June 1990

[17] CCITT Draft Recommendation I.363, "B-ISDN ATM Adaption Layer (AAL) Specification, June 1990

Experience of Modelling and Implementing a Quality of Service Management System

Uffe Harksen (UH Datacom Consultants, Denmark),
Andreas Mann (IBM, Germany), George Pavlou (University College London)

ABSTRACT

This paper presents the experience from modelling and implementing a Telecommunications Management Network (TMN) Quality of Service (QoS) management system, based closely on the principles of the evolving TMN standard [1]. The management information models (Managed Objects-MOs) for each TMN layer, their relationships and the modelling decisions are presented, along with an implementation based strictly on real OSI management services and systems management functions. All management capabilities are expressed as managed object properties and all interactions are based on hierarchical manager-agent relationships of TMN Operations Systems (OSs). The full system development cycle, our experience from this modelling and implementation work and our conclusions are presented.

1. INTRODUCTION

In future Integrated Broadband Communication Networks (IBCNs) there will be complex provider and user schemes. Lower level service providers will sell their services to higher level service providers, the latter being their users. At the top of this layered service hierarchy, end users will be using underlying services. The TMN prototype system models three "stratas" [2] of service providers:

- a basic bandwidth provider, providing point-to-point links
- an ATM connection provider, providing ATM connections
- a service provider offering higher level services i.e. teleservices.

The teleservices span from telephony to file transfer, interactive data transfer, video broadcasting and conferencing etc. End users make use of these services which results in a layered service hierarchy consisting ultimately of four layers. The prototype management functions cover the top three layers (users, teleservices and ATM connections), basic bandwidth management being outside its scope. The service management functions in particular cover the teleservice and user layers and focus on Quality of Service (QoS) management. The evolving TMN standards [1] are based on the OSI Management model, enabling the management of resources (and thus services) through abstractions of them known as Managed Objects(MOs). A layered logical decomposition is suggested which recommends a similar structure for the associated management information models. The service management part of the prototype has been developed to conform closely to the emerging TMN architecture and uses real OSI management protocols and systems management functions (event reporting, logging etc.). A hierarchical structure has been adopted for the service TMN management information model (MOs) and all the management capabilities regarding the associated resources (services, users etc.) are expressed as managed object properties. The service TMN Operations Systems (OSs) i.e. service management applications, are organised in a logical layered architecture and all the interactions between them are according to hierarchical manager-agent relationships.

110

In this paper the approach to the modelling, design and development of this TMN service management system is explained, highlighting models and sources of influence and explaining important design decisions and experiences gained through this process. Its realisation gives insight into the problems of structuring TMN applications and systems. It must be noted that in the absence of a real IBC environment, the network, services and users are simulated but the actual QoS management functions are based on real OSI management services. The rest of the paper has the following structure: an overview of the system development cycle is given. The sources of input to the specification and design process are explained. Issues regarding the implementation architecture of the above model are then presented. Finally the experience from modelling, designing and implementing this system is presented along with our conclusions.

2. THE SYSTEM DEVELOPMENT CYCLE

Before examining in detail each of the various stages involved in modelling and implementing such a service management system, a first look at the full system development cycle will identify these and their relationships. Figure 1 gives an overview of the system development cycle.

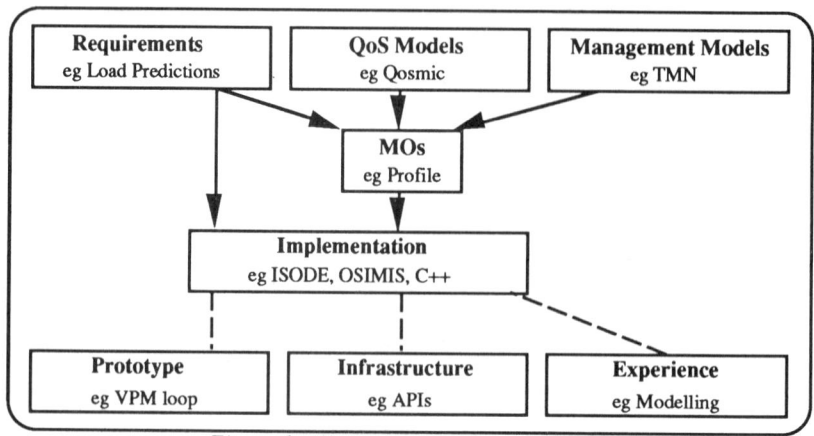

Figure 1 - The system development cycle

The first source of input is the requirements specification. This is translated to the desired functionality of the service management applications, these being the handling of QoS complaints for service degradation and the generation of load predictions as explained later in detail. The application context is thus service management, focusing at Quality of Service as opposed for example to customer administration or other service management functional areas. This brings up two important issues, Management and Quality of Service. Management and QoS models have being considered both by standardisation bodies and other projects and together with the functional requirements they should be combined to provide the management solution, allowing for necessary deviations/extensions specific to the problem in hand. Bearing in mind the conformance to the evolving TMN standards and the OSI management model, the target is to produce the prototype QoS service management information model in accordance with the TMN logical layered architecture. This will specify in effect all the capabilities related to the managed entities and will provide a framework for the development of management applications. The latter will use the information model (MOs) for their awareness and implementation functions according to the NETMAN ADI model [3], while the decision functions will be dictated by the system requirements (management policies).

Then in the implementation stage the information model together with the management policies should be translated to an implementation architecture. According to the Open Distributed

Processing (ODP) viewpoints [4],the system model from the information viewpoint should be mapped onto an engineering model. Computational aspects should also be examined to provide the management policies and achieve the translation between adjacent layers of the information model. Finally, different technologies could be used to realise the system (technology viewpoint). The implemented system will be installed in the testbed and cooperate with the rest of the prototype. This cooperation assumes a common understanding of the management model, the communications infrastructure and the specific management scenarios. The final outcome is not only the service management part of the prototype. Parts of the system that were considered generic and implemented that way may be reused in future applications in a similar context. As an example, the infrastructure for converting the management information model into a concrete MIB implementation and providing access to it i.e. the management platform, can be reused by any management applications following the same model. This is better facilitated if suitable technology, e.g. an object oriented one, has been used for the implementation. More important, the experience of modelling the system in such a way may be used in future attempts to organise Management Information Bases (MIB) in a hierarchical fashion.

3. INPUT TO THE SPECIFICATION AND DESIGN PROCESS

3.1 The Requirements

The main objective of the complete prototype management system is to optimise the utilisation of the underlying ATM network, maintaining at the same time the quality of service to the levels negotiated with the service users. Service management can be viewed as a part of a control system, which should contain monitoring as well as active control functions. A complete service management system should contain applications to serve planning functions on all management levels. In line with the QOSMIC timeline model [5], these functions may be classified in three categories, service level management, applications that support long term strategic planning of new services and phasing out of old services; user level management, applications that support the general management of service subscribers e.g. covering contract negotiation and user individual QoS monitoring and billing, call level management and applications that monitor and verify the QoS of individual calls and provide active control functions (see also section 3 in [5]). The prototype focuses on the following types of service management applications: monitoring and presenting service user behaviour over time and providing load predictions to the network TMN, monitoring and logging the provided QoS levels and alarm generation in the case of degradation below certain thresholds, and monitoring lower strata QoS and reporting to lower strata service providers.

3.2 The Management Models

The prototype management model is based on the following models and principles:
- the Telecommunications Management Network standards [1]
- the OSI Reference Model [6] and in particular the Management Model [7], [8], Service [9] and Protocol [10]
- the B-ISDN Reference Model [11]
- the ATMOSPHERIC Stratified Reference Model [2]
- the QOSMIC timeline model [5].

The analysis of the input from these models showed that the design of a service management model for a layered services system requires a multidimensional model covering:
- the service layering, as proposed in [2] and [5]
- the timeline concept as proposed in [5]
- the protocol layering from [6]
- the TMN layering as proposed in [1].

A generic service management model for layered B-ISDN communication services complying to these dimensions was developed and is presented in [12]. This model provides a generic way of relating MIBs of stratified management systems and the distribution of managed objects in the different management layers. The model comprises a generic data flow model and a global conceptual information model. The dataflow model is based on the timeline one, which structures QoS management functions according to the following timelines:

- the service timeline, covering functions and information related to a service over its lifetime, scale being in the order of years
- the customer-provider timeline, covering functions and information related to a customer-provider engagement over its lifetime, scale being in the order weeks or months
- the call timeline, covering functions and information related to a single service call, time scale being in the order of seconds or minutes.

3.3 The Quality of Service Model

Before being able to measure the quality of service, a model is needed to define what quality of service is. For the purpose of this system, the following Network Quality of Service (NetQoS) parameters were considered as providing an indication of the "network" quality of service i.e. QoS at the level of the ATM connection provider:

- Network Connection Establishment Delay (N-CED)
- Network Connection Signal Delay (N-CSD), the average end-to-end delay through the network
- Network Connection Signal Delay Variation (N-CSV), the maximum variation of the N-CSD
- Network Connection Disconnect Delay (N-CDD)
- Network Cell Loss Rate (N-CLR), equivalent to the Bit Error Rate (BER)
- Network Connection Signal Quality (N-CSQ), a unitless number between zero and 1, where 1 denotes the perfect quality of service
- Network Connection Establishment Probability (N-CEP).

For each teleservice, the equivalent Service Quality of Service(SrvQoS) parameters can be calculated based on the NetQoS ones provided by the ATM network. This requires the development of different calculation models for each teleservice, as the effect of e.g. cell losses will have different effect on a PCM-telephone and a Bulk File Transfer service.

As an example, in the PCM-telephone case the cell loss rate measured at the entry point into the ATM-network (N-CLR) will be mapped directly to an equivalent service one (S-CLR), indicating that there is no intermediate adaptation or compensation for cell losses. In the Bulk File Transfer case, the S-CLR will always be zero, as there will be an intermediate cell loss detection and retransmission mechanism (e.g. the OSI FTAM), which in reality transforms cell losses to increased end-to-end delay (S-CSD) and reduces the effective mean bandwidth. As a second example, a video conference which is based on 6 or more ATM-network connections e.g. two for voice and 4 for the picture transfer. The calculation of the service connection establishment delay and probability (S-CED and S-CEP) is very complex and depends heavily on the connection establishment algorithm.

4. THE SERVICE MANAGEMENT INFORMATION MODEL

The service management information model is the most important step towards the service management prototype as it specifies completely the management capabilities of the system. A managed object model has been developed for the service management information base, complying to the service and management models and principles described above. In particular, it is layered according to the TMN logical layered hierarchy, the managed objects of the higher layers providing increased abstraction of managed resources. This layering is complete as interactions between adjacent only layers are allowed. At the time of the management information base specification for this prototype, there were no complete/detailed models for MIBs related to service management. The prototype MIB has been developed based partly on material from the following sources:

- the OSI generic management information model [13]
- the TMN generic management information model [14]
- the RACE Common Functional Specification (CFS) documents, mainly [15]
- the generic service MIB model described in [12]
- the Network Management Forum (NMF) object library [16]

In the following two subsections the information model is discussed from two different points of view. Section 4.1 explains the classes of MOs arranged in the inheritance tree while section 4.2 discusses the instantiation of MOs, that is the containment hierarchy.

4.1 The Inheritance Hierarchy

All MOs defined and implemented in the prototype are shown in figure 2. This figure can be divided vertically and horizontally revealing different aspects of the information model. Going through the inheritance hierarchy from left to right, three groups of MOs can be identified as follows.

The three leftmost branches starting with actor, address and service are mainly derived from [15], [16]. MOs in this branch identify "resources" with respect to service management. It is obvious that some representation of service, service provider, user is needed. Some of them are further refined. As the main focus of the prototype lies on on-line performance management aspects such as planning, accounting etc. are not reflected in the MOs properly. This is for effort reasons only, the general concept of these MOs has proven suitable.

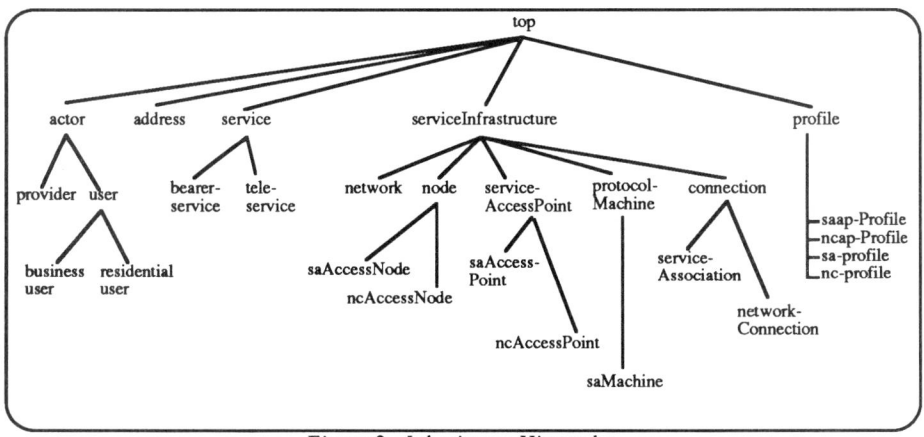

Figure 2 - Inheritance Hierarchy

The middle branch starting with serviceInfrastructure is mainly derived from [13] [14] where network, node(to some extent similar to entity), serviceAccessPoint, protocolMachine and connection are defined. MO serviceInfrastructure is used to group all resources belonging to one service provider. It enables a service provider to distinguish resources required for different offered services. When these MOs are instantiated a provider has to define carefully for what service this resource is actually needed. This MO was found helpful in structuring the layered service architecture.

The right most branch starting with profile contains all supporting MOs. Some of them such as system, discriminator, log etc. are not shown. These support all kinds of management applications and are defined commonly by OSI and CCITT [17]. In this branch figure 2 shows a specific MO called profile only. Profiles are needed to store various kinds of information, e.g. how many network connections are needed for one service, what are the QoS requirements that are guaranteed to a user.

Going through the inheritance hierarchy from top to bottom, two different layers can be identified:

- The uppermost MOs of this figure are part of a "generic layer". MOs are independent of a service and are refined for a bearer service offering pure network connections from A to B, for a teleservice offering what is called a service association realising the transmission and computation facilities for a video conference. Most of the MOs in the generic layer such as provider, user, service, network, node, connection, profile can be found in some standards or documents going to become standards
- MOs at the leaves of the inheritance tree are service specific. As mentioned above, a generic connection is refined to a networkConnection modelling the transmission facility a bearer service provider offers to its users, that is to providers of teleservices. On the other hand, a generic connection is refined to a so called serviceAssociation modelling the facilities (comprising for example transmission, adaptation, error correction) a teleservice provider offers to its users, the end user. Similar examples hold for the refinement of node, serviceAccessPoint and protocolMachine. For effort reasons only, some were not further refined e.g. network, provider.

This concept of layering the inheritance tree into a generic and a service specific part was found very helpful. The separation of generic and specific parts in the inheritance hierarchy was recommended by other authors very recently [18]. The reuse of the service management part of the prototype and its service information model in other projects is not restricted by principal design decisions.

4.2. The Containment Hierarchy

All classes of MOs instantiated in the prototype are shown in figure 3. This shows the three levels of the service management system realised in the prototype and is a refinement of figure 4 in section 5.1. The functional decomposition regarding the management information trees is discussed. The lowest level (SSA - Service Simulator Agent) implements a mediation function translating the service simulator language into a standard conformant one, e.g. defining MOs, allowing CMIP communication. The communication between service simulator (the real resource) and SSA is private matter, communication between SSA and other managers is conformant to system management standards [6], [7], [10].

The middle level (NELSM - Network Element Level Service Manager) manages the SSA and translates its information model into its own [19], realising a so called information conversion function. Among others, it implements a very important data reduction function, aggregating and compressing information over time. While SSA operates on real time data, NELSM aggregates information over some aggregation interval that may be changed by other management applications.

The upper level (SLSM Service Level Service Manager) manages the NELSM and performs all kinds of administration work. Among others it aggregates and compresses information further. This condensed information may then be used (not implemented in the current prototype) by other management applications e.g. to decide if a service should be stopped or the QoS values guaranteed to each user can be improved.

Again, figure 3 can be divided vertically and horizontally revealing different aspects of the information model. Going through the figure from bottom to top three groups of MOs can be identified as follows.

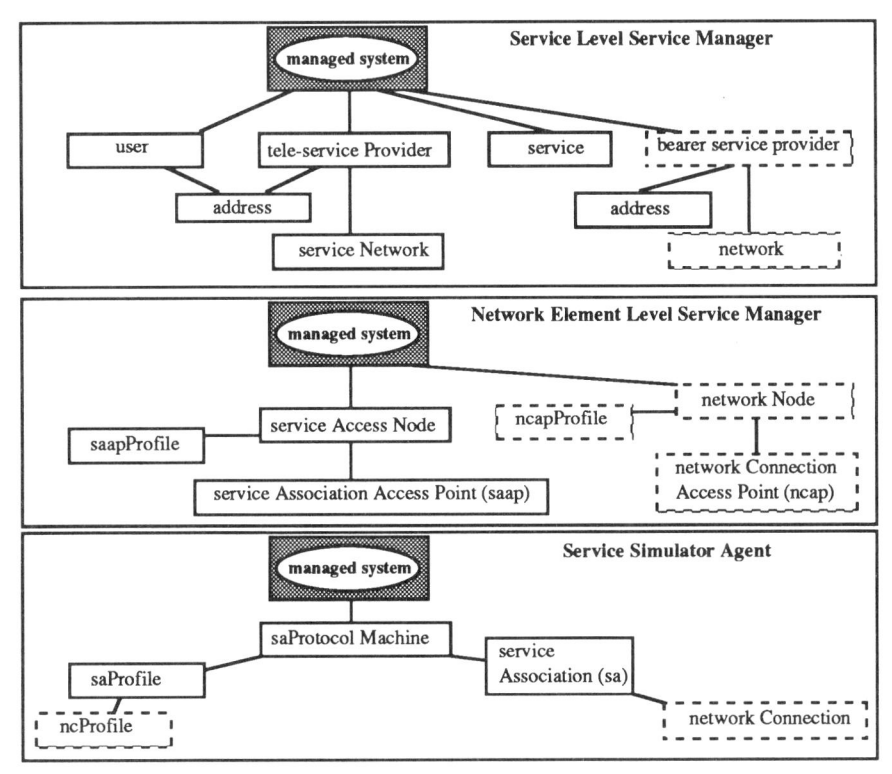

Figure 3 - Containment Hierarchy

MOs instantiated at the SSA level represent mainly information out of the call timeline [5]. This is especially true for serviceAssociation and networkConnection. These MOs are created dynamically as often as some user requests a service. Thus, MOs at this level have a short lifetime compared to other ones such as serviceAssociationAccessPoint. These MOs are updated very often by the real resource, the service simulator. Attributes of these MOs change values frequently and notifications may be emitted quite often. The number of MOs to be handled at this level is huge and varies significantly over time.

MOs instantiated at the NELSM level represent mainly information out of the user timeline [5]. A serviceAssociationAccessPoint (SAAP) exists as long as a user subscribes to a service. Together with the MO SA-Profile (instantiated in SSA) it reflects the contract between end user and teleservice provider and contains information e.g. about the QoS the provider guarantees to deliver to this user.

MOs instantiated at the SLSM level represent mainly information out of the service timeline [5]. MOs as service and provider objects exist for as long as a service is offered. The users using this service may differ over time, but the service remains. Each service has default QoS values stored in the MO SAAP-Profile. If in some contract more stringent QoS values are agreed between user and provider, these values will be stored in MO SA-Profile as discussed above.

Going through the figure from left to right two different types of MOs can be distinguished:
- MOs drawn with solid lines represent resources controlled by the service manager, user, serviceAssociationAccessPoint, serviceAssociation being typical examples. For example a service management application can decide to base a service association on 4 instead of 3 network connections (say) in order to increase the efficiency of a high-speed FTAM service. Whenever a user asks for this service the corresponding SA-ProtocolMachine has to get the required network connections from the network provider in order to set up the requested service association
- MOs drawn with dashed line represent resources provided by the underlying service provider, the network provider. NetworkNode, networkConnection are typical examples. These resources are controlled by the network provider. The service manager needs a representation of them in order to measure delivered QoS and to express information to the network provider. For example, it makes no sense the service manager telling the network provider that 1000 (say) service associations of type A are expected to be set up during the next hour. Instead the network provider needs to know the estimated number of network connections for the next hour in order to provide enough transmission capacity in time. This makes sense as the network provider does not know how many network connections make up one service association.

5. IMPLEMENTATION ASPECTS

5.1 The Overall Engineering Model

The management information model described was translated to a set of hierarchically organised TMN operations systems. This overall organisation of the TMN prototype, including both service and basic bandwidth management functions is shown in figure 4. The architectural model described applies only to the service management functions, referred to for simplicity as "service manager". The basic bandwidth management functions are similarly referred to as "traffic manager". Last but not least, the IBCN network, the services and their users are simulated and referred to collectively as the "simulator".

The network and services are found in the Network Element layer of the proposed TMN layered hierarchy, which comprises the managed elements. It should be noted that managed elements can be either physical or logical and services are logical managed elements manipulated by the service management functions. Services at that level exist physically at the customers premises and like other network elements (nodes etc.) they are outside the TMN. They should be managed by functions in the Network Element Management layer which is in the TMN.

Access to the network elements in a real system will occur mostly through non-fully compliant management protocol stacks because of the complexity and overheads involved in realising these by simple elements. Though some services may be able to support them, translation functions (mediation devices) will be needed for some others to provide an object-oriented view according to the OSI management model. In the case of this prototype, the unit implementing the simulated service functions had been developed before the rest of the service management system and used a simple message passing mechanism as its management hooks. The Service Simulator Agent (SSA) is thus exactly a mediation function box between the latter and the rest of the service management system.

In the Network Element Management layer, there is an operations system realising an information conversion function which translates the lower level information model to its own through aggregation and compression functions. It is also at that level that the complaints regarding poor quality of service are produced and sent to the bandwidth provider's management system (Link Bandwidth Manager OS) to help optimise the usage of the network resources. This operation system is known as the Network Element Level Service Manager (NELSM).

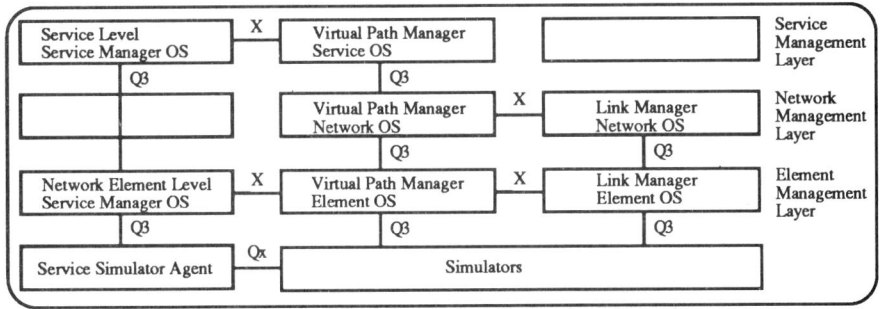

Figure 4 - Hierarchically Organised TMN Operations Systems

Finally, at the Service Level of the TMN hierarchy there is another operations system which has its own information model as described in section 4.2 and is responsible for taking a global view at the service network and producing load predictions which are used by another part of the bandwidth provider's management system (Virtual Path Bandwidth Manager OS) to optimise the allocation of bandwidth to virtual paths. This operation system is known as the Service Level Service Manager (SLSM).

It is mentioned that in the prototype there are no service management functions in the Network Level of the TMN hierarchy. This is a very subtle and debatable point as is the allocation of the management functionality to the layers suggested by the example TMN layered architecture [1].This decision is in line with the particular example architecture but more important is the use of such a hierarchical layered decomposition rather than the debate of which label is put on each layer. The overall engineering model for the service management part of the prototype thus consists of four different units, each implemented by a separate process in operating systems terms: Service Simulator, SSA, NELSM and SLSM.

5.2 The Communications Infrastructure

There are two aspects in the communications infrastructure used in the service management prototype: the communication within the TMN and the communication with the managed elements (services) outside the TMN. The communication within the TMN is based on the use of the OSI management service and protocol CMIS/P realised over a full interoperable OSI stack. These are q3 reference points realised as Q3 interfaces over CMIS/P. There is always a hierarchical manager to agent relationship between the management units involved (OSs and mediation function blocks), with units higher in the hierarchy acting in a manager role with respect to the units of the level below which act as management agents in that instance of communication. This recursive relationship terminates at the lowest level where real resources (services in this case) are managed. Units at the same (peer) level may act in both roles with respect to each other: the basic bandwidth provider's management units to which the NELSM and SLSM send their information may be thought of as such.

The communication between the mediation function block (SSA) and the managed services (ServiceSimulator) is realised in a proprietary fashion which is whatever the service boxes support as management hooks. In our case, this is a simple message passing facility requiring no presentation services and is implemented using the Internet Transmission Control Protocol (TCP). This is a qx reference point which is converted to a Qx interface though TCP.

The implementation of CMIS/P is realised by the OSIMIS management platform [20] which in turn uses the ISODE OSI protocol stack [21]. The latter allows interoperable services over different transport service stacks: TP0 over X.25, TP4 over CLNP or even TP0 over TCP/IP using the RFC 1006 method [22]. The latter is mainly used in this experiment which mostly operates on a local area network environment. The communication with the rest of the system (basic bandwidth management) does not use fully conformant Q3 interfaces as the rest of the system uses a proprietary management platform based on TCP. Were it using the same management platform, communication at the Network Element Management level (QoS complaints) would be naturally realised as CMIS event reports while communication at the Service Management level (load predictions) could be modelled either as event reports or as information stored in the service network managed object using aspects of the workload monitoring systems management function [17], which any interesting party could periodically read.

5.3 The Management Platform

Using CMIS/P is only part of the problem of exploiting the powerful features of the OSI management model. The difficult part is how to convert the formal specification of management information into managed objects in software and also how to provide necessary OSI systems management functions such as event reporting, logging, workload monitoring, access control etc. The solution to this comes from the genericity and object orientation of the OSI management model: if the generic features of the latter are implemented carefully, they can then hide a large part of the complexity of using CMIS. The systems management functions can also be implemented in a generic way and they could be "hidden" from managed object implementers. This is exactly what is provided by the Generic Managed System (GMS) of the OSIMIS platform which provides an object-oriented Application Program Interface (in C++) for implementing managed objects with minimal complexity. It allows implementers to concentrate on the intelligence built into the managed objects and the communication with the associated resource rather than the mechanics of CMIS access and the complexity of the systems management functions. It should be added that complex but powerful aspects of CMIS such as scoping and filtering are completely hidden. Support is also provided for periodical polling functions and communication to the associated real resources through arbitrary communications mechanisms. This is what has been used to convert the information models for each layer into operations systems with agent capabilities (SSA, NELSM, SLSM).

An additional problem though comes from using CMIS in a manager role to access subordinate management information models e.g. SLSM to NELSM etc. Though this is not as complex as the agent part, a special object-oriented API was developed which made transparent the use of remote management information bases. This is known as the Shadow MIB (SMIB) method and allowed implementers to concentrate on the management policies rather than the mechanics of CMIS access. This infrastructure is now an integral part of the OSIMIS platform. The object-oriented implementation architecture of the generic operations system is shown in figure 5. Each such system is implemented as one process in operating system terms and handles an MIB instance (a management information tree). It offers an agent interface to peer and higher levels and acts as a manager itself with respect to peer or lower information models which it uses to implement its management policies. It may also translate lower information models into its own providing higher levels of abstraction.

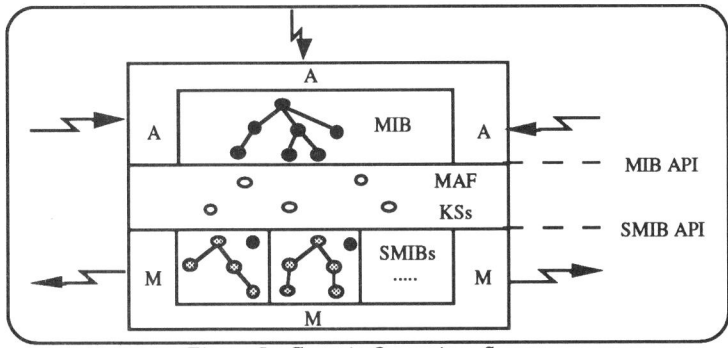

Figure 5 - Generic Operations System

6. CONCLUSIONS

The approach and the experiences from modelling and implementing a Telecommunications Management Network (TMN) Quality of Service (QoS) management system were presented in this paper. The management system was implemented based closely on the evolving management standards from CCITT as well as ISO. As the prototype is finalised and running we can draw some main conclusions.

In the process of developing the service management information model we chose the approach first to establish a global conceptual information model and then, during the development of the separate MIBs for the specific service management layers, we assigned objects to each of these layers based on performance considerations and other practical issues. We found that this approach ensured a generic, consistent and yet reasonably simple information model which, combined with the timeline model, provided a useful platform for various service management applications. The concept of layering was found to be very helpful in handling the complexity of management applications. But, we found the use of such a hierarchical layered decomposition more important than the debate on how to label each layer. Furthermore, we developed the "Shadow MIB" concept to support a more transparent and global view of the MIB from an applications point of view. But we still do not believe that it is realistic to establish a perfectly transparent global MIB based on the current management protocols. For the future we envisage using the service management part of the prototype in a real environment and in other projects to further validate the value of our concepts.

7. ACKNOWLEDGEMENTS

This work was supported by the Commission of the European Community under the RACE program. The authors wish to thank the partners of the RACE NEMESYS project.

8. REFERENCES

[1] CCITT Draft Rec. M.3010, "Principles for a Telecommunication Management Network (TMN)", Report R 28, Nov. 1991.

[2] Tuanhaduong et al : "Stratified Reference Model, An Open Architecture Approach for B-ISDN", XIII International Switching Symposium, Stockholm 1990.

[3] Plagemann, S., Turner, T. : "Telecommunications Management Conceptual Models", Sixth RACE TMN Conference, Madeira, September 1992.

[4] Geihs, K., Mann, A : "ODP Viewpoints of IBCN Service Management", IFIP/IEEE Int. Workshop on Dist. Systems: Operations & Management, Munich, Oct., 1992.

[5]   Cochrane, D. : "Quality of Service Mappings", Sixth RACE TMN Conference, Madeira, September 1992.

[6]   ISO/IS 7498, Information Technology - Open Systems Interconnection-Basic Reference Model, ISO/IEC IS 7498, 1984.

[7]   Information technology - Open Systems Interconnection Systems management overview, ISO/IEC IS 10040, Aug. 1991.

[8]   Information technology Structure of Management Information, Part 1: Management Information Model, ISO/IEC IS 10165-1, Aug. 1991.

[9]   Information technology - Open Systems Interconnection Common management information service definition, Version 2, ISO/IEC IS 9595, July 1991.

[10] Information technology - Open Systems Interconnection Common management information protocol specification, Version 2, ISO/IEC IS 9596, July 1991.

[11] CCITT Draft Rec. I.321, "B-ISDN Protocol Reference Model" Rev. June 1990.

[12] Harksen, U. : "Service Modelling in NEMESYS", Proc. 5th RACE TMN Conf., London, Nov. 20-22, 1991.

[13] Information technology Structure of Management Information Part 5: Generic Management Information, ISO/IEC DIS 10165-5, Feb. 1992.

[14] CCITT Draft Rec. M.3100, Generic network information model Working Party IV/3, Report R28, Nov. 1991.

[15] RACE Common Functional Specifications : "Telecommunications Management Objects", CFS H550, Dec. 1991.

[16] Forum Library of Managed Object Classes, Name Bindings and Attributes OSI Network Management Forum 006, Issue 1.1, June 1990.

[17] Information technology - Open Systems Interconnection - Systems Management, Various parts (mainly 5 and 6), ISO/IEC IS 10164, Jan. 1992.

[18] Schott, B., Clemm, A., Hollberg, U. : "An ISO/OSI based Approach for Modelling Heterogeneous Networks", Proc. of the 4th Int. Conf. on Inf. Networks and Data Comm., INDC-92, Espoo, Finland, March 16-19, 1992.

[19] Mann, A., Pavlou, G. :"Quality of Service Management in IBC: An OSI Management Based Prototype, Proc. 5th RACE TMN Conf., London, Nov. 20-22, 1991.

[20] Knight, G., et al : "Experience of Implementing Network Management Facilities" 2nd IFIP Conference, Integrated Network Management, Washington, April 1991.

[21] Rose, M. T. : "The ISO DE User's Manual, U. Delaware, 1990.

[22] ISO transport services on top of the TCP: Version 3 RFC 1006

The Management of Telecommunications Networks
R. Smith, E. H. Mamdani, J. G. Callaghan (Editors)
© Ellis Horwood 1992

Quality of Service Mappings

Don Cochrane (Dowty, UK)

1. INTRODUCTION

As services become more important, so the users' expectations grow regarding the quality of those services that is delivered to them. This is especially thought to be true for the future IBC services (Integrated Broadband Communications). In parallel with this, the concept of Quality of Service has grown in importance. This started as a vague concept that it was something to do with how satisfied users were but couldn't be measured and has rapidly evolved to a clearly identified and scoped subject over the last three years. This paper presents:

- the basic concepts of performance
- the relationship between some terms in common use
- the relationship of QoS to service methodologies
- the Timeline Model
- the Brick Model
- QoS and the Object Oriented paradigm and Open Distributed Processing (ODP)
- the Language of QoS.

2. BASICS OF QoS AND SYSTEM PERFORMANCE

2.1 QoS and System Performance

There are two classes of performance parameter relevant to the IBC: QoS parameters and System Performance (SP) parameters. The description of QoS and other terms developed in [1] is:

" QoS is a set of user-perceivable attributes of that which makes a service what it is. It is expressed in user-understandable language and manifests itself as a number of parameters, all of which have either subjective or objective values. Objective values are defined and measured in terms of parameters appropriate to the particular service concerned, and which are customer-verifiable. Subjective values are defined and estimated by the provider in terms of the opinion of the customers of the service, collected by means of user surveys."

System Performance (SP). Any provider, of any service at any level of an IBC, be it at the network level or any of a stack of nested value added services (the multi-service environment) can measure quantitatively a number of parameters which indicate how well the service is performing. System Performance is the general term for all such parameters. It is assumed that System Performance comprises provider-measurable parameters that contribute to QoS as seen by the user. Network Performance (NP) is the specialised case of SP which applies to the Bearer Service of a telecommunications service. SP and QoS are related but are not the same. The user cannot perceive SP at all. The provider (acting strictly as a provider) cannot measure QoS but can infer it from SP values and a mapping function to QoS. This is illustrated in Figure 1 where the two directions of viewing quality can be seen as opposite at service boundaries. A frequent method of avoiding this problem of non-visibility of QoS to Providers is to also act as customers and so to experience the QoS.

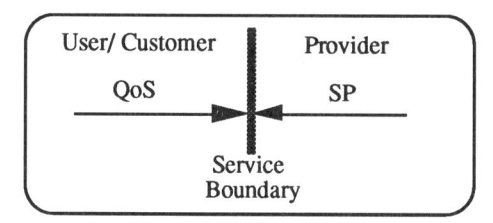

Figure 1 - QoS & SP Relationship

SPs are always objective and can be measured. QoS attributes can be either subjective or objective. An example of objective QoS is information error rate while an example of a subjective QoS is picture quality on a TV screen. Subjective QoS can be approached by such methods as sampling of users. A pragmatic (but accurate) rule for distinguishing QoS from SP is that if a parameter is visible and measurable by the provider then it is an SP. If the user can measure it then it's a QoS. Clearly there are a number of parameters that can be measured equally well by both actors; these are the common QoS parameters that have a 1 to 1 mapping to corresponding SPs [1]. SPs are measurable throughout the IBC. The destination of all these measurements is the TMN. It is the TMN that can perform concatenation of the performances of the Network Elements to form SPs and from there infer QoS.

2.2 QoS and SP Relationships

It tends to be a characteristic of the subject of QoS that there is a misuse of terms and the relationship between terms is often unclear. This section is intended as a clarification of the relationships between terms used in the QoS field. QoS is something that requires to be planned before use of a service. It is also something measured or inferred during the operation of a service. These two aspects of QoS may be categorised as being *Planned* and *Achieved*. *Planned* QoS is the static QoS that is the goal and may be set down in a contract or other document. *Achieved* QoS is that which is delivered to a user and can be perceived by that user. QoS is either Objective or Subjective in nature. Merely because current technology does not permit quantification of a particular subjective QoS today is no reason to ignore it.

It is also reasonable to assume that a Service (or Network) Provider will plan levels of SP (*Planned SP*) and measure the actual SP in the network (*Measured SP*). In addition, CCITT have introduced the concept of Degree of Satisfaction (DoS) [2], and ETSI have described User Opinion (UO), both of which can be used to provide feedback to the service and network providers. The relationship between all these terms is shown in figure 2. This diagram is primarily used (in its current form) during the timescale of a single call (as opposed to an entire customer contract period or the entire life of a service).

The figure contains a number of parts with relationships between them indicated by a number of arrows. The parts are:
- SP. This is a unit containing the two items *Planned SP* and *Measured SP*. These are grouped together as they are both SPs. *Planned SP* is the set of SP parameter values that a Provider intends to maintain, believing that the maintenance of these levels will assure an adequate QoS level for the Customer. This is shown as being derived from a Contract. *Measured SP* is the set of SP values for the same parameters as measured by the provider. There may be differences, either good or bad with the planned values.

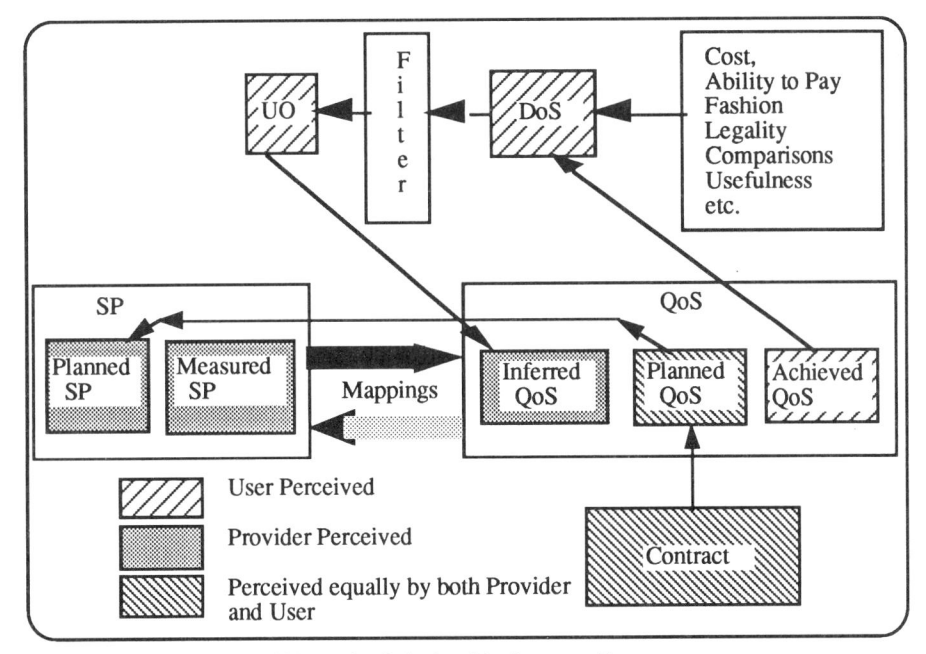

Figure 2 - Relationship Between Terms

- To the right of these is a unit entitled QoS. As for SP there are contained items; this time they are *Planned QoS*, which is common to both Provider and Customer and is derived from a Contract, *Inferred QoS* which is the provider's view of QoS as inferred from SPs and from User Surveys, and *Achieved QoS* which is the QoS as perceived by the User, i.e. what is usually meant by QoS. The providers and users views of these can be seen from figure 3 where Planned QoS can be seen equally well by both actors and each has their own view of the QoS as actually delivered; Inferred QoS for the provider and Achieved QoS for the User. The scopes of the views of the provider and User are shown explicitly in figure 3.

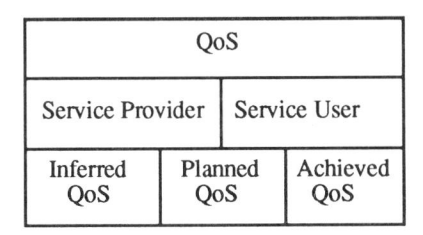

QoS		
Service Provider		Service User
Inferred QoS	Planned QoS	Achieved QoS

Figure 3 - Actors Views of Planned QoS etc.

At seen in figure 2 there are some external influences which affect *DoS*. QoS as perceived by the user is one factor in DoS but is by no means the only one. Also, as seen in figure 2, there is a *Filter* and *UO*. These and the arrows between them indicate that DoS is modified by a *Filter* when examined by means of surveys. The filter is necessary because it is a common human characteristic that when people are asked, in a survey, of their opinion about something, they might not give a true reflection of their DoS. The answers to questions can be affected by the phrasing of the question, for example, or that users dislike causing possible offence by being honest. Alternatively, a recent, isolated bad experience may result in a poor report.

The arrows indicate major interactions as follows:

- *Cost etc to DoS* : This arrow indicates that DoS is affected by many more effects than just QoS. Cost is clearly not a QoS; although important to both customer and provider. (If cost were a QoS then a free service would be without quality.) Legality means that however good a service is, if the user has it by means of e.g. stolen equipment, the QoS may still be high but the user may be less than totally satisfied by the risk of detection. The list of factors given is not exhaustive
- *Achieved QoS to DoS* : This arrow indicates that one of the constituents of DoS is QoS as perceived by the user
- *DoS via Filter to UO* : This indicates that DoS is affected by mental and social filters on its way to being gathered by Providers via User Surveys as UO
- *UO to Inferred QoS* : User Opinion is one of the ways that Providers have of estimating QoS. UO is an input to the overall picture of QoS that Providers have
- *Measured SP to Inferred QoS via Mapping* : These are the SP to QoS and QoS to SP mappings. The provider can measure SP. This can be translated into the QoS that the User experiences. This too is an input to Inferred QoS. In the middle is a unit marked *Mappings*. The mappings in the two directions are inherently different: the solid arrow indicates that the SP to QoS mapping is deterministic but the shaded arrow indicates that the reverse mapping is not, i.e. there are a number of sets of SPs that can provide a given QoS. As an example of this, a given user File Transfer throughput can be achieved equally well by either a combination of a very high data rate with a high error rate (and thus many retransmissions) or else a lower data rate with a much lower error rate
- *Planned QoS to Planned SP via Mapping* : The Provider, when setting up the service or changing its performance, can plan the QoS to be offered (Planned QoS). In order to convert this into SP parameters which the provider can measure directly and control, a further QoS to SP mapping is required.

3. THE TIMELINE MODEL

When considering such different QoS parameters as information error rate and accuracy of billing, it becomes very difficult to mentally reconcile that these are both QoSs. Clearly they are both QoS parameters but there is a fundamental difference between them. As more QoS parameters are examined, this area of difficulty becomes worse. In order to solve this problem, the Timeline Model [3] is introduced. See figure 4.

Timeline	Stage		
Per Call	Access	Information Transfer	Disengagement
Per Customer	Subscription	Service Use	Contract Cancellation
Per Service Infrastructure	Provision	Operation	Discontinuation

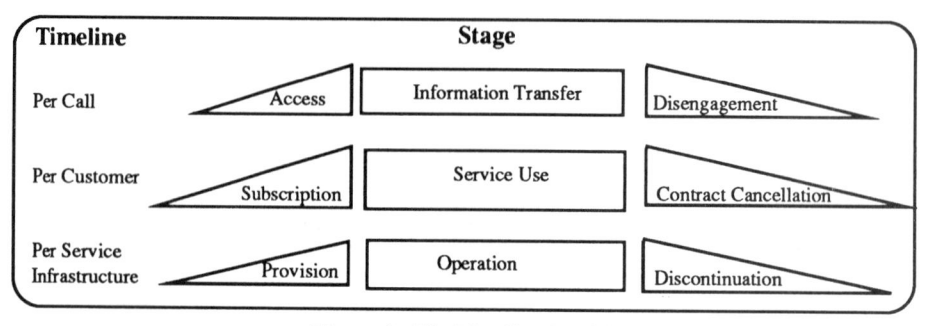

Figure 4 - The Timeline Model

This model shows that there is a hierarchy of timescales involved in the interactions between a customer (user) and a service provider. The benefit that this model brings is to put such concepts as QoS of Billing and Customer Services alongside such QoS parameters as Bit Error Rate without the mental gymnastics required to accommodate these dissimilar concepts. There are three Timelines in the Model; Per Call, Per Customer and Per Service Infrastructure.

3.1 Per Call Timeline

The top horizontal timeline of the diagram represents a single call. Here the user of the service (not necessarily the customer) is involved. Using the concepts of CCITT Recommendation E.810, it is meaningful to examine the stages of a single use of a service in chronological phases. Thus a single call has an *Access phase*, an *Information Transfer phase* and a *Disengagement phase*. These are very characteristic of a connection-oriented architecture. It is assumed that the basic structure of service of the IBC is connection oriented, i.e. possessing all of these three call stages, as opposed to a connectionless mode of operation.

The *Access phase* covers the user indicating that use of the service is required, identifying to the provider the destination of the call and negotiating any parameters of the call (bandwidth, reverse charging, Closed User Group etc.), the provider either making the connection or failing to do so (incorrect number, network congestion, equipment failure etc.) and the recipient indicating call acceptance or rejection.

Information Transfer covers the period from the destination user accepting the call until either party indicates that disengagement is required.

Disengagement concludes the process, generating and storing a charging record for that call. In the ISDN, a user's terminal could have a number of calls in progress simultaneously.

Each of these phases has its own QoS and SP parameters.

3.2 Per Customer Contract Timeline

A single call occupies only a small portion of the time period during which a customer subscribes to a service (of the order of seconds or minutes) but there will be many such calls and possibly in parallel. This relationship is shown by the middle horizontal timeline. Note that this timeline applies to both the customer and the user (who may be either the same or different persons).This middle horizontal timeline of figure 4, which covers the provision of a service to a single customer, itself has three phases in a similar manner to the Per Call timeline. These are:

- *Subscription*, e.g. connecting the user to the local loop
- *Service Use* which is the normal in-service state where a service is available to the user on demand
- *Contract Cancellation* when a user moves premises or the contract between provider and Customer is otherwise ended. This timescale spans days, months or years.

Examples of QoS for the middle timeline would be parameters such as:
- Time to provide service
- Time to repair totally failed service
- Accuracy of directory service entry for subscriber
- Accuracy for provider acknowledging bill payment (i.e. claiming that a bill was unpaid and discontinuing service, when the bill had actually been paid on time)
- Training of users
- Help Desk availability.

3.3 Service Infrastructure Timeline

The lowest horizontal timeline of figure 4 involves Customers but not users (unless they are the same person). It has a still longer timescale than the Customer Timeline and may be up to tens of years. This is the entire lifetime of a service, from Provision (initial planning and installation of the necessary equipment to provider the service), through Operation (within which there will be many sequential and parallel instances of the middle timeline) and finally Discontinuation (when the service is replaced or obsoleted e.g. UK Inland Telegrams).

The Operation period of the entire service has QoS and SPs, which would not be applicable to the network operations staff but would be of interest to the Business Planners. An example for this timeline could be:

- Service "image" (A subjective QoS as it is perceived by users but cannot be precisely quantified beyond "professional", "They do not strain themselves to help you", etc.).

4. RELATING THE THEORY TO THE REAL WORLD — MANAGEMENT METHODOLOGY FRAMEWORK

4.1 Need for a Methodology

The complete set of actions required to conceive, plan, provision, introduce and run an IBC service are many. QoS is a non-trivial adjunct to many of them. In order to make this situation manageable, a methodology for handling QoS in any stage of Service Operation is required. As the stages of service operation are different, so it becomes impossible for a single methodology to cope. The different phases of operation must be identified and an appropriate methodology specified.

4.2 The QOSMIC Cube of Service Management Methodologies

In order to make manageable all of the different methodologies for the different stages of a service required, a framework is necessary. Three dimensions have been identified as sufficient for the framework required to put Service and QoS Methodologies in context.

These are identified as follows:

- Call, Customer and Service Infrastructure. The Timeline model for understanding QoS divides the chronological perspectives for QoS into three nested timescales; Call, Customer and Service Infrastructure, thus providing the first axis of the framework.

- Introduce, Active and Terminate. QoS is relevant to the three phases of each of the three timelines. These phases are generically called Introduce, Active and Terminate in figure 5 thus identifying and classifying the second axis. These specific terms are used henceforth, as given in the Timeline figure 4.

- Provider and Customer. In addition, we can identify the two actors implied in any service operation: the Provider and the Customer of the service. Without both of these, there is no service. A Service Management Methodology Framework which is intended to be relevant to QoS (as this one is) must include the Customer because QoS is a Customer-oriented concept. The relationship between Customer and Provider can be complex as any Customer can also be a Provider of a higher level service and vice versa at lower layers. Consider further that Customers can receive services from more than one Provider.

We can therefore take these three axes and produce a cube model showing all these interactions shown as figure 5 (taken from [3]). At each intersection of these axes we identify a potential service methodology, each of which has its Quality of Service perspective.

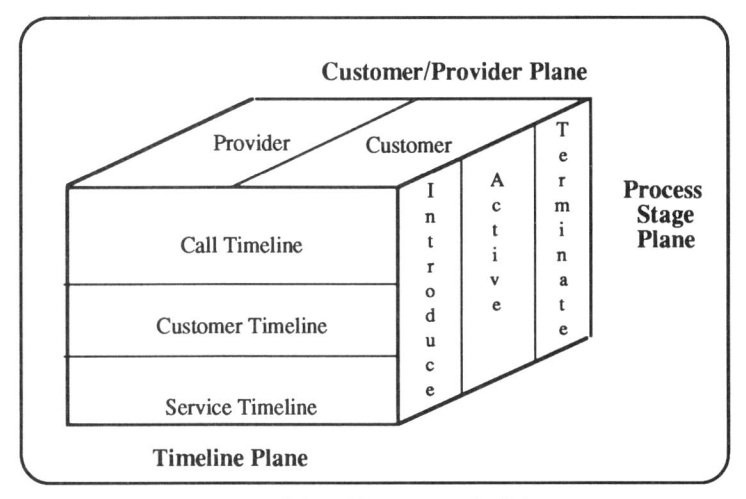

Figure 5 - Cube of Service Methodologies

The main features of this cube are:
- Adherence to the Timeline Model. This permits non-Per Call Timeline aspects such as Billing and Help desk to be considered and also anticipates the provider monitoring QoS over multiple calls or multiple customers as well as on a per-call, per customer basis
- Facilitates the dual Customer/Provider role of a service provider, receiving QoS and delivering QoS to an upper layer. The methodology thus unifies the Provider and Customer views of QoS and so can be used, selectively, by the Service Provider and by the Service Customer. It thus accommodates the multiple hierarchical service concept of IBC
- The methodology is set within a broad perspective covering the whole range of activities required during the introduction, monitoring, use and termination of a service. This ensures that all QoS related matters are considered within the methodology.

In the cube, we use the terms of the Timeline Model.

The Service Timeline concept has to be viewed from both Provider and Customer's viewpoints.

The Provider's view of the Service Timeline is that it is his livelihood and is thus the Business of the Provider. From the Customer's viewpoint the Service Timeline can be regarded as the Enterprise Timeline.

Similarly, the Customer Contract Timeline can be expanded to include the perspective of both Customer and Provider. Hence the term 'Contract' in the Timeline name. This reflects the fact that the aspect of this view common to both parties is that there is a contract involved.

For the Per Call timeline, the service of the call is the commodity which is being bought and sold by Customer and Provider respectively.

In order to examine all the intersections of the faces of the Cube, a mapping from three dimensions to one dimension is necessary to produce a linear list of entries. The intersections of the axes of the cube can be expanded in different ways depending on the order in which the axes are taken. An expansion and development of the methodologies is given in [3].

5. THE BRICK MODEL

IBC is seen as being a wide area multi-service environment in which there are both multiple providers of basic services (bearer services) as well as a large number of Value Added Service Providers. Services as seen by users will be assembled from different service components from different suppliers. Thus there is interworking between services both on an upper layer to lower layer basis and on a peer to peer basis. These different services will both receive and provide QoS on the bases of peer and higher-layer/lower-layer (in functionality terms) relationships. The Brick Wall Model helps us understand the relationships between the services.

There are three aspects to the Brick Wall model:
- It allows the examination of each single service `Brick' as a component in the provision of the final QoS as perceived by the end user
- It relates the performance of the service as measured by the provider to received QoS and the functionality of the service
- It allows the composite effect of the QoS of multiple services on the QoS of the end user to be scrutinised whatever the configuration (relationship) between the services which make up the final service.

The service `Brick' mentioned above is a concept which includes both OSI layers (e.g. Transport, Session etc.) and also CCITT communications concepts such as the ATM Adaptation Layer (AAL) etc. It can also be used to refer to the service provided by a physical object such as a switching node. By means of this wide scope of the meaning of the term, it is possible to reconcile the OSI and CCITT views of communications, i.e. functional layers separated by Service Access Points versus physical units separated by reference points. The aim of the Single Brick is to illustrate the propagation of QoS through a service layer.

The brick, has two possible classes of QoS input and two possible classes of QoS output as shown in figure 6.

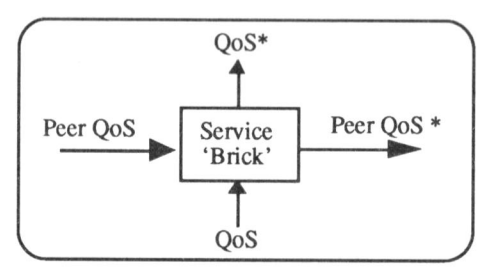

Figure 6 - The QOSMIC Brick

All inputs and outputs may not always be present due to the fact that services do not necessarily operate in an environment with peers. There may be many occurrences of a single QoS parameter from a single service, e.g. a distribution service.

Inputs:
- Peer QoS is that QoS received from Peer providers, i.e. providers at the same OSI level and providing a similar service to the provider in question. e.g. two national networks are peers.
  - QoS from lower level providers. These are providers which make an underlying service available that the provider in question needs to function. e.g. an e-mail service requires an X.25 service below it and an Adaptation Layer (AAL) requires an Asynchronous Transfer Mode (ATM) layer.

Outputs:
- Peer QoS* is the QoS provided by a provider to its peers. This is their received Peer QoS. Note that this may be the same Brick that is providing Peer QoS to this Brick, i.e. the relationship is mutual as in the case of both ends of the Transport Layer.
- QoS* is the QoS provided to the overlaid services, i.e. it is their received QoS from a lower layer.

Having described a single Brick, we can consider the effect of building a 'Brick Wall' where multiple services are stacked alongside and on top of each other. There are conversions from SP to QoS and vice versa as the quality of the underlying Brick is modified, changed and extended by subsequent Bricks. By means of this model, the effect of all components of a service can be taken into account.

6. QOS AND THE OBJECT-ORIENTED PARADIGM.

A simple diagrammatic representation of Computational ODP (Open Distributed Processing) Objects is shown in figure 7, where object A is offering two interfaces, providing access to services M and S. Object B accesses service S and object C accesses service M. An interpretation of the diagram is that interface S is a Service as provided to the User by a Provider object A whilst M is the Management interface of object A to some management entity C. Objects A and C and the interface M between them are thus in the Provider's domain whilst B is in the User's domain.

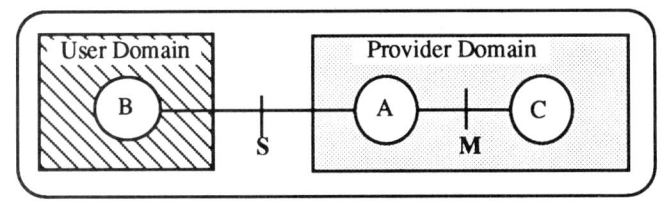

Figure 7 - Configured Objects

In QoS and SP terms, interface S, between the Users' and the Provider's domains exhibits QoS as perceived by B, whilst interface M is in the Provider's domain and reports the SP of the service from A to B across S.

In CCITT parlance, Reference Points are described between functional blocks. They may be made manifest by means of interfaces. Such interfaces have a number of parts:
- The M-part (message part) which contains the syntax and semantics of the information to be passed.
- The P-part or Protocol part which is concerned with carrying the information (M-part) from one functional block to the other. The P-part can be changed in different implementations without affecting the M-part at all.
- The Q-part or QoS part. This can be implemented in different ways depending on the implementation of the M-part.
- The S-part, or Security Part. This is being studied elsewhere.
- There may be other parts defined.

The question then arises of their implementation and modelling forms. In TMN, it is taken as an axiom that the object-oriented paradigm will be used for this purpose. Thus the M-part consists of an area of shared management knowledge in the form of managed objects and the messages directed to and from these.

The P-part in an OSI-conformant environment is CMIP (Common Management Information Protocol). The ODP aspect of the P-part is that this is part of the Technology Viewpoint. CMIP, when combined with its underlying layers of protocol, is extremely complex and processor intensive. The nature of some of these supporting protocols (e.g. Transport) forbids their use at throughput speeds above about 10 Mbit/s due to the time wasted waiting for small windows of data blocks to be acknowledged. In this model, it is extremely reasonable and simple to substitute a lighter protocol without affecting the M-part.

The form of the Q-part in this environment then arises. There are two possible approaches which are non-exclusive.
- All objects which are concerned with QoS have one or more attributes to register QoS
- There are specific QoS Data or parameter objects.

The form taken will depend on the source and purpose of that QoS. The categories of QoS attribute possible are:
- A record of the QoS being received from a peer or underlying object (Received QoS)
- An analysis or history of 1 above (Historical QoS)
- The QoS contracted for across the interface concerned (Planned QoS)
- A QoS value inferred by mapping from an SP parameter (Inferred QoS).

In addition, there will be attributes for SP parameters which are not mapped to QoS parameters but are only of interest to the Provider. An example of such SPs is the amount of buffer space available, etc which does not interest the Customer directly until they run very short. The Provider is interested because they ultimately affect the resiliency of the QoS, equipping levels, etc.

Received QoS would appear to be a candidate for QoS Managed Objects (MOs), whilst Inferred QoS would be most suited to be an attribute (along with measured SP parameters). However, Historical QoS and Planned QoS could be implemented in either manner.

7. THE LANGUAGE OF QoS

QoS is a user-oriented concept. It is described as being perceived directly only by users. Therefore it is an absolute axiom that QoS must be expressed in terms which are both meaningful and relevant to the user. This means:
- meaningful implies that the user can perceive the QoS in the unit and under the conditions stated. For example, it is common to state the SP of data error or data loss in terms of a number of lost or errored cells out of a total number of cells. End Users have no visibility of cells at all so any QoS statement referring to cells is totally meaningless to them. However, as a user can be another layer, e.g. AAL is a user of ATM and would understand cells and cells would be meaningful in this context. For human users or applications, a more meaningful unit must be chosen (usually a unit that is less directly visible or meaningful to the provider.)
- relevant implies that quoting a QoS in terms that the user cannot ever hope to verify is irrelevant. For example, stating that there will be one Prematurely Terminated Call in 100 million is meaningless to a domestic user; how many single users will make that many calls in a lifetime? Again, the QoS must be rethought to be expressed in a way that is relevant to the user. In this case, a domestic user should never experience Premature Termination.

The language appropriate to any given situation is dependent on a number of parameters. The identified parameters are User Type and Timeline, i.e. who the user is and what he/she is doing on which Timeline.

Different users have different expectation of QoS. They also differ in the type of language that is meaningful. The expressing of a QoS in terms of e.g. events per 1000 calls may be quite meaningful to a Network Manager in a large corporation but is pretty useless to an elderly person living alone making, at most two calls a week. Neither of these users can understand a rate expressed in terms of events per 100 000 000 calls.

These are many different types of people who have special needs; varying from elderly people with a general slowing down of faculties to people with particular faculties inoperative due to disease or accident. They have special needs for the requirements on the actual quality offered to them and thus in the specification of the Language of QoS applicable to them. For example, aurally handicapped persons who are using videotelephony need a higher standard of ability of the video transmission/presentation system than may be usual in order to support the use of sign language. These needs affect the language of QoS in that this parameter must be specified and in a language relevant to the application of sending sign language. This may well not be very meaningful to a non-aurally handicapped person.

The second axis contains the timelines. These are progressively less sensitive to the actual user application as the timescale passes from Call to Service infrastructure.

8. ACKNOWLEDGEMENTS

The concepts presented in this paper are based on work performed by the RACE I project QOSMIC during its three year duration. The author is grateful to colleagues both inside and outside QOSMIC, for the help and inspiration received.

9. REFERENCES

[1] RACE Project QOSMIC Deliverable 1.3C : "QoS and Performance Relationships" 82/KT/LM/DS/B/013/b1.

[2] CCITT Rec. E.800 : "Quality of Service and Dependability Vocabulary", Blue Book, 1988.

[3] RACE Project QOSMIC Deliverable 1.4C : "QoS Verification Methodologies", 82/DOW/DSC/DS/B/014/b1.

IV - Implementing the TMN

The Management of Telecommunications Networks
R. Smith, E. H. Mamdani, J. G. Callaghan (Editors)
© Ellis Horwood 1992

An Architecture and other Key Results of Experimental Development of Network and Customer Administration Systems

C. Bernard Hurley (Broadcom, Ireland)

ABSTRACT

Telecommunications management support systems of the near future will be highly complex systems. Developing such systems will require the backing of a concrete and focused approach that only recommendations from experience can provide. This paper describes the key results of experimental development of network and customer administration systems. The descriptions of the system requirements, the information modelling approach, and a complete system architecture are forerunners of necessary and important recommendations to emerge at the end of 1992.

1. BACKGROUND

Consider the impact of this opening statement: management of the imminent broadband telecommunications networks and services will be supported by powerful and complex computer systems. Powerful in this context embraces helpful, useful, efficient, task simplifying, integrated and other commendable system attributes. Complexity derives from network scale, service variety, customer plenitude, and a heterogeneous, distributed and evolving domain. That the systems are powerful will make them desirable, and hence a reality. That they will be complex is unavoidable, but the factual nature of the opening statement presents one obvious impact: how these systems are to be developed and subsequently maintained needs addressing.

At this point, one can expect shouts from various factions of the software engineering world, such as the following intentionally vague comments: "Use the X methodology"; "Follow a structured design technique"; "Be object oriented"; or even "Employ an experienced large scale system development consultant company." In fairness, certain groups will have important contributions to make. However, to cope with the potential scale, complexity, and longevity of a network management system, a more concrete and focused approach is required.

The goal of the ADVANCE project was to produce recommendations on how future Network and Customer Administration Systems (NCASs) should be designed and developed. To put the term administration in context, it can be interpreted as referring to the near real-time but non-time critical, or frequently off-line activities of network management. For example it includes accounting, planning and provisioning. The project has therefore tackled the indicated issues which impact from the opening statement of this paper. That the project has been focused on its task, and produced something of concrete usefulness is without doubt. The approach of defining problem areas within NCAS development, and subsequently prototyping solutions using advanced technologies led to the recommendations which will shortly emerge in the final report [1].

This paper provides a succinct description of several key result areas which are directly related to the forthcoming recommendations. The importance and relevance of these results are guaranteed by the focus on a particular aspect of network management (i.e. network and customer administration), and the hands-on experience of prototype development.

2. INTRODUCTION

This paper discusses several useful and usable concepts for NCAS design and development, with emphasis being placed on an NCAS architecture. The concepts can be grouped in three streams, namely (i) requirements, (ii) overall design, and (iii) specific component design. They are listed in their respective groupings:

Group (i)	Group (ii)	Group (iii)
Requirements,	Model Based Management,	Management Applications,
	The NCAS Architecture,	Computing Platforms,
	Network and Service Model,	User Interfaces,
		Modelling Languages,
		OBSIL.

Following this introduction, individual sections of this paper will address each area in turn. In order to avoid duplication, where an area is documented in detail elsewhere, a reference to the source is provided with a very brief description of the area. Prior to doing this, it is useful to consider how these particular areas emerged, and how they inter-relate.

A number of factors influenced the emergence of these concepts, and shaped their nature. Firstly, emphasis was placed on the highly desirable attributes of near-future, broadband NCASs. These required that NCASs be open, distributed, flexible, expandable, re-usable and integrated. Secondly, the project's objectives demanded that use of Advanced Information Processing (AIP) technologies be emphasised where plausible. Therefore, the state of the art computing disciplines such as object orientation, distributed systems, information modelling principles, knowledge based systems and man-machine interfaces were brought to bear on this telecommunications management task. Thirdly and finally, the project adopted the approach of combining theoretical analyses of AIP techniques with practical applications of the techniques in the prototype development of a partial NCAS. Therefore, the results that emerged did so on the basis of experience in each of the specification, design and implementation phases of system development. This latter point has an important impact on how the reader should view these concepts. It cannot be over emphasised that experience has led to these concepts being promoted as worthwhile.

Regarding the inter-relation of the concept areas, the prime characteristic that they have in common is that they are each what is termed a key result area of ADVANCE. The combination of all key results form the basis of conclusions and recommendations on NCAS design and development. Requirements are simply the requirements of an NCAS which need satisfying in the specification and design. Model based management is a term used to describe the philosophy of viewing virtually all aspects of the managed domain in a conceptual model representation, thus enabling the NCAS to be designed in terms of functional and information models embodying behaviour and specified interaction. This has benefits for both system developer and system end-user, one of which is the common view of the system with which these traditionally disjoint parties are provided.

The NCAS architecture, incorporating model based management principles, is the basis for the design of any NCAS. An architecture is required because a once off system development is not at issue. Rather, as a result of the desired open, de-regulated nature of broadband communications [2], very many NCASs will be implemented. Each will require sufficient commonalty to enable interaction and integration. The architecture is thus the basis for ensuring this commonalty. For similar reasons, a mutual, common understanding of a network and service model will be required, at least at specification and design level.

Moving from the more general design support to specific component design, concepts for the design of management applications, underlying computing platforms, and man-machine interfaces for user access are described. Finally, two languages developed to fulfil needs which arose during experimentation are discussed. One language was used for an information model description and implementation, and the other was used for system interaction.

2.1 Critical Business Problems

Thus far, the content of this paper has been introduced from a technical development viewpoint. This is correct considering the source of the information was a project with a technology focus. However, the project was not blind to the critical business problems which an NCAS should support. These include the ability to:

- Enable a Telecommunications Administration (TA) to administer effectively the provision and use of multimedia communication and management information
- Reduce the telecommunication services introduction delays that arise from the current functional separation of network elements and operations systems
- Minimise the redundancy of data, functionality, and physical computing equipment which exist between network elements and operations systems, and between different types of operations systems
- Enable a rapid response to a customer requiring a service or service adjustments, with the associated reduction in service deployment delays
- Cope with increasingly complex and sophisticated management requirements from customers wishing to exert management control over their use of public multimedia services.

These problems are presented here to help illustrate later how the technical approach of the overall design concepts (group (ii)), and in particular the architecture, enable the support of the problems.

3. REQUIREMENTS

It is not proposed that a comprehensive set of requirements be presented in this section, but instead the approach adopted to compiling the requirements is described. The reasoning is that in many respects this entire project was about requirements, and therefore subsequent sections of this paper will address requirements on certain NCAS aspects. Furthermore, the project never intended producing a full set of functional requirements - a sister project (NETMAN) has addressed such issues [3].

Two strands characterise the approach to requirements compilation. The first strand included theoretical studies early in the project which focused on identifying NCAS user roles and their associated requirements. This resulted in a source of functional requirements for later use and elaboration. Other studies addressed how NCAS might benefit from the use of advanced technologies, and the requirements an NCAS would have on such technologies. Coupled with these studies were the analyses of implementation experiments which were carried out in middle and later stages of the project. Thus it is reasonable to predict that a coherent set of requirements will emerge from this strand of investigation.

The second strand focused on producing a computational view of the NCAS requirements. It adopted the Open Distributed Processing (ODP) abstract requirements model for distributed computing [4]. Briefly, ODP looks at requirements from five viewpoints: enterprise, information, computation, engineering, and technology. Within each viewpoint, requirements for each of seven aspects can be identified, namely storage, process, user, communication, identification, management, and security. By adopting the model, a comprehensive matrix of requirements can be compiled. This project began investigating the ODP approach at a relatively late stage, but has found the model to be useful. Naturally, it is but one model that can be adopted. The important point found with its use is that a common understanding of the viewpoints is achieved with respect to the system under question - in this case the NCAS.

4. MODEL BASED MANAGEMENT

Model based management describes the situation where the functional/processing components of the NCAS are supported in their task by a conceptual model of the world of discourse [5]. The model (i) defines all shared management information, (ii) maintains the consistency and correctness of the information, and (iii) provides a single means of access to it. The model is a logical repository of information, and is thus implementation independent.

Pursuance of the model based management approach within the development of an NCAS is motivated by several facts. The use of models by network management standards and related bodies [6] lend credence and desirability to the approach, and has also influenced the use of the modelling approach by other related projects. However, the ability of conceptual modelling theory to meet, among other needs, the shared information requirements of an NCAS played a large part in adopting this approach.

Conceptual modelling theory dictates that the model provide an intuitive conceptual representation of the real world through the close mapping between real world entities and entities and concepts in the model.

A conceptual model improves the definition and understanding of a consistent set of terminology, i.e. it defines a domain of discourse. Its objective is to capture as many of the real world semantics as are necessary to properly represent real world entities and concepts (e.g. information and behaviour). Modelling concepts should be applied consistently throughout the model, thus ensuring the semantic consistency of the model. Syntactic consistency should be ensured through the use of templates and an appropriate modelling language. Precise syntactic and semantic definitions aid understanding and usability of the model, and help to ensure that it is sound and complete.

The benefits of such a model include the provision of a common view of the system to both developer and end-user (though the former will be more detailed), the support of a common means of access to shared information requirements of NCAS functional elements, and the ability to represent all aspects of administration, whether they be the administrated entity, the NCAS itself or other external entities in a single coherent manner.

The impact of adopting the model based management approach is visible in the following sections of this paper. The NCAS architecture contains a specific component as a direct result of embodying model based management. Section 6 on the network and service model gives a description of the mentioned NCAS component. Other sections describe how NCAS components rely on or support the model at the centre of the model based management approach. Section 10 describes a modelling language developed and used by the project.

5. THE NCAS ARCHITECTURE

Why have an architecture? Consider the alternative to an architecture which would be to produce a single, standard NCAS specification, which, whilst being more specific, would also be excessively prescriptive. It would in effect lead to all NCAS implementations being very similar, thus obliviating the desired openness and choice [2]. On the other hand, an architecture is the basis for ensuring commonalty between NCAS developments, whilst allowing each to remain unique. Therefore, a system architecture is required in order to avoid being in conflict with the desired open, de-regulated nature of broadband communications, whilst still providing necessary system structure. Thus is evident the general desirability of an architecture. More specifically, the NCAS architecture incorporates the model based management principles central to this project. The architecture also imposes a tremendously

useful structure on the NCAS. Furthermore, since its inception, the NCAS architecture has been imbued with several technical objectives which ensure the critical business problems outlined in the introduction are addressed. The technical objectives are to:
- Enable the rapid introduction of telecommunication services and management services through providing a basis for application modularity, reusability and portability
- Enable the integration of management applications through their execution on a common, open, distributed processing platform, which masks the complexities of facility heterogeneity and data and process distribution
- Separate the telecommunication transport technology dependent aspects of switching and transmission equipment from the standardisable management and service related aspects through the specification of generic models, interfaces and NCAS common services
- Enable the integration of sophisticated user-interface methods (e.g. graphics, audio, intelligent interfaces) with management services of the NCAS
- Enable Customer Network Management (CNM) through providing customer access to NCAS services.

A surfeit of "why?" should be avoided. It is time to abandon reasons for the architecture in favour of describing it as portrayed in figure 1.

The NCAS architecture comprises five distinct building blocks, namely (i) Management Applications (MAs), (ii) the Common Information Model (CIM), (iii) Common Services (CSs - also called Servers), (iv) the NCAS Resources, and (v) the Distributed Processing Support Platform (DPSP). These building blocks reside within the NCAS domain, as do the NCAS users (though the latter are not depicted in the diagram). However, the telecommunications network itself remains external to the NCAS domain, in line with current standards [7].

The management applications and the CIM together provide the operational management functionality of the NCAS. In essence, these entities define and form the management capability of the NCAS. The management applications can be viewed as the front-end of that functionality - they are autonomous software components each of which provides a specific management service/facility. Comprising an arbitrary number of software modules, a management application may be invoked either through a Man-Machine Interface, or in the case of automated applications, through the distributed processing support platform. Management applications are capable of interacting with one another, thereby allowing the offering and use of each others services.

The CIM is an object oriented, information model, structuring all information required by management applications. It is logically centralised within an NCAS. Thus, to the management applications the CIM acts as a source and sink for information, which is accessible through a single, well-defined and well-structured interface. Information object definitions of the CIM include information attributes and, more innovatively, behavioural relationships. Furthermore, the CIM contains NCAS functionality which is either (i) common to several management applications, or (ii) intrinsically coupled to an information object within the model (behaviour is embodied in this manner). Functionality of type (ii) can be used to ensure the consistency of the CIM. This is achieved by the propagation throughout the model of operations which have an impact, due to object inter-relationships, on more than just the originally targeted object.

Because the CIM logically contains both the system information and functionality of type (i), it can be viewed as the back-end of the operational management functionality of the NCAS, thus completing the front-end and back-end view of the NCAS user functionality.

In order to explain adequately the role of NCAS common services, it is necessary to first define the meaning of NCAS resources. NCAS resources include management system components such as file systems, databases, host processors and NCAS user terminals. They also include facilities to access network equipment such as exchanges, service modules and tariff tables. The network equipment is typically multivendor sourced and has been accumulated over time, generally involving a large capital investment. It is thus unavoidably heterogeneous in nature. Furthermore, it is topographically distributed due to the nature of telecommunications. The NCAS resources have inherited similar heterogeneous and distributed characteristics.

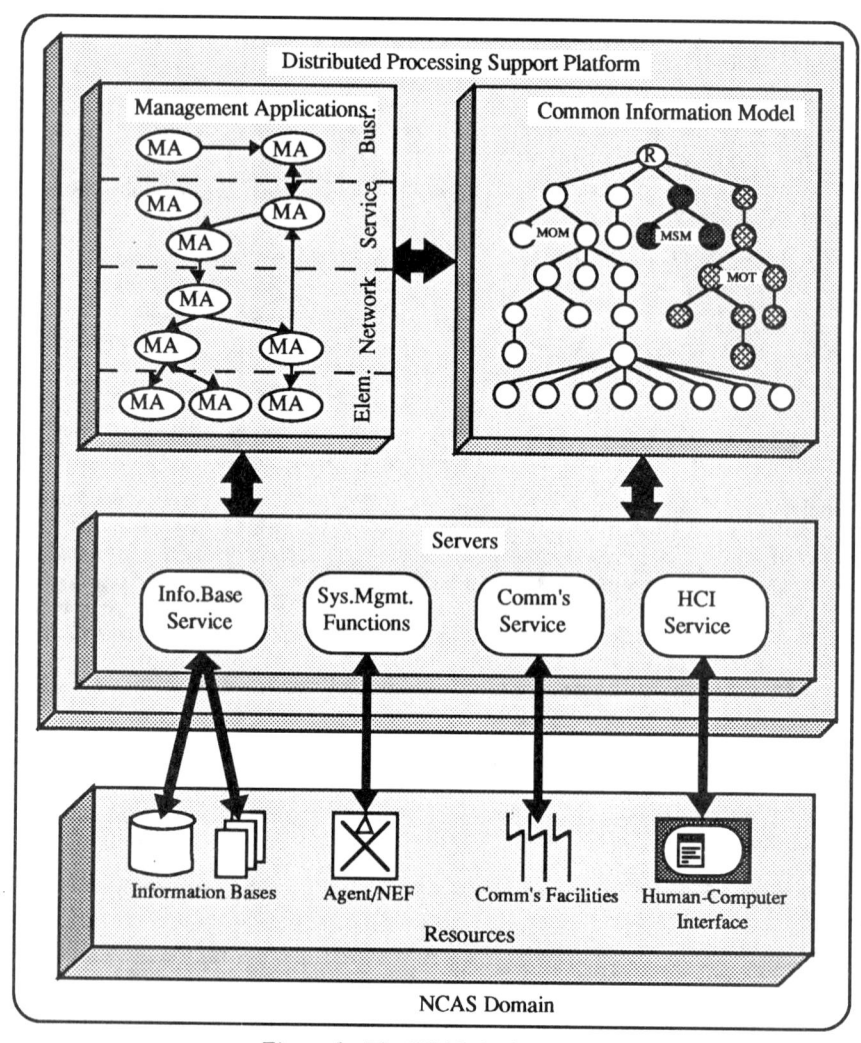

Figure 1 - The NCAS Architecture

Consider now the following scenario an NCAS just comprises management applications (which is too frequently close to the situation of systems currently in use). Two immediate observations could be made: management applications would exist as large, monolithic, unmaintainable and fault prone entities, which interacted (if indeed they were capable of interaction) with tremendous difficulty on the part of the developer and inefficiency on the part of the software components; and secondly, functional and data duplication and their associated redundancies would be endemic in the resultant administration system. Justification for the CIM as described above is immediately visible, in that it enables the avoidance of the twin redundancies of functional and data duplication. However, the combined challenges of heterogeneity and distribution of NCAS resources could continue to make management applications and the CIM very complex in their realisation.

To mask these complexities, common services (or servers as they are frequently called) support both the management applications and the CIM. The common services are tasked with the hiding of complexities brought about due to heterogeneity and distribution of NCAS resources [8]. They may communicate directly with the relevant resources in a totally proprietary manner (this direct, common service to resource communication is indicated by the long, double headed arrows in figure 1) or conform to TMN standards [7]. The common services have little association with the provision of the management functionality of NCAS, apart from their support of the management applications and CIM. Four server groupings have been identified within the architecture, each of which support the same message interface as the CIM. The groupings are (i) an Information Base Service which provides a consistent interface for transparent access to persistent information stored in a variety of formats, (ii) System Management Functions which are provided in conjunction with agent or network management facilities residing within network elements, (iii) a Communications Service which provides a single, consistent interface to local and wide area communication facilities [9], and (iv) a Human-Computer Interaction Service which provides user interface facilities to support information presentation and exchange in an appropriate format (visual/audio, textual/graphical, etc.) to human users. Specific servers are further described in sections 8 and 9 of this paper.

As stated earlier, the interaction between common services and NCAS resources is taken as provided. However, in order to support both the modular nature of the architecture and the internally distributed nature of the architecture building blocks, one must consider the implications and requirements that this creates on distributed processing. The distributed processing support platform addresses these implications and fulfils such requirements. It provides a supporting communications medium, visible as the block underlying the management application, the CIM and server blocks. A communications mechanism is specifically depicted as the thick, short, double-headed arrows in the distributed processing support platform of figure 1. This communications mechanism supports the messaging interface which has been twice alluded to thus far in the architecture description, with reference to the CIM and the servers. The NCAS architecture advocates the use of one interface language for communication between management applications, the CIM and the common services. This language is further described in section 11 on the Object Based System Interaction Language, OBSIL.

In concluding this overall description it is useful to consider the overall structure of the NCAS architecture. A multilevel structure is evident from figure 1, with the management applications and CIM on the top level. Underneath and in support of these are the common services, which reside above the resources. An orthogonal two level structure exists with the management applications, CIM and common services residing over the distributed processing support platform. Though the resources do not reside over the distributed processing support platform, all the building blocks reside within the PV2 domain. Further structuring is apparent within the CIM and management application building blocks. This is described in sections 6 and 7 respectively.

6. NETWORK AND SERVICE MODEL

A detailed description of the network and service model is given in [10], whilst some general information is provided here to complete the architecture description. ETSI, in a recent Strategic Review Committee report on Public Networks to the ETSI Technical Assembly [11], recommends that their standardisation work pay closer attention to the aims of achieving a consistent network description model, and an enhanced service description model. This NCAS research has for some time recognised the vital importance of not only these two models, but for a complete, modular information model, which provides a logically centralised source and sink for NCAS management information. This innovative concept, introduced in the architecture description as the CIM, has its roots in the model based management approach. It also stems from the Management Information Base (MIB) concept of ISO OSI [12], but advocates some significant enhancements, in particular the incorporation of behaviour within the information base.

Being a logical model, it follows that any subdivisions within the model are also logical rather than physical. Three logical divisions are identified within the CIM in order to improve comprehension of the model and to highlight object groupings which can potentially share common requirements and behaviour. Figure 1 portrays these within the CIM block and they are described in the next paragraph. Whilst the CIM is depicted as being structured hierarchically, this merely portrays the possible naming capabilities embodied within the model definition. The model design (i.e. the model based management approach) incorporates further relationships which are more likely to be net structured.

The Managed Object Model (MOM) equates closely with ISO OSI MIB objects, being the information objects required for the management of networks, services and customers within a single TMN domain.

The Model of Other TMNs (MOT) contains objects representing management information of external TMNs, which may have additional security requirements, and will have special communication requirements.

The third division is the Management System Model (MSM) which is best understood when viewed from the context of managing the NCAS itself. Being a complex, multicomponent, distributed processing system, NCAS will need to be managed. Thus, by incorporating an MSM, the same system paradigms that enable the management of networks, services and customers can be applied to the management of NCAS. The MSM therefore contains representation of system components such as management applications, servers, and other related objects.

Information objects of the network and service model, or CIM, are not restricted to being data records, but through fully embracing object oriented concepts can exhibit behaviour, which is realised through object functionality. Three forms of behaviour have been identified, namely (i) the CIM and its objects automatically mimicking the behaviour of the managed resources (e.g. reflecting the status of network elements), (ii) the CIM and its objects executing operations received from management applications and (iii) the CIM and its objects ensuring that changes made by the first two forms of behaviour are consistent with rules held within the model.

At this stage, adequate information on the network and service model has been provided to assist in the further understanding of the NCAS architecture, and to reveal how the CIM supports the task of the next key result area, the management application implementations.

7. MANAGEMENT APPLICATION IMPLEMENTATIONS

A detailed description of the management application implementations is given in [13], whilst an appropriate overview is provided here. Referring once again to figure 1, it is evident that management applications are layered according to BT's Co-operative Networking Architecture for Management (CNA-M) proposal[14], business, service, network and network element management layers. Strictly speaking, the precise functional nature of the management applications is outside the scope of this research, other than that they provide network and customer administration services to the end-user. However, management applications as components of the NCAS architecture required investigation to elicit their requirements and impact on other architecture components. To this end, the project adopted a service provisioning scenario within which several management applications were developed for experimental purposes. The functional design of the applications stressed the experimentation with advanced information processing techniques such as object orientation, knowledge based systems, and man-machine interfaces as an integral part of the application functionality. Furthermore, the applications utilised to the full the capabilities of the CIM and servers, thus exploiting the NCAS architecture. Migration of application functionality into the CIM was tested where appropriate. The stepwise development and latter stage integration of the MAs, though difficult, lead the project to conclude that the model based management approach and associated NCAS architecture led to easier integration of the applications and a flexible and expandable system. Several other important conclusions emerged also, and are presented in [13].

8. COMPUTING PLATFORMS

The NCAS architecture identifies the need for a distributed processing support platform. Moreover, it is frequently the case that NCAS common service component functionality is available in "off the shelf" packages of computer ware (e.g. distributed databases, user interface management systems, etc.). The term computing platform refers to an entity which provides both distributed processing support and other generic, reusable computer system components. Within the context of the NCAS architecture of figure 1, the computing platform embodies both the distributed processing support platform block and the common services block. From the earlier descriptions of the role of common services and the distributed processing support platform, it follows that the task of the computing platform is to mask both heterogeneity and distribution. This project developed a prototype computing platform to serve the needs of both the network and service model, and the management applications, both of which have been described above. The development included a delivery mechanism with a send/receive interface (an implementation of the distributed processing support platform), an 'other information base server' providing relational storage of object oriented information (a service within the architecture's information base server), a model updating service which maintains consistency between the CIM and the NCAS resources (a service within the system management functions), and support for both the modelling and system interaction languages described in sections 10 and 11 respectively. The project has contributed extensively to co-operative research on computing platforms, and this is further documented in [15].

9. USER INTERFACES

Any management system is merely a management support tool, to assist a human or organisation of humans to perform their management task more effectively. In fact the key elements in telecommunications management systems are the management personnel who use the management support tools to carry out a management activity. Thus, the nature of the user interface is of prime importance in an NCAS as it acts as the conduit through which the human manager can avail of the NCAS as a management support tool. The more successfully the user interface integrates the management personnel and the management system, the more effective will be the overall system.

In harmony with the importance of user interfaces, their development has attained the level of providing powerful user interface packages for customisation by a system developer. The services such packages provide are made available in the architecture through the human-computer interaction service.

Development of the user interface design system (UIDS) focused on viewing the human-computer interaction service as a terminal based interface facility. However this would not be sufficient were a fully user-centred methodology (as opposed to processing-centred which emphasises the machines capabilities) to be adopted. In such a case, many other forms of physical interface need to be considered, including paper based, wall mounted (i.e. many users simultaneously), and video based (e.g. security of remote stations). Both [16] and [17] provide more detailed results of the user interface area of investigation.

10. MODELLING LANGUAGE

In order to support an entity such as the CIM, or indeed to test its practicality, some means of defining and realising the model must be available. To accomplish both the support and testing aspects, a modelling language was defined. Termed a Managed Object DEfinition Language, MODEL, it effectively enables the definition of the CIM and its subsequent realisation in an object oriented deductive database form. The network and service model that largely comprise the CIM have been defined and developed in MODEL. To do justice to the language in the space available would be not only impossible but pointless as [5] deals with the language in detail.

11. SYSTEM INTERACTION LANGUAGE

The modular structure of the NCAS architecture identified the need for defining interfaces to satisfy the system interaction requirements of the NCAS. Furthermore, the role of the distributed processing support platform has been identified as providing a communications medium and mechanism to support distribution and system interaction. Thus was born the Object Based System Interaction Language, OBSIL. It combines the expandable usefulness of object oriented operations with the powerful expressiveness of relational querying capabilities. OBSIL addressed and satisfied the interaction requirements of the NCAS architecture building blocks under the NCAS experimenters' control. It is described in detail in [18].

12. CONCLUSIONS

Adopting the model based management approach has proven useful in the research and experimental development of network and customer administration systems. The subsequent form of the NCAS architecture emerged to a large degree independently of other related architectures, yet remains comparable to such architectures, whilst embodying several innovations. The NCAS architecture was found not just useful, but necessary in the development of the partial NCAS, and its subsequent integration. It is considered a concrete and focusing basis for NCAS development.

In total, nine key result areas of experimental development of NCASs have been identified and, within space limitations, described. It is considered an important step towards integrated management of broadband telecommunications networks and services that these results are now emerging.

13. ACKNOWLEDGEMENTS

The author wishes to acknowledge the contributions that a large number of researchers have made to the ADVANCE project since its beginning in 1988. It is due to their substantial endeavour that these important results are now beginning to emerge.

14. REFERENCES

[1] ADVANCE92. : The Final Report of RACE Project ADVANCE (R1009). ADVANCE Deliverable number 4, due for release 31 December, 1992.

[2] Commission of the European Communities :"Towards a Dynamic European Economy: Green Paper on the Development of the Common Market for Telecommunications Services and Equipment", Com (87)290, 30 June 1987.

[3] Walles, A. : "Functional Description of Network Management", Sixth RACE TMN Conference, Madeira, September 1992.

[4] ISO/IEC JTC1/SC21/WG7. Basic Reference Model of Open Distributed Processing, Nederlands Narmalisatie-Instituut, 1990.

[5] Manning, K., and Spencer, D. : "Model Based Network Management", Proc. 4th RACE TMN Conf., Dublin, Ireland, November 1990.

[6] OSI/Network Management Forum. Forum Library of Managed Objects, Classes, Name Bindings and Attributes, Forum 006, GDMO DIS Draft, 1991.

[7] CCITT M.3000 etc. standards which define the separation between management and network

[8] Hurley, B., and Gardiner, P. :"Resolving the Heterogeneity and Distribution Aspects of Communications and Data in IBCN TMN Systems" Proc. 4th RACE TMN Conf., Dublin, Ireland, November 1990.

[9] Hurley, C.B. : "A Communications Handler - Background, Design and Implementation", M.Sc. Thesis, Trinity College, Dublin, Ireland, February 1991.

[10] Manning, K., Sparks, E. : "Service and Network Model Implementation", Sixth RACE TMN Conference, Madeira, September 1992.

[11] ETSI (European Telecommunications Standards Institute). Strategic Review Committee on Public Networks Report to the Technical Assembly. March 1992.

[12] ISO (International Standards Organisation). Information Processing Systems - Open Systems Interconnection - Basic Reference Model - Part 4: Management Framework. ISO IS 7498-4, Geneva, 1989.

[13] Da Silva, L. et al : "Implementation of Management Applications for Network & Customer Administration Systems", Sixth RACE TMN Conference, Madeira, September 1992.

[14] British Telecom Plc, "Management Reference Model: Management Layers and Domains", Co-operative Network Architecture, Vol. 7, Infrastructure, PA 0036:Part 2:1.00, 1991

[15] Wade, V. et al : "Experience Designing TMN Computing Platforms for Contrasting TMN Management Applications", Sixth RACE TMN Conference, Madeira, September 1992.

[16] Meroni, M., and Morris, C. : "A Methodology for Developing NCAS User Interfaces", Proc. 5th RACE TMN Conf., London, England, 1991.

[17] Mandich, N., Belleli, T. : "HCI Considerations in TMN Systems", Sixth RACE TMN Conference, Madeira, September 1992.

[18] ADVANCE92. Object Based System Interaction Language, OBSIL, User Guide 4.0. RACE Project 1009 Internal Document, 1992.

The Management of Telecommunications Networks
R. Smith, E. H. Mamdani, J. G. Callaghan (Editors)
© Ellis Horwood 1992

Viewpoints on Traffic and Quality of Service Management in Telecommunication Management Networks

Pravin Patel, Alex Galis (Dowty, UK), George Pavlou (University College London, UK),
Kurt Geihs (Johann Wolfgang Goethe-University, Germany)

ABSTRACT

This paper describes the modelling of an experimental Traffic and Quality of Service (QoS) Telecommunications Management Network (TMN) system that manages a simulated Integrated Broadband Communications (IBC) network based on Asynchronous Transfer Mode (ATM) technology. The TMN system of the IBC network is based on the emerging TMN standards [1] and TMN Architecture [2]. The modelling techniques used are based on the emerging Open Distributed Processing (ODP) [3] viewpoints and the OSI/Network Management Forum (OSI/NMF)[4] perspectives. The entire experimental system [5], from the IBC network (a simulator) to the TMN system was modelled using the ODP viewpoints, while the OSI/NMF perspectives were used to model and implement the QoS management functions of the TMN system.

1. INTRODUCTION

IBC networks of the future will offer a multitude of end user services ranging from existing voice and data services to new services such as video telephony. Traffic and QoS management functions are an important part of TMN as attempts are made to obtain the maximum throughput from the networks whilst maintaining the QoS provided to the end users. The Traffic and QoS management functions of these networks will be very complex. This complexity arises from distribution of the network, services and management and the different types of services offered on the same network. The experiment implements a subset of the issues involved in Traffic and QoS management for IBC networks. The experiment consists of a number of TMN emulators which interact with a simulated IBC Network. To understand the system being modelled, a knowledge of the issues involved in the Traffic and Quality of Service management is a prerequisite [6].

TMN systems are very large and complex distributed systems to which ODP and OSI/NM viewpoints are applicable. OSI/NM perspectives are specific to management systems and could be mapped directly onto the ODP viewpoints [2]. Section 2 is a short introduction to the experimental work. Section 3 is an introduction to the ODP viewpoints and OSI/NMF perspectives. Section 4 and section 5 describe the use of ODP viewpoints and OSI/NMF perspectives in the NEMESYS experiment. Our conclusions on this subject are outlined in section 6.

2. NEMESYS EXPERIMENT

A description of Traffic and QoS management for IBC networks was introduced in [6] and expanded in [7]. These issues have also been addressed by the CCITT in the M and I series recommendations [1], [8]. The functions are detailed in [9] and RACE Common Functional Specifications [10]. The experiment case study and design are documented in [5] and [11]. The following are the main subsystems of the experiment :

IBC Network Simulators

The simulated IBC network consists of a user simulator, a service simulator and a network simulator. The network simulator establishes and simulates ATM connections across the simulated network. The network simulator supplies the management system with information about the state of the connections and calls made in the network. The service simulator informs the management of any QoS degradation and user call profiles. The simulators implement management decisions.

Traffic Management System

This manages the network simulator. The traffic management system controls the routing and bandwidth allocated to virtual paths (VP). The Call Acceptance Function (CAF) parameters are computed by the traffic manager. These are required by the network simulator for the acceptance of a new connection onto the network. The CAF parameters are computed by the Call Acceptance Management (CAM).

Service Management System

This monitors the service simulator. From the simulator data it computes the QoS as perceived by the users and computes the traffic load predictions for the network. The QoS reports and the load predictions are send to the traffic manager for action on the network simulator.

3. MODELLING TECHNIQUES

The modelling techniques used in the NEMESYS experiment are the ODP viewpoints[3] and OSI/NMF perspectives [4]. The OSI/NMF perspectives are directly mappable into the ODP viewpoints and are specific to management systems [2]. The ODP viewpoints are applicable to any distributed system.

3.1 ODP Viewpoints

ODP addresses the underlying architecture for the development, management and operation of distributed information systems. To this end, the ISO ODP group have generated a Basic Reference model of ODP in five parts: Overview, Descriptive model, Prescriptive model, User model and the Architectural semantics [3].

In an attempt to deal with the full complexity of a distributed system, the ODP standards consider a system from different viewpoints, each of which is chosen to reflect one set of design concerns. Each viewpoint represents a different abstraction of the original distributed system, and is only concerned with the issues from that particular viewpoint, all other issues being ignored. The viewpoints are Enterprise, Computational, Information, Engineering and Technology. Each of these viewpoints has aspects, these being Process, Storage, User, Communications, Identification, Management and Security.

3.2 OSI/NMF Perspectives

The OSI/NMF is specific to distributed network management systems. It has a general architectural framework for addressing interoperable issues and a number of different perspectives are used for modelling the entire distributed management system. Each perspective describes a different abstraction of some aspects of the general model, its major components and their interactions. Collectively, the different perspectives address the total problem of interoperable network management. The five perspectives are Enterprise, Single Managed Object, Managed Object Relationships, Logical Distribution and Physical Distribution Perspectives. These have been defined in the OSI/NMF Forum Architecture document [4].

4. THE USE OF OPEN DISTRIBUTED PROCESSING VIEWPOINTS IN THE NEMESYS EXPERIMENT

4.1 Approach

The experiment was implemented in three stages. The first stage involved an understanding and decomposing of the complete experimental system using the ODP viewpoints. The second stage involved the analysis of the experiment components using traditional analysis and design techniques and the emerging object oriented analysis and design methods. The third stage involved the development of the system.

The approach undertaken for the first stage was a pictorial analysis of the experiment using ODP viewpoints. A number of diagrams represented each ODP viewpoint. These diagrams were further refined to reduce the level of abstraction and put more detail into the design of the experiment. These viewpoints were revisited after the second stage to reflect the accuracy of the experiment. In all cases the viewpoint models are compliant with the TMN implementation architecture defined in [12].

4.2 Enterprise Viewpoint

This addresses the legal, administrative, legislative and business concerns of an enterprise. The enterprise being modelled is an IBC network and its management. The derived model mainly addresses the administrative and ownership concerns. However this viewpoint is vital and necessary for development of the other viewpoints since it forms the basis from which other viewpoints develop. The experiment based on an IBC network and its management is modelled from three views in the enterprise viewpoint:

- The Network Enterprise model: This shows administrative and business concerns of the network and its management

- The Organisation Enterprise model: This shows that the experiment components are organised in relation to the network enterprise and is derived from the Network Enterprise model

- The Human Interfaces model: In any human enterprise, the involvement of the people is of vital concern to the operation of the enterprise. This model shows the relationship of the IBC network enterprise to human users

The Telecommunications Network Enterprise Model

Figure 1 represents the ATM Network Enterprise model mapped onto the experiment. The successive stages required to arrive at this model are described in the NEMESYS Experiment design [11]. The physical network is derived from the work of the CCITT study group XVIII [8].

The role of providers, suppliers, ownership and administrators is still under review by the legislative bodies in most countries. For example, in a deregulated telecommunications market the virtual network would most likely be controlled by value added suppliers, whilst in a regulated market all the control is by the public telecommunication supplier. This conflict was resolved by the Network Enterprise being owned by one organisation but the control and management of segments of the IBC network being done by departments within the organisation.

Figure 1 also shows the system interactions with the human user.

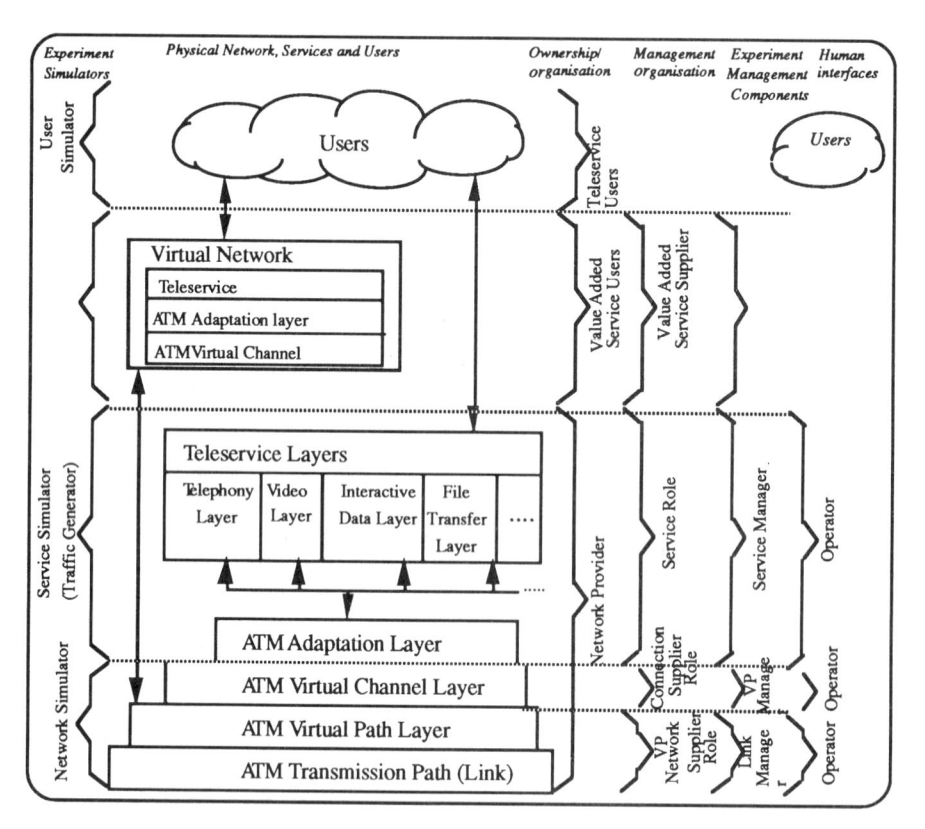

Figure 1 - Telecommunications Network Enterprise model mapped onto NEMESYS Experiment

4.3 Information Viewpoint

Information is interpreted as data flows between the components of the system. It is created and stored at various places. This includes the Management Information Base (MIB) as defined in [2] and flow of information as events which may/may not trigger management activity and/or network reconfiguration. The location of the information is necessary in this viewpoint. The viewpoint reflects this by showing the logical level in which the information is stored. This complies with the CNA-M levels defined in [2].

The information viewpoint is modelled from the following areas of concern:

- Event Interfaces: This is information exchange between functional components. The standards bodies have stressed the importance of the interfaces between TMN Operational Systems [1], this viewpoint shows compliance with the q and x reference points of the TMN architecture[2]

- Management Information Base (MIB): The Manager Information Base (MIB) is based on the CMIS standard [13]. This is the repository of all management information in the TMN system

- Data files. Since there are on-line and off-line components in this experiment, the interfaces between these are through files.

The Event Interfaces model

The event interfaces of the information viewpoint are shown in figure 2. This displays potential event interfaces between the simulators and the Management Applications (MAs) and inter-MA event flows. A MA is the smallest amount of TMN functionality that can be assigned to an operating system process. In practice, several MAs are assigned to each process. This shows one TMN system composed of three components which interact with each other via the q3 reference points. The interface between the network (simulators) and the TMN systems is via the qx reference point.

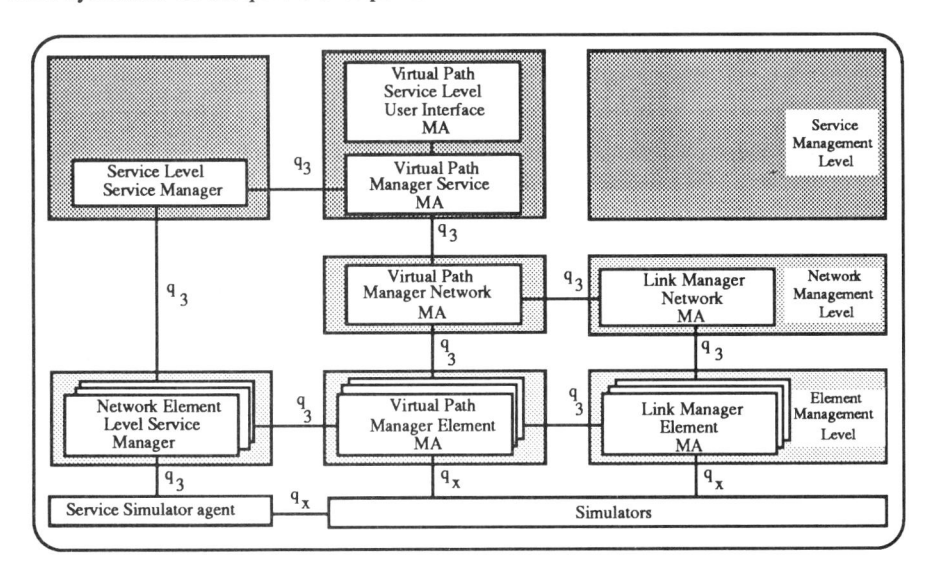

Figure 2 - Reference Configuration for the NEMESYS Experiment

Management Information Base (MIB) Model

Figure 3 shows the MIB of the NEMESYS experiment. Details of the Managed Objects (MO) can be obtained from [5]. This figure shows the structure and types of MO that have been implemented in the experiment. Each level has shadow MOs of the level below except for the Service Simulator agent (SSA). The SSA is a mediation function which translates simulator events into managed objects and management actions on the managed objects into events for the service simulator. Details of the managed objects are given in [11].

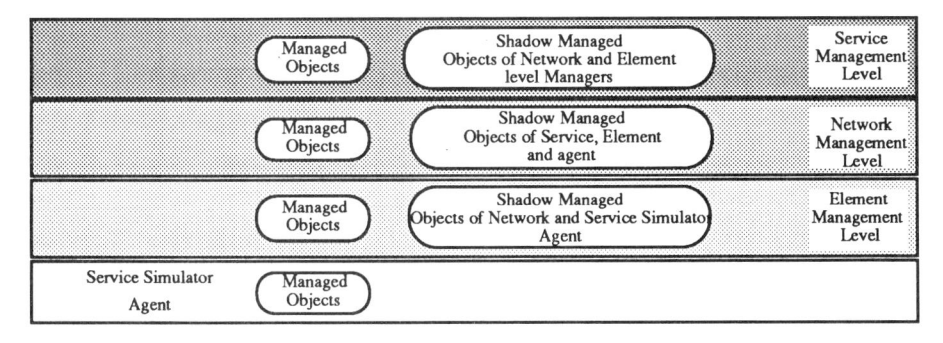

Figure 3 - Management Information Base (MIB) model

4.4 Computational Viewpoint

This viewpoint lays the foundation for the development of the system. Figure 4 shows the experiment computational model; it also shows logical interaction of the events and the information flow. The experiment has a number of functional blocks:

- Simulators Functional blocks: These are the components of the simulator and the functional performed in these blocks

- Experiment Platform Functional blocks: These are the components of the experiment control which provides the distribution framework for the other components

- Service and Traffic Management Functional blocks.

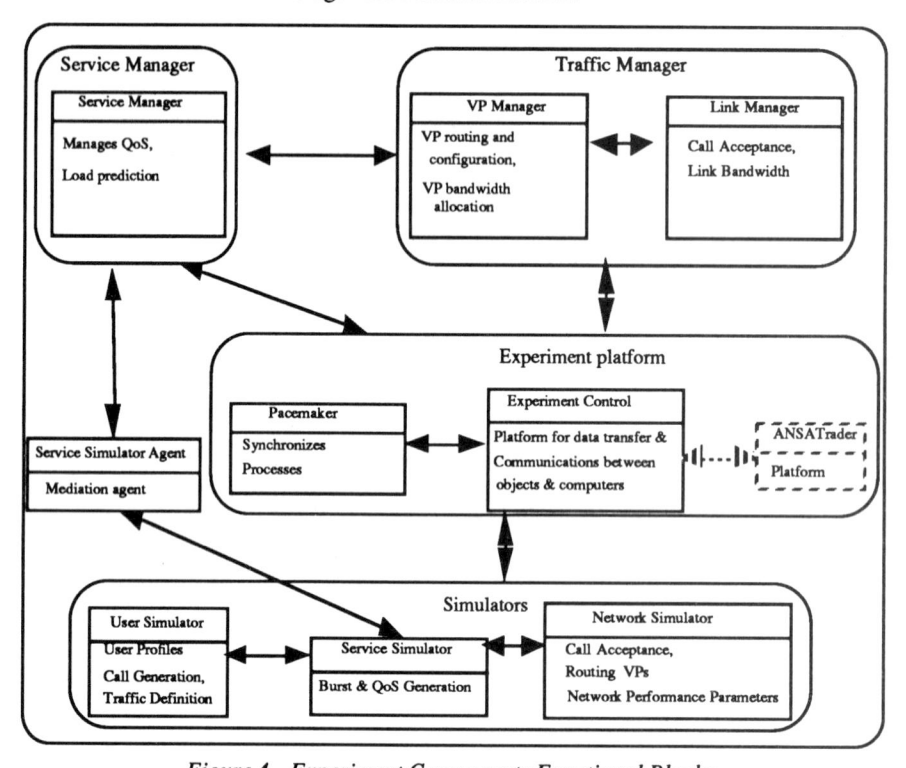

Figure 4 - Experiment Components Functional Blocks

4.5. Engineering Viewpoint

This is the physical implementation of the computation viewpoint onto an information processing system. The experiment is engineered by mapping experiment functional blocks to the operating system processes. The techniques used in the experiment for the implementation are part of this viewpoint. The mapping of the functional blocks to the processes is achieved by a configuration file, in most cases there is a one to one mapping between the functional block and a process. Figure 5 shows the engineering techniques where used in the experiment.

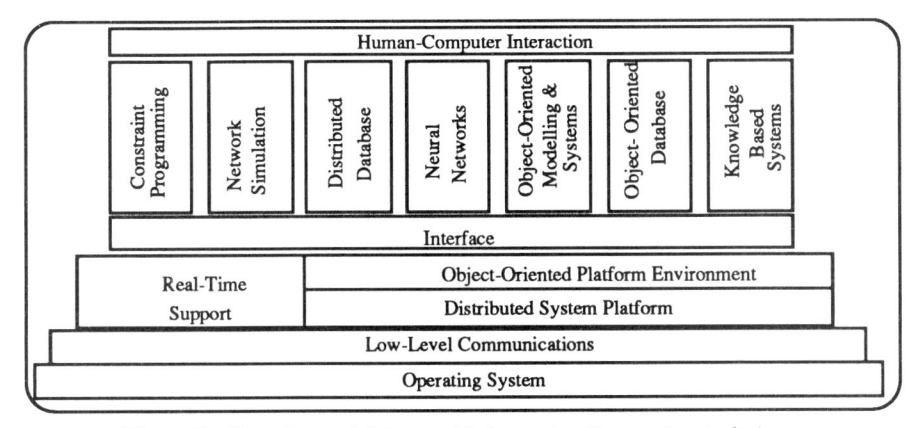

Figure 5 - Experimental Advanced Information Processing techniques

4.6 Technology Viewpoint

The hardware and the software tools used in experimental are highlighted in this section. The tools used to implement the engineering techniques used shown in figure 6.

KEY

UDP= User Datagram Protocol
RPC= Remote Procedure Call
CMIP= Common Management Information Protocol
TCP/IP= Transmission Control Protocol/ Internet Protocol
GMS/SMIB= Generic Management System/ Shadow Management Information Base
MAIB= Management Application Information Base

( ) = Engineering Tools.

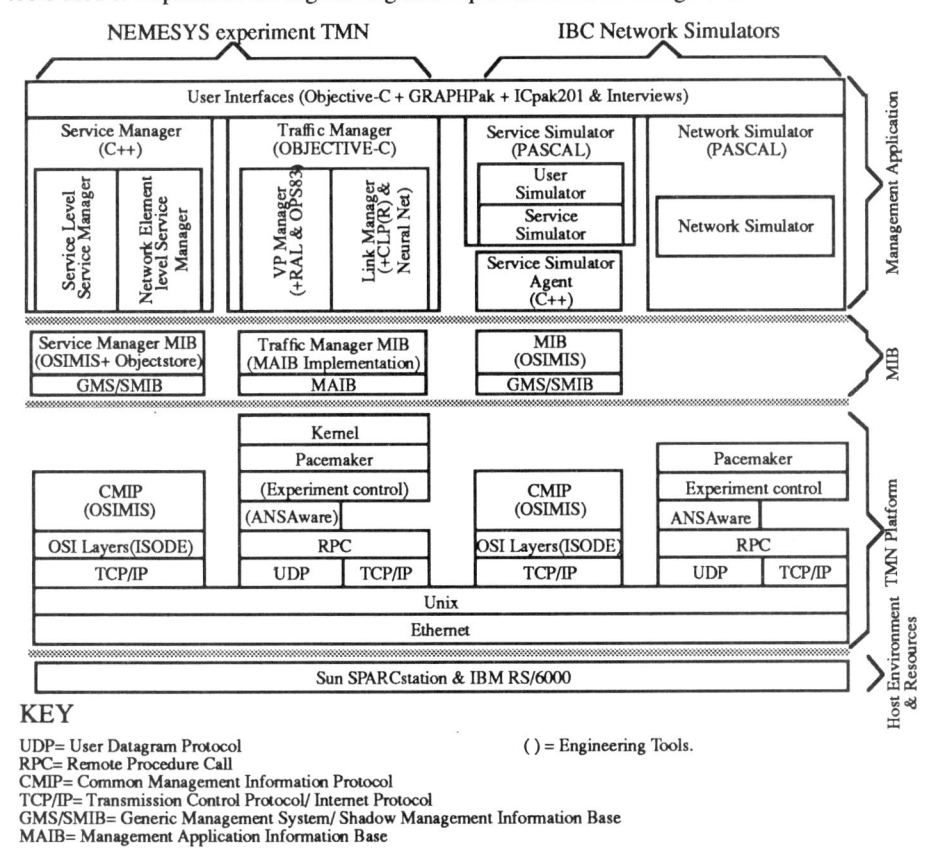

Figure 6 - Experiment Technologies and their relationship to the TMN Layers

5. MODELLING THE SERVICE MANAGEMENT FUNCTIONS USING OSI/NM PERSPECTIVES

The ODP viewpoints provide a powerful tool for modelling distributed systems of any nature. TMN is a distributed application of specific nature and it is based on an object-oriented framework. The management of physical or logical real resources is enabled through abstractions of them known as Managed Objects (MO), accessed via a management protocol [14]. Management applications in agent roles support these objects while applications in manager roles access them in order to implement management policies.

The OSI Network Management Forum (OSI/NMF) [4] has defined five different perspectives, each modelling a separate aspect of the management system and collectively addressing the problem of distributed interoperable network management. The service management functions of the experimental TMN system were modelled using these perspectives and also implemented using real interoperable management services and protocols.

5.1 The Enterprise Perspective

The enterprise perspective of network management is concerned with user requirements, policies and the broadest level of interoperability modelling. Interoperability is achieved using the interoperable interface across a management network, in this case the TMN.

In the NEMESYS experiment there is one TMN owned by one organisation but control and management of the IBC network and services may be distributed within that organisation. Teleservices for example are offered by providers having no control over the management of the underlying IBC network. Instead, they should interact with the ATM network providers to introduce and withdraw services, point-out problems and provide statistics that may serve for a better planning at the network level.

These interactions should occur over interoperable interfaces and aspects such as interoperability, shared conceptual management information schema and interworking policies become important. Interoperability is achieved by using the OSI management protocol CMIS/P over a full OSI protocol stack. This is adequate for the management applications of the teleservice provider as there exists a common understanding of the management needs and information model. An interworking policy must be agreed which leads to a shared conceptual management schema.

In this case, the teleservice suppliers provide complaints on network connections offering poor quality of service according to previously agreed thresholds and they also process utilisation data to provide load predictions for the future. This is the agreed interworking policy and these interactions are naturally realised as event reports emitted from the related managed objects: QoS complaints are emitted from network connection objects receiving a poor service and load predictions from a "network" managed object as understood by the teleservice provider. The structure of the management information exchanged and its semantics must be commonly understood and that is what is achieved through the shared conceptual schema.

5.2 The Single Managed Object Perspective

Management systems exchange information modelled in terms of managed objects. The properties of a managed object are defined in an abstract form [14] and this specification together with the management access protocol CMIS/P [13], [15] identify uniquely the interoperable interface.

The managed objects are modelled according to the management application needs. As a first step after the requirements specification, the managed objects should be identified. This is a difficult task as all management capabilities should be expressed as managed object properties. Extensibility provisions to management needs should also be made and careful modelling may minimise the interactions between management applications.

From this perspective, the necessary managed object classes need to be identified. This includes their attributes, supported actions and emitted notifications. Operations on attributes and actions may trigger interactions to the associated real resource: it is through this modelling that management capabilities will be achieved, enabling monitoring through event reporting and reading of attribute values and control (intrusive management) through setting attribute values and actions.

In the case of service management, the managed object classes that may be identified relate to service associations (calls), network connections, service providers, service users, networks, services, performance profiles etc.

5.3 The Managed Object Relationships Perspective

Managed objects do not exist in isolation. In fact, relationships such as inheritance and containment are fundamental to the existence of managed objects: managed object classes come into existence through registration to a global inheritance tree while managed object instances are named through containment relationships in the Management Information Tree (MIT) [14]. .Information and behaviour may be inherited through generic objects and also alternative behaviour may be supported through polymorphism. For example, the classes teleservice and bearer-service are refinements of the generic class service. With containment, managed object instances acquire unique names through the relative names of their containing objects. Containment relationships are specified through name bindings which tell which class can be the superior and which attribute names the particular class in that instance. There are also other relationships such as IS-CONNECTED-TO, BACKS-UP, USES which may be expressed by special attributes or even special relationship managed objects. The inheritance and containment trees for Quality of Service Management are shown in Figures 7 and 8.

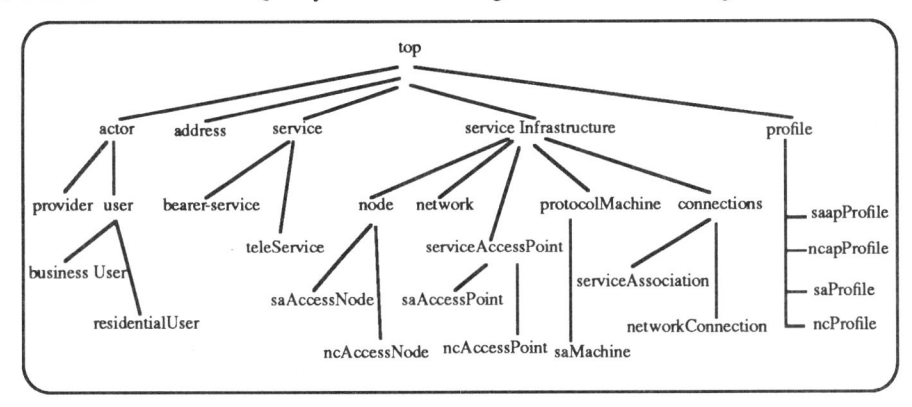

Figure 7 - Managed object relationships perspective- Inheritance Hierarchy

5.4 The Logical Distribution Perspective

The relationships between managed objects do not consider at all their distribution. The global QoS containment tree shown in figure 8 could be distributed physically across many interoperating management applications across the network. This physical distribution is decided according to the management needs and the authority relationships.

In this case, the relationships are hierarchical as the management model complies closely to the TMN Logical Layered Architecture [2]. There are three layers in this architecture regarding QoS management. The Network Element layer which in this case comprises the actual teleservices, these being the "elements" of service management. Monitoring and controlling these may be done in an non standard way, but an object-oriented view can be presented to higher layers through a mediation function (SSA - Service Simulation Agent as services and the IBC network are simulated). Objects at this level comprise service associations, the underlying network connections and associated profiles.

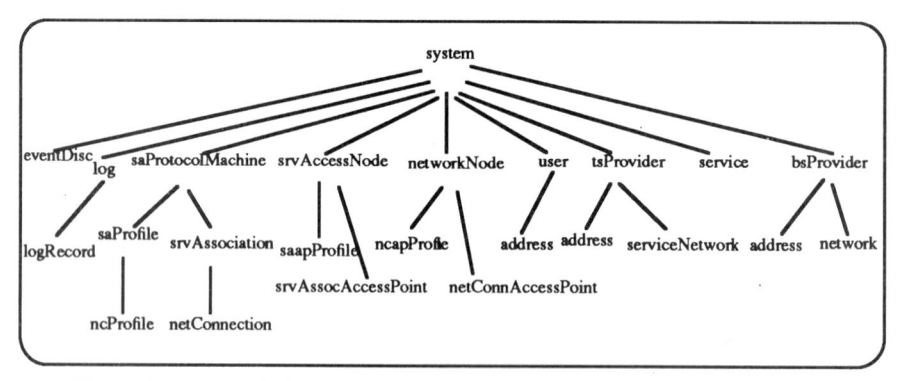

Figure 8 - Managed object relationships perspective- Containment Hierarchy

The Network Element Management layer manipulates managed objects at the network element layer and creates higher level abstractions. These are service and network access points and associated profiles. It is also responsible for producing the quality of service complaints when the monitored QoS decreases and for aggregating utilisation information at each access point. The latter is used by the layer above in order to evaluate patterns of usage in the service network.

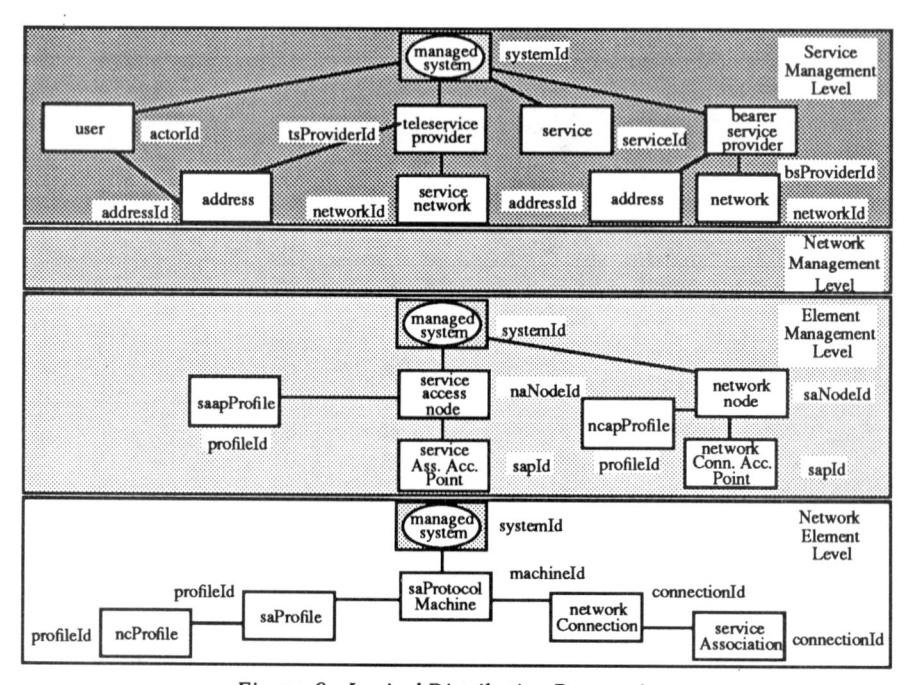

Figure 9 - Logical Distribution Perspective

Finally, the Service Layer has a global view of the service network in terms of service providers, services and users. A part of this information is made available to peer level functions of the network providers. Load predictions derived from the patterns of usage are sent to assist their management functions. The management schema is exported and the management knowledge is shared. This logical distribution is presented in Figure 9, the containment relationships and the managed object classes at each layer being shown.

5.5 The Physical Distribution Perspective

The physical distribution covers the mechanisms, structures and rules when the cooperating applications are realised by a physically distributed set of information processing systems. The logically layered management information model may be realised as a set of distributed applications (figure 10). At the lowest layer, services will exist at the customers premises and there may be many mediation function boxes. The actual services and users are simulated by one application, the service and user simulator. There is also one only mediation function block, the SSA.

The same applies to the other two layers where there is one only Operations System in each layer: the Network Element Level Service Manager (NELSM) and the Service Level Service Manager (SLSM). In a real system, there would probably exist more operations systems in each layer and relationships would also be peer-to-peer as opposed to strictly hierarchical.

Interoperable interfaces and protocols used are also important. The Q3 and X interfaces are realised as managed object interactions through CMIS/P and a full OSI stack. The Qx interface to the actual services could be non standard: in the experiment case, it is a TCP-based message passing mechanism between the Service Simulator and the SSA.

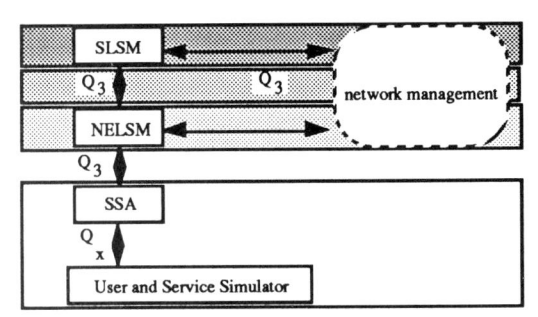

Figure 10 - Physical distribution perspective

6. CONCLUSION

Broadband Networks and Open Distributed Systems are technologies which significantly change the way information is processed and disseminated. Broadband Network Management Systems will be linked to form large distributed systems, which will be inherently heterogeneous. Management Systems will have to be open in order to facilitate interaction and cooperation between themselves. Open Distributed Processing sets up the framework within which this cooperation will be feasible.

This paper presents the applications of the ODP and OSI/NM frameworks to the design of Traffic and Quality of Service Management Systems for Integrated Broadband Communications Networks. It argues that the ODP and OSI/NM viewpoints represent a significant modelling tool in the analysis of the problem domain as well in the design of a solution. The usability of the ODP modelling in TMN requires automation tools which will assist in the development and the formalisation of the models and their consistency.

Viewing the system from five different viewpoints produces five different abstractions which must be consistent with each other. Maintaining the consistency when developing five different models of the same system is one of the main issues to be clarified in the future.

7. ACKNOWLEDGEMENTS

The work described in this paper was carried out in the RACE NEMESYS project as part of the investigation and evaluation of the role of Advanced Information Processing (AIP) techniques in Traffic and Quality of Service (QoS) management for IBC networks.

8. REFERENCES

[1] CCITT Draft Recommendations M.3010 - "Principles for a telecommunications management network," November 1991

[2] GUIDELINE Deliverable ME8 : "TMN Implementation Architecture", 03/DOW/SAR/DS/B/012/b3, RACE Project R1003 GUIDELINE, March 1992.

[3] ISO/IEC JTC1/SC21/WG7: Basic Reference model of Open Distributed Processing. Parts 1 to Part 5.

[4] OSI/Network Management Forum Architecture- January 1990.

[5] NEMESYS Deliverable 9 : "NEMESYS Experiment Case Study Description", 05/ICS/GS/DS/C/019/A1 September 1991.

[6] Patel, P., Griffin, P. : "Traffic Management for IBC Networks" - Proceeding of the 5th RACE TMN Conference. 20th-22nd November 1991.

[7] Gentilhomme, A. et al : "The Use of AIP Techniques in Traffic and Quality of Service Management Systems", Sixth RACE TMN Conference, Madeira, September 1992.

[8] CCITT COMXVIII Recommendation I.610 : "OAM Principles of the B-ISDN access."

[9] RACE project GUIDELINE Deliverable ME 7 : "Initial Report on the ITF Common Case Study", 03/BCM/ITF/DS/C/011/b1, February 1992

[10] RACE Common Functional Specifications - RACE Central Office, January 1992.

[11] NEMESYS Deliverable 10 : "Experiment 3 Design", 05/DOW/SAR/DS/B/024/A1 May 1992.

[12] Callaghan, J. G. et al : "TMN Architecture", Sixth RACE TMN Conference, Madeira, September 1992.

[13] ISO/IS 9595 - Information Technology - Open Systems Interconnection - Common Management Information Service Definition 1991

[14] ISO/IS 10165-4 Information Technology- Structure of Management Information: Guidelines for Definition of Managed Objects, 1991

[15] ISO/IS 9596 - Information Technology -Open Systems Interconnection - Common Management Information Protocol Specification 1991

[16] ISO/IS 10165-4 Information Technology - Structure of Management Information - Part 1: Management Information Model 1991

The Management of Telecommunications Networks
R. Smith, E. H. Mamdani, J. G. Callaghan (Editors)
© Ellis Horwood 1992

Implementation of Management Applications for Network & Customer Administration Systems

Luis da Silva (CET, Portugal), P. T. Wright (Roke Manor Research, UK)
J. Girvin (GPT, UK), F. Carvalho (CET, Portugal)

ABSTRACT

This paper reports on the definition and issues concerning the Provisioning task within Network and Customer Administration Systems (NCAS) and describes the results of the use of novel approaches for the implementation of an integrated set of Management Applications (MA). These include the use of object orientation in the software life cycle, the use of Advanced Information Processing (AIP) techniques such as Knowledge Based Systems (KBS) in order to produce more modular, re-usable and capable management applications. Also, the use of Man-Machine Interface (MMI) tools and methodologies, and the adoption of a user centred MMI design approach in order to provide consistency across NCAS operative interfaces, are described.

1. INTRODUCTION

This paper describes one aspect of the work carried out within the RACE project ADVANCE. Within the scope of ADVANCE project activities, two areas had been subject of particular attention:

- investigation of an architecture which can be used as a baseline for the development of an NCAS system
- prototyping a specific example of such an NCAS system in order to produce concrete results as an aid to prove or disprove some of the concepts being developed within the investigation work packages.

This paper concentrates on the investigations and prototyping of Management Application (MA) functionality within an NCAS system, particularly focusing on how the work with respect to application functionality has been biased towards:

- exercising and demonstrating the main concepts of the ADVANCE NCAS logical architecture [1]
- investigating and demonstrating the applicability of the use of AIP techniques for the realisation of certain MAs.

2. SCENARIO DESCRIPTION

In order to exercise the architectural components of the second ADVANCE prototyping phase, an integrated set of Management Applications has been chosen which is involved with all aspects of the NCAS function, i.e. customer, service and network administration. A Management Application scenario has been devised which begins with the interaction between the NCAS and an external customer (or potential customer) who may have a requirement for a new service, or a complaint or query about an existing service. The example adopted in the MA scenario is a large banking business with a number of geographically distributed sites, each with varying service needs. Overall, this customer will require a number of multi-media services,

provided in a number of configurations between a number of sites. The analysis is split into four main functions:
- interaction with the customer & elicitation of requirements
- analysis and construction of the service details
- analysis of the proposed service impact on the network status in order to determine what network alterations may need to be undertaken in order to provide the service
- analysis of service costs and scheduling implementation.

Figure 1 shows the applications that are being prototyped. The breakdown of application functionality was undertaken for pragmatic reasons to permit project partners to investigate and develop an easily identifiable function within the MA scenario. This breakdown also permits the investigation of the complexities of the interaction between separate NCAS application functions. At the time of developing the MA scenario, there was little detailed information available from the standards bodies such as CCITT and ETSI.

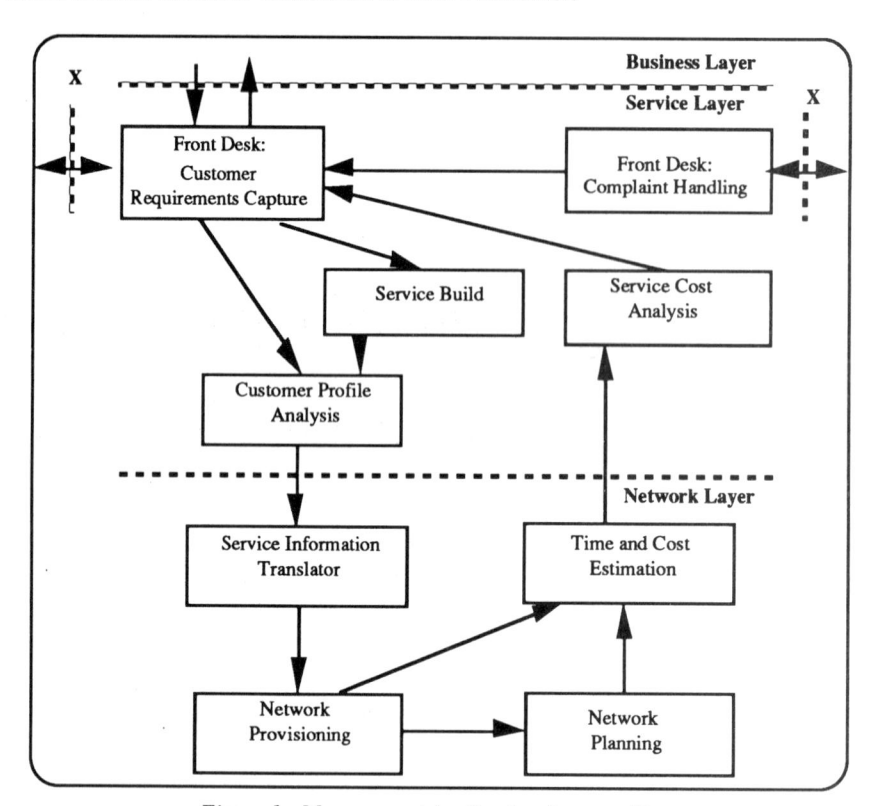

Figure 1 - Management Application Interworking

The focus of the scenario is the Customer Requirements Capture function (which is part of the overall Front Desk operation). Two types of customer input to the Customer Requirements Capture are being investigated:

Customer Request. The banking customer decides on the need for a new service or the upgrading of an existing service. The customer interacts with the Service Provider TMN via the Front Desk: Customer Requirements Capture function in order to extract all the necessary

requirements that will permit a detailed analysis by the Service Management Layer. Information that will be required from the customer will include:
- current services (if not known by the Service Provider TMN)
- any requirements for additional supplementary services
- geographical information
- special billing requirements, security requirements
- end user terminal information
- predicted usage details.

Customer Complaint. Alternatively, users within the customer organisation may also log complaints with some Customer Management function when they have particular problems with a service. A complaint handling function within the Customer Management Layer can interact with its peer applications within the Service Provider TMN through the X interface (as defined in M.3010 [2]). Correlation of users complaints might involve four outcomes:
- fault detection, logging and analysis
- performance problems, logging and analysis
- customer having an incorrectly specified service and therefore a need for this service to be replaced or upgraded
- service provider having incorrectly specified and implemented the service and therefore there is a need for this service to be replaced or upgraded.

Once the customer facing function has the necessary information, stored in the form of Customer Requests, it is necessary to perform a detailed analysis of this information at the Service Management Layer in order to produce a "quote" to the customer of the service cost.

Two particular functions have been investigated at the Service layer to analyse the customer needs. The first of these functions is concerned with the construction of the new service specification from the technical detail gained from the customer. The service is constructed from a basic set of service components. The second function deals with the information from the customer regarding the usage of the service and produces a detailed analysis of the various requirements that must be met by the network to support the service. The requirements may include:
- information processing (data, video, audio etc.)
- information storage (data, video, audio etc.)
- encryption
- signalling
- local & remote communications.

Hence, a detailed description of the overall service impact will be presented to the Network Layer as a basis for analysis of the network support requirements. Service requirements passed down to the Network Layer functions are network independent in nature, i.e. a service specification deals with a purely service view of the requirements. Within the Network Layer, the first task is to translate service specific information into network specific information. One particularly important sub task involved here is to identify network and network elements that will be affected by the proposed service in order to perform an initial analysis of the impact on the different technology networks. Within the ADVANCE prototyping work, an ATM based network is being investigated.

Analysis of a particular network type is broken down into a number of sub-functions. The two main sub-functions deal with provisioning and planning of the network. The impact of introducing the service for a customer may have little or no impact on the network. In such cases some simple provisioning of the network may be sufficient to support this extra network traffic, e.g. reconfiguration of routing tables. If provisioning actions cannot be made to support this traffic it will be necessary to undertake some more complicated network restructuring which will entail planning changes to the network, e.g. increasing the capacity of links and/or nodes or even introducing new links and/or nodes into the network. This is quite a complicated task and it is possible that several options could result, each with different consequences, for example:

- cost
- time scales
- extra capacity to introduce other planned services more quickly.

Therefore a number of Network Change Plan options can result from Network Planning. It must be noted that the scenario assumes that the provisioning and planning functions only concentrate on the local access networks as the effect of the extra traffic on the backbone network will result in a minimal increase in traffic over that network.

The required network reconfiguration details, whether from the Provisioning or Planning functions, are passed on to an estimation function in order to analyse the Network Change Plans to determine an overall change plan in terms of time scales and relative costing of the proposed changes. This information is necessary for functions at the Business Layer that need to determine whether the predicted revenue from particular services will cover the network and running costs and give sufficient return for profits. Network Layer costing information is passed up to the Service Layer where a similar function produces a Service Costing Analysis for the proposed service based on customer size, present customer usage, tariffing schemes etc. Hence costing information from the Network and Service Layers is calculated and stored for use by the Business Layer.

3. NETWORK PLANNING MA DESCRIPTION

3.1 Introduction

Section 2 described the overall functionality of the MA scenario, however in order to provide the reader with a more detailed view of the work done in this area, one of the MAs specified and prototyped, is described in more detail. This MA is the Network Planning application. The Network Planning MA resides in the Network Layer (as defined in BT's CNA model [3]) and presents a set of generic characteristics, such as its inherent complexity, interface requirements with the Common Information Model (CIM) [4], interaction requirements with the other MAs and with the user. These aspects present an interesting and complete case study not only for AIP techniques investigation, but also for the evaluation of the ADVANCE NCAS logical architecture.

3.2 Relevant Concepts

The service requirements passed down to the Network layer, which are network independent in nature, are translated by the Service Information Translator MA into network specific requirements. This is accomplished by specifying the appropriate bearer services that are used to transport the service, and the set of requirements not directly mapped into pre-existing bearer services and referred to as non-bearer service information.

In the ATM scenario, the Network Provisioning and Planning MAs are presented with a request to establish a set of trails (Virtual Path Connections) from one source NAP (network access point) to a sink NAP. The set of trails (opposed to a single one) is due to the fact that each bearer service related traffic may flow across the network using different Virtual Path Links and Virtual Path Switches [5]. The NAP represents the aggregation of customers of a given geographical area. In the context of the network scenario, they correspond to the concept of Network Access Points instantiating the connection point between CPNs and the ATM based public network. Figure 2 represents the ATM Virtual Path and Virtual Channel levels, which are important concepts of the ATM network scenario.

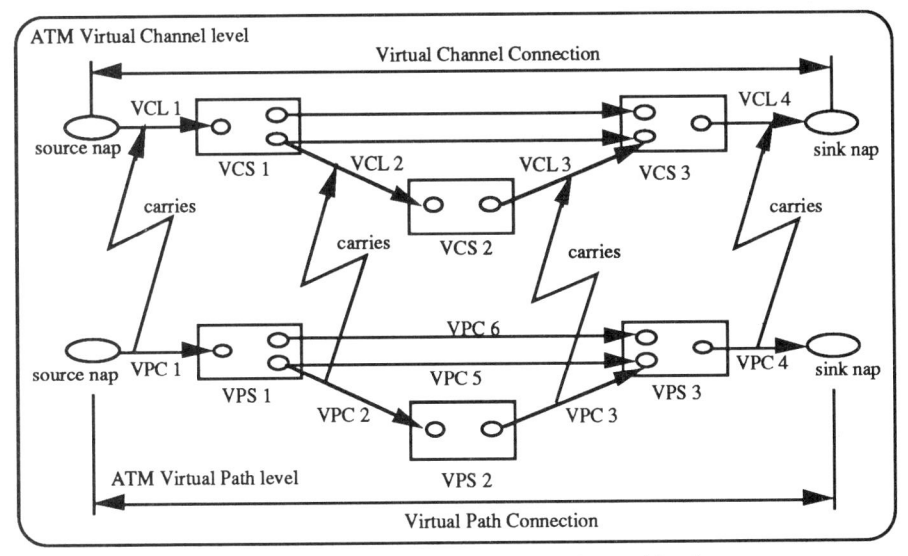

Figure 2 - The ATM Virtual Path and Channel levels

3.3 Functional Breakdown

Figure 3 depicts the functional decomposition of the Network Planning MA. It comprises the following modules: Coordination, Requirements Establishment, Network Analysis, Plan Development, Plan Reporting and Private Object Base. The diagram also shows the Common Information Model (CIM) [4], an important component of the ADVANCE NCAS logical architecture with which the MAs are conformant.

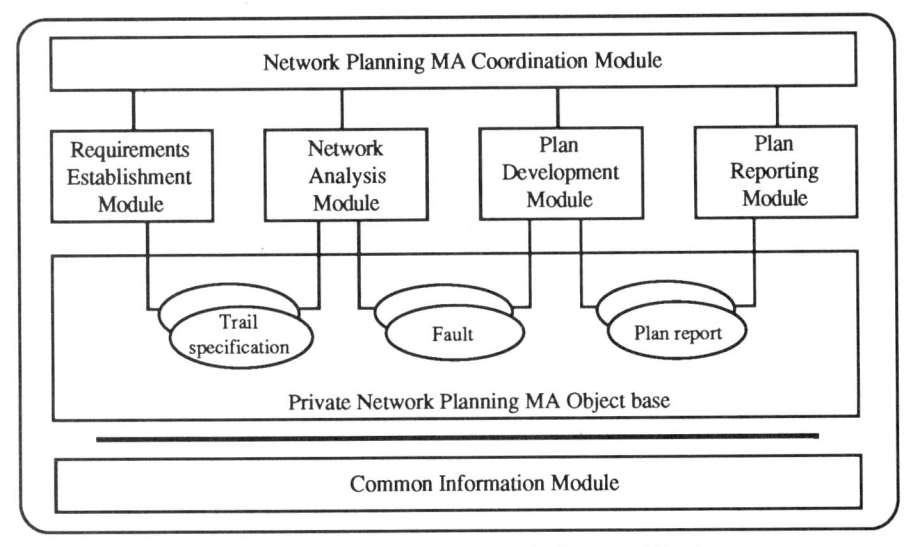

Figure 3 - Network Planning MA: Architectural Design

Within the framework of the experiment, this Planning MA is triggered by the reception of Network Provisioning Reports coming from the Network Provisioning MA, and reports back its proposal plan to the Time & Cost Estimation MA in the form of Network Planning Reports. Each module is described in some detail below:

Network Planning MA Coordination Module

This module is responsible for the correct execution of the Network Planning MA. It implements the overall functionality of the MA, calling all the other modules and controlling their proper execution. The planning process is not a sequential one, several iterations may be needed, which are triggered and controlled by this module.

Requirements Establishment Module

The Requirements Establishment module deals with the elaboration of the requirements that the service provisioning process, through the Network Feasibility Request, indirectly poses to the Planning MA. This module retrieves from the CIM the Network Provisioning Reports, which have been generated by the execution of the Network Provisioning MA, and processes them. The main objective of this module execution is the generation of the requirements for the ATM based network capability planning.

The Network Provisioning Reports provide a number of Trails and a number of Paths for each Trail, together with the failed network elements on these Paths. Based on the information obtained from the Network Provisioning MA, the Requirements Establishment module instantiates in the Private Object Base, corresponding Trail Specifications and Path Specifications, which will be used by the Network analysis module.

Network Analysis Module

This module is responsible for the task of network analysis for provisioning purposes. It aims at testing the ability of the network to sustain the specified trails, as these are identified or calculated in the Requirements Establishment module. Since the connectivity between the two NAPs of a trail has been examined (resulting in paths identification), each path is traced in order to determine the reasons for its deficiency to sustain the trail.

These reasons, which, among others, may indicate insufficient capacity in a link or switch, or lack of special features in a network element, are given in the form of Fault objects that are stored in the Private Object Base for later retrieval by other modules.

In order to evaluate the total traffic figures at each link, this module needs information on the bandwidth associated with each service. This information is retrieved from the CIM.

Plan Development Module

The Plan Development module deals with the task of changing the network configuration either by reconfiguration procedures or by upgrade. A decision making process is employed to find out the appropriate method for overcoming faults. This means that the network deficiencies, which have been generated during the Network Analysis module execution, are analysed and, according to this analysis certain planning related actions are determined. The methods and functions available to carry out these actions follow:

Trails Reconfiguration and Node Reconfiguration

The two methods attempt to satisfy the requirements for provisioning by rerouting some of the already existing virtual channel connections (VCCs) in order to free bandwidth on a specific physical link (Trails Reconfiguration), or attempts to reassign the available exchange capacity to appropriate port-groups (Node Reconfiguration).

Node Upgrade and Link Upgrade

If the node reconfiguration does not succeed, the node upgrade methods and the link upgrade methods are attempted. These methods aim at adding new ports at specific port-groups in order to increase the available capacity of the exchange (Node Upgrade) and at increasing the available capacity of the physical links (Link Upgrade).

Plan Reporting Module

This module deals with the generation of Network Planning Reports, which outline what proposals for network configuration changes the Network Planning MA is to report. This Network Planning Report is stored in the CIM to be retrieved, at a later date, and analysed by the Time and Cost Estimation MA.

The Network Planning addresses the problem of planning the network from the technical point of view but the validation part is the responsibility of the Time & Cost Estimation MA.

3.4 Design Aspects

The design of the Network Planning modules followed the Object Oriented paradigm. The OOSD (Object Oriented Structured Design) methodology [6] was used in the design representation, and the SPOKE [7] Object Oriented environment was the tool used in the implementation of the modules. KBS techniques were used in the design of the Plan Development module, addressing in particular the Decision making process which selects and invokes the appropriate planning method (or combination of methods).

A MMI has been included allowing user interaction with the modules mainly for demonstration purposes. This MMI was implemented using SPOKE's PanelBuilder tool, which is a screen layout editor.

4. IMPLEMENTATIONS ISSUES

A number of AIP techniques have been used during the development of the different MAs. Specifically the Object Oriented paradigm was extensively used in all the MAs. The reason for choosing object-orientation is that it seems a very promising candidate for advancing the state of developing and maintaining of software systems in operating organisations. This extensive use of Object-orientation in the MAs for PV2 also followed the conclusions and recommendations from the first phase of prototyping of the ADVANCE project [8].

Three OO programming languages were used in the development of the MAs, namely SPOKE, ART-IM [9], and C++. Within the project there was great concern on the use of an OO design methodology. A survey was undertaken which identified the OOSD as the best candidate to be used in the MA specification activity. The OOSD methodology was extensively followed in the specification and design of most of the MAs.

Although all the MAs have implemented User Interfaces, the majority of them aim to be used only for stand alone demonstration purposes. It was in the MAs of the Service Layer and especially in the Front Desk MA where the MMI techniques were particularly important. The OpenLook GUIDE toolkit [10] and the ADVANCE MMI guidelines [11], produced by a special MMI activity within the ADVANCE project, were used in the design of the Front Desk MA User Interface. This User Interface has the special feature of using the "intelligent questionnaire" concept as the underlying model rather than the GUI (Graphical User Interface) computer window system.

Knowledge Based System techniques were used in the Front Desk, the SIT, the Network Planning and the T&C MAs. SPOKE Engine and the ART-IM KBS were used in the implementation, based on a rule based approach, of the Decision Making processes in those MAs.

The MAs were designed and implemented taking into consideration the fact that the common management information they need to retrieve or update for the accomplishment of their functionality was in the CIM - Common Information Model. This component of the NCAS architecture contains not only the shared management information, but also shared functionality.

The MAs interact with the CIM using an interaction language called OBSIL [12] (Object Based System Interaction Language) which is based on the ISO CMIS (Common Management Information Service) message specification, but allows for much greater flexibility in the retrieval of information from the CIM. This permits the MAs to pose very powerful, selective queries to the CIM, and leave the CIM to perform the selection and processing of the MA requirements. An example of such a query is shown below:

"give me all transit exchanges in a certain geographic area which are currently blocked"

Hence, a good deal of complex information processing is abstracted away from the MAs and into the CIM. Ultimately, this could lead to low level MA functionality being processed in the CIM, e.g. route finding, etc. The CIM is also used in the support of MA interaction. The MAs prototyped do not interact directly with each other. All the interactions are made through the CIM, which provides a notification based mechanism that insures that the correct MA will be invoked to accomplish any information processing needed.

The inter-TMN interaction aspects were also investigated by this activity. The interaction between a Customer Management System and the service layer of a Network Provider Management System was specified and prototyped. The correspondent communications functionality was modelled in manager, agent and managed objects terms, using CMIS messages for communication.

5. THE REPRESENTATIVE MA

As a result of the work described in Sections 2, 3 and 4 of this paper, it has been possible to abstract away from the specific requirements of these management applications, in order to develop the more general requirements of management application functions which have been identified as necessary to support the overall functionality of the NCAS system. The requirements of the management applications can be sub-classified into: functional requirements, and non-functional requirements.

5.1 Functional Requirements

As stated above, the MA scenario concentrates mainly on service provisioning within the NCAS. However, the ADVANCE project has previously looked in some depth at the application functionality of a typical NCAS [13]. The functional areas described below can be seen to map mainly onto the Configuration and Accounting Management areas of the OSI FCAPS [14] model. Application functionality is partitioned in the following way:

Customer Administration

Functional requirements of customer administration systems include the selling of services to customers including customer queries and customer requirements, creation and maintenance of customer accounts including charging and billing and the customer service desk function which must perform such tasks as advice to customers, handling customer complaints, service modification and customer disconnection and security checking.

Service Administration

Functional requirements of service administration are concerned primarily with pre-service functions such as long term analysis, service planning and service specification and negotiation with the customer. Also in the context of these requirements are the in-service functions such as the provision, modification and removal of services, providing service information to the end-user, other administrations and the business and network management layers and the monitoring of services and restoration of fixed services.

Network Administration

Functional requirements of network administration may include correlation of network information, planning the future of the network based on traffic, fault and performance statistics, monitoring network behaviour and implementation of configuration changes.

5.2 Non Functional Requirements

The non-functional requirements of NCAS management applications has been sub-classified into the following aspects (roughly in line with the aspects of the ISO ODP model [15]):

Information Requirements

Information requirements of an NCAS management application can be categorised into three separate perspectives:
- *owned information*. This information may be highly dependent upon the task of a particular management application and is kept under the responsibility of that application
- *shared information*. Information held within the NCAS but used by a number of applications. Requirements include concurrent access, transparency of information etc.
- *external information*. Information on managed resources either in another part of the TMN or from another TMN.

Security Requirements

Management applications provide the window into the NCAS through which information can be exchanged by humans and/or machines. Therefore, management applications provide an obvious mechanism through which the security of the NCAS may be threatened. Security requirements of NCAS applications include the following:
- an authentication facility for verification and acceptance of a user
- an authorisation facility to permit restrictions on the access of users to the NCAS resources via one or more NCAS applications
- an association function to ensure the security over the transfer of information to/from applications within the NCAS.

Design Requirements

The design of an NCAS Management Application requires specific NCAS knowledge. The application developer must take into consideration all the NCAS features, constraints and facilities. Information and presentation may vary depending on whether the application is dealing with customers, services or networks. A user centred approach to the design takes into account the type and experience of the user and the requirements upon the application in terms of performance required, frequency of usage of the application, importance of the application to both the users overall job role and the integrity of the NCAS, etc.

User Interface Requirements

Two particular aspects have been identified with respect to user interface (UI) requirements for NCAS management applications:
- the ergonomics of the user interface, i.e. those particularly associated with the psycho-physical characteristics of the interface, particularly the requirements of NCAS applications with respect to use of colour, size of fonts, display of complex information on the screen, etc.
- conversational aspects of the UI, i.e. how the user and the interface carry on a dialogue.

Processing Requirements

An NCAS management application may be viewed as a number of threads of processing with each thread having its own processing requirements. Such processing of an NCAS management application must consider the following as necessary considerations:
- reliability (i.e. the extent to which a process is expected to perform its intended function with the required precision)
- accuracy
- cost-effectiveness, i.e. the amount of resources consumed to achieve the task
- completeness, the degree to which a full implementation of the application function has been achieved
- determinism, the ability of the application to process the inputs received by it.

6. CONCLUSIONS

The investigation and experimentation work on management applications has produced a greater understanding of both the functional and non-functional requirements of the NCAS as an integral part of the TMN. Prototyping work has shown how NCAS management applications can operate as part of an NCAS based upon the model-based management concept which has been adopted within the ADVANCE project.

The concept of the Common Information Model (CIM) is central to the NCAS architecture. Common information regarding customers, services and networks is accessed via the CIM which is responsible for maintaining the correctness and consistency of this information. Management Applications have been shown to operate within the model based management concept. All application interaction with the other resources within the NCAS (including other applications) is performed via the CIM. This approach lends itself very neatly to the concepts that the NCAS should be an open, transparent, distributed, easily extendible system. The CIM allows application designers to concern themselves only with local application information. Common, shared information is accessed via a standardised access language (OBSIL) based upon and extending the CMIS concept. Maintenance, security and availability of this information is not the responsibility of the applications.

Also investigated were the possibilities of migration of functionality into the CIM, where it can be identified that application functionality, at a coarser level, share application sub-functions, e.g. route-finding. However, the trade-off between decreased duplication of both functionality and data must be weighed against the increasing size and complexity of the CIM and the number of entities wishing to access the CIM and their potential frequency of access. The benefit of such an approach would allow the application designer to browse through CIM functionality in order to build part or all of their application from generic building blocks located within the CIM.

The design of Management Applications has used an AIP approach in specification and design in order to assess the applicability of such techniques to the future development of NCAS systems in particular and TMN systems in general. All of the applications have been specified and developed using an object oriented approach in order to assess the suitability of object orientation to the design and development of complex, distributed systems such as an NCAS. Wherever possible AIP techniques such as KBS have been used as a means of automating as much of the NCAS functionality as possible. As the relationship between customers, services and networks become increasingly more complicated with the ever more diverse nature of telecommunications services and networks, the need to manage these vast amounts of information in terms of analysing performance, potential network problems, predicting expansion plans, etc., provide network management tasks that are particularly well suited to the adoption of KBS and use of embedded knowledge in expert systems.

Object orientation has also been investigated as an alternative to the more conventional software engineering approaches used in today's present systems. Object orientation provides a more natural way of decomposing the design process, allowing for the more natural grouping of an entities data and functionality. Object orientation provides a more natural means of producing modularity in the design, which permits the designer a better means of being able to initially decompose the task. It also permits a better means of maintenance of the application software and the distinct modular entities permit for simpler code re-use for upgrades, enhancements, etc., by the Management Application maintainer.

The use of user interface design and building tools and adoption of basic MMI guidelines has assisted the application design, particularly with respect to the production of standardising the 'look and feel' of the application user interfaces to the NCAS operators. A further development of this work would be the investigation of an MMI service which would allow applications requiring similar user interface functionality to share common MMI resources. Also, the adoption of a user-centred design of the user friendly interfaces will lead to the production of user interfaces tailored to meet the needs of the NCAS operative rather than the developers perception of those needs.

7. REFERENCES

[1] Hurley, C. B. : "An Architecture and other Key Results of Experimental Development of Network and Customer Administration Systems", Sixth RACE TMN Conference, Madeira, September 1992.

[2] CCITT Draft Rec. M.3010 "Principles for a telecommunications management network (TMN)"; Q23/IV December 91.

[3] BT CNA (Co-operative Networking Architecture) Management / Network Modelling, PA0003- Part 1, DEC 91.

[4] Manning, K., Sparks, E. : "Service and Network Model Implementation", Sixth RACE TMN Conference, Madeira, September 1992.

[5] CCITT Draft Rec. I.311 "B-ISDN General Network Aspects", Melbourne, Dec. 91.

[6] Wasserman, A. et al : "Concepts of Object-Oriented Structured Design", Technical Report, IDE.

[7] "SPOKE, Object Oriented Programming Environment", V3.1.2 Alcatel ISR / Alcatel Alsthom Research

[8] R1009 ADVANCE, Deliverable 2: "Report on Prototype Version 1", July 90.

[9] " ART-IM Reference Manual", Inference Corporation, 1991

[10] OpenWindow Developers Guide 1.1", User Manual, June 1990.

[11] R1009 ADVANCE "MMI General Guidelines", Internal Doc. ADSYD056, June 91.

[12] Azmoodeh, M., Shomali, R. : "OBSIL: An object based system interaction language" in RACE-TMN Object-Oriented Modelling SIG proceedings, June 1991.

[13] R1009 ADVANCE, Deliverable : "Network and Customer Administration System Model", 09/BCM/LMI/DS/B/017/B1, July 1990.

[14] ISO - International Organisation for Standardisation; Open Systems Interconnection - Basic Reference Model - Part 4: Management Framework, ISO/IEC 7498-4:1989

[15] ISO - International Organisation for Standardisation; Basic Reference Model of Open Distributed Processing", ISO/IEC JTC1/SC21/WG7 N308, Oct. 1990.

V - Experimental Results

The Management of Telecommunications Networks
R. Smith, E. H. Mamdani, J. G. Callaghan (Editors)
© Ellis Horwood 1992

Virtual Path And Call Acceptance Management for ATM Networks

Pravin Patel, Peter Loader (Dowty, UK), Alain Gach (IBM, France),
Verna Friesen (Institute of Computer Science, Crete, Greece),
Christian Kaas-Petersen (Copenhagen Telephone Company, Denmark),
Charles Mialaret (Gsi Erli, France)

ABSTRACT

This paper presents the traffic management functions for a multi-service Integrated Broadband Communications (IBC) network using Asynchronous Transfer Mode (ATM) technology. The specific areas of traffic management which are considered are aspects of Virtual Path Management (VPM) and Call Acceptance Management (CAM). This paper describes the VPM and CAM of an experimental traffic management emulator which manages a simulated ATM network. The role of traffic management for ATM networks is outlined. Details of the experimental management components are presented. In addition experimental results available at the time of writing are outlined.

1. INTRODUCTION

The work described in this paper was conducted in a project whose aim was the evaluation of Advanced Information Processing (AIP) techniques in the context of traffic & Quality of Service (QoS) management. This project developed a series of traffic and QoS management emulators as a means to perform this evaluation. Although the experimental emulators were only a means to an end they implemented realistic traffic and QoS management functionality. This paper gives an overview of the traffic management aspects of that functionality. The reader may refer to [1] in this volume for an overall ODP-centred description of the experimental system developed. The reader may also refer to [2] for an elaboration of the service and QoS management and [3] for AIP evaluations in the experiment. These papers describe separate but related aspects of this work. Further details of the experiment specification and design can be found in [4] and [5].

Requirements of Traffic Management

The main requirement of traffic management is to optimise the use of the network so that the network resources are used as efficiently as possible whilst ensuring that the QoS experienced by the service users is not degraded. Because the IBC network is intended to transport many different services it must be able to cope with a multitude of connection types with differing characteristics including: bandwidth requirements at connection establishment, QoS requirements, and user behaviour. Many of these parameters are unknown today and will be unknown or changing during the operation of such networks. In addition ATM technology involves the concept of Virtual Paths for flexibility of both call acceptance and routing. This means that control functions in the network are not efficient if they are fixed. New traffic management functions are required to control/manage dynamically the operating parameters of the network in response to changing requirements. An important distinction is the difference between control functions that operate within the network itself and the management of these control functions. When configured, the network can operate on its own using the control functions although it may use resources inefficiently. Management aims to

optimise resource utilisation and does so by altering the control parameters in the network. This paper describes
- The structure of the experimental performance management emulator
- Detailed traffic management functionality including Virtual path and routing management where user behaviour changes throughout the day and Call acceptance management where the attributes of the calls are not well known.

1.1 OVERVIEW OF TRAFFIC MANAGEMENT FUNCTIONS

Control functions of routing, call acceptance and switching functions are performed automatically in the network. Management functions attempt to optimise network performance by setting the parameters of control functions. This section gives a synopsis of the network functions which are impacted by management actions. This is followed by the management of these functions.

ATM Network Traffic Functions

The technology recommended within RACE and CCITT for implementing IBC networks is ATM [6]. Information in ATM networks is transported in fixed length cells. Cells are identified on a link (ATM transmission path) by their Virtual Channel Identifier (VCI) and Virtual Path Identifier (VPI). Cells belonging to a particular connection are allocated to a virtual channel and groups of virtual channels are allocated to a virtual path which in turn are grouped onto a link. Virtual paths form a logical overlay network on the links [6],[7].

In this paper "Virtual Paths" (VP) are virtual path connections and "connections" are virtual channel connections. VPs will usually span more than one link. Cells belonging to a particular VP will be routed between links (along which their VP traverses) at VP cross-connect nodes. Cells belonging to particular virtual channels can be switched between virtual paths at virtual channel switches. Also in this paper, a call is defined as an end-to-end service provided to a user/customer of the network. A call may consist of several connections in the network. These connections have certain common characteristics and can be grouped into Classes of Service (CoS). Within each CoS, the random distributions and parameters of the traffic are similar, as are the performance targets (cell loss rate).

A unique traffic management problem in ATM is the management of the VPs. VP configuration allows the network behaviour to be changed dynamically, since all the traffic is transmitted via the VPs. This problem is unique to ATM for the following reasons:
- VPs form flexible logical networks
- Bandwidth allocated to VPs can be altered dynamically
- VP routing tables and route selection can be altered to meet changing traffic requirements.

Another basic traffic management problem is the admission control of new call requests onto an ATM network. In this paper this is referred to as "Call Acceptance". It is a problem of particular importance in ATM networks for several reasons:
- ATM networks put few constraints on traffic sources, which can have throughputs varying with time, filling up to the source or the link capacity, unless some policing mechanism is used (e.g. leaky bucket)
- ATM networks have few capabilities for reducing instantaneous throughput of traffic sources. In fact, due to the large throughput considered, any reaction may be too slow to avoid congestion efficiently
- Contracts between network users and operators will include QoS commitments. This implies that the network operator must be able to monitor and control network performance and hence influence QoS.

Traffic Management Structure

The experiment is composed of a number of management emulators which interact with a simulated network [1], [4], [5]. The management emulators being:
- The service managers which are responsible for the management of the QoS received by the users and the load prediction calculations derived from user behaviour. Details of this component is described in [2]
- The Virtual Path Manager (VPM) which is responsible for VP management
- The link manager which is responsible for the management of the link traffic. It has two main functions: Call Acceptance Management and Link Bandwidth Management.

The VPM and the Link manager together form the traffic manager. Figure 1 shows the overall structure of the experiment and the relationships between the functional blocks.

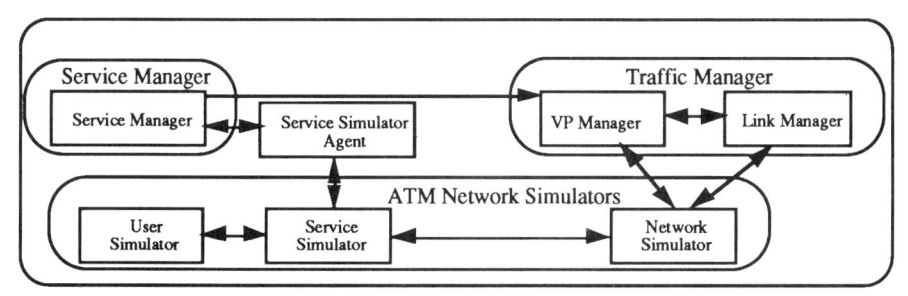

Figure 1 - Experiment structure

Traffic Management Functions

The traffic managers continually monitor the performance of the network and modify its parameters so that the network is used efficiently and users do not suffer degraded QoS. The VPM is responsible for creating and modifying the VP routing tables and their bandwidth. CAM on the other hand is responsible for identifying the bandwidth required by connection types and for identifying the bandwidth of multiplexed connections. Further details of traffic management functions can be obtained from [8]. Traffic management attempts to influence the following:
- The strategy employed to accept calls in the network at the nodes. The algorithm used by the Call Acceptance Function can be altered by the traffic manager
- The bandwidth allocated to a connection in the network. Algorithms will be used to allocate the bandwidth to VPs, and to connections on the virtual paths during the call establishment phase. These algorithms can be managed from the traffic manager
- Routing table generation and the route selection strategies can be managed by the traffic manager. This is particularly useful for congestion control and load balancing as well as re-routing during network element failures.

2. VIRTUAL PATH MANAGER

2.1 Introduction

The role of the VPM is to change the VP bandwidth and alter the routing topology of the VP network. This is done dynamically to cater for changes in the bandwidth requirements from the connection service supplier. This section discusses the issues involved in changing the bandwidth and the routing of VPs. The objective of the VPM is to maximise VP utilisation without degrading the QoS experienced by its customers. The VPM cannot work in isolation. It needs information from other management components to make sensible decisions on the bandwidth requirements for providing adequate service to the connection supplier. The service manager provides load predictions of service usage on the network. This gives details of the number of connection requirements for each source/destination/CoS for the network [2] and is used to predict the maximum and minimum VP bandwidth requirements on the network. The link manager computes the call acceptance parameters which detail the traffic mix (number per CoS) which can be accepted for multiplexing on a particular VP. This is used to calculate the traffic throughput that is possible on the VPs. The network must inform the VPM when a call is successful and on which VP the connections have been made. This is necessary to maintain a network wide traffic profile for VP management. Once the VPM has made decisions on the Routing and/or Bandwidth changes it must transmit that change to the Network. Figure 2 shows the VPM and its interactions.

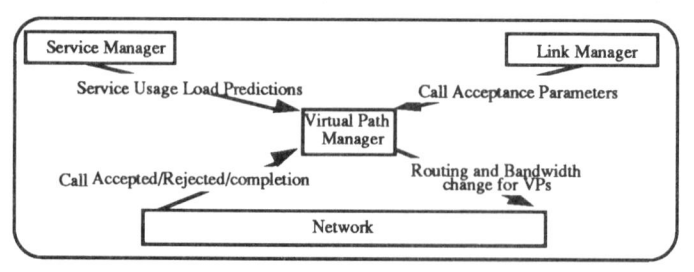

Figure 2 - VPM Interactions with other management and network components

The VPM manages the following aspects of traffic management:
- the amount of spare and allocated bandwidth of the VP
- the Classes of Service (CoS) which are to be allowed onto the VP network
- the topology of the overlay network to cope with failure, load balancing etc.
- implementation of bandwidth changes on VPs at element level.

The VPM functionality is distributed, with the changes to the network being conducted at the element level and the computation of the routing done at the network level whilst the load prediction processing is done at the service level. Figure 3 shows the distributed management of VPM functions according to the CNA-M layering discussed in [9].

Layering	Functions
Service Level	Mapping Load Prediction from Service Manager into VP bandwidth requirements. Network wide Class of Service management. Network wide Target Quality of Service management.
Network Level	Routing and VP network topology Configuration management
Element Level	Link Bandwidth Management. Bandwidth changes to the VPs. Efficiency calculations

Figure 3 - Distribution of functions for the VPM

Details of the management functions/algorithms at the different levels are described below.

2.2. Service Level Functions

The main goal is to compute network wide strategies. The main VPM functions are:
- Computation of the load between source and destination pairs
- Computation of target QoS expected from the VPs. The Target QoS is the level of connection rejections per CoS that can be accepted by the VPM
- Computation of CoSs acceptable in the network.

Efficiency calculations computed at element level are logged at the service level for determining the effect of management decisions.

Mapping Load Predictions From Service Manager into VP Bandwidth

Load predictions about service usage gives information about the maximum and minimum number of connections per CoS per Source/Destination node pair. This information needs to be converted into the bandwidth requirements at the network level. This is achieved by converting the load predictions into bandwidth requirements by multiplying the mean/peak bandwidth usage per CoS. This bandwidth is summed for each source/destination pair and VP bandwidth requirements calculated using the routing tables.

Network Wide Class of Service Management

Load predictions together with the efficiency computations are used to compute whether or not a given CoS is handled by a VP or not.

Network wide Quality of Service Management

Using the computed efficiency values, the management system makes decisions about setting targets for the inferred QoS expected from the network.

2.3 Network Level Functions

Routing and VP Network Topology Configuration Management

The major network level function is Routing And Configuration (RAC). Its main purpose is to set individual targets for the amount of spare bandwidth that the element level will try to reserve for VPs. RAC tries to conform to targets set by the service level. These specify the spare bandwidth to be reserved between each source and destination node for each CoS in a connection and the maximum spare bandwidth to be allocated to any connection, and the maximum number of VPs, to be used between a source and destination node.

A simple algorithm is used to implement RAC. The algorithm attempts to concentrate spare bandwidth on VPs serving high priority routes, and so reduce the number of available VPs and routing table entries that are actually in use. For each combination of source, destination node and class of service, the algorithm uses the routing table entries recursively to discover the routes from source to destination, in priority order. It allocates as much of the spare bandwidth as it can to each route discovered, updating the spare bandwidth required for each connection along the route, and marking the routing table entries that are needed. When all of the information from the service level has been processed, the updated targets for the spare VP bandwidth and routing are sent to the element level. This algorithm's handling of overload conditions has not yet been optimised. One possible improvement would be to scan through the CoSs in priority order instead of numerical order. This would result in the routing of low priority traffic being selectively disabled under overload conditions. Another possible improvement would be to provide feedback to the service level when its targets cannot be met. However, despite these shortcomings, the algorithm provides an improvement of RAC function which is robust enough to allow service level algorithms to be compared.

2.4 Element Level Functions

At the Element level one is always trying to ensure that there is always just sufficient spare bandwidth to cope with the multiplexed peak traffic. This involves monitoring the load from the network and analysing if there is sufficient bandwidth to meet the demands of the usage. If not, the bandwidth of the VPs which are causing the problems is increased. If there is more than sufficient spare bandwidth then the bandwidth is decreased.

Link Bandwidth Management (LBM)

The VPs exist on the links (transmission paths). The links can only accommodate a fixed amount of bandwidth. The sum total VP bandwidth usage must not exceed the bandwidth of the link, otherwise the link will lose cells. During VP bandwidth allocation LBM must ensure that the sum total of the allocated bandwidth on the VPs does not exceed the link bandwidth. Another consideration when allocating bandwidth for VPs on links is that a VP must have the same bandwidth across all the links it is traversing. To ensure this, the bandwidth is not allocated to the links until the management is certain that the required bandwidth is available on all the links that the VP is traversing. Finally, VP bandwidth cannot be deallocated if that bandwidth is currently being utilised. Hence the management must ensure that the bandwidth is only deallocated when it is absolutely certain that the released bandwidth is not necessary.

Bandwidth Changes to VPs

A simple algorithm for continuously requesting for the required bandwidth from the LBM until the request is satisfied has been implemented. More sophisticated algorithms could be employed however they were deemed unnecessary for the experiment.

Efficiency Calculations

Efficiency is a measurement of how effectively the network is being utilised. The experimental measure of the resource usage is the number of connections per CoS per Source/destination pair. The measure of the poor QoS is the number of connection rejections per CoS per Source/Destination pair. Every time poor QoS is realised the efficiency is decremented. Hence the aim is to keep the efficiency value as high as possible. This efficiency takes into account poor service to existing users when the CAM parameters disallow calls due to poor multiplexing of connections which in turn has an effect on efficiency. The efficiency calculations are as follows.

efficiency = No. of cells * Unit cell cost + peak usage * duration * unit duration cost

where the peak usage is the maximum bandwidth allocated to the connection. Every time a connection fails the efficiency is decremented by the predicted usage of that connection. The unit costs are normalised values for particular CoSs.

3. CALL ACCEPTANCE MANAGEMENT (CAM)

3.1 Description of Techniques and Experimental Results

Call Acceptance is divided into two parts, the Call Acceptance Function (CAF) and the Call Acceptance Management (CAM). The CAF is performed in the network, and makes on-line decisions on whether a call request should be accepted or rejected. It is normally a simple and fast process, able to respond quickly to user requests. The CAM is performed at the network management level. Its task is to compute and refine call acceptance parameters, and download them to the CAF. It is an iterative process where the goal is to stabilise the CAF parameters so that the users experience good QoS. The most common performance measure used in the CAM functions are the cell loss probability and the congestion probability (the probability of exceeding the link capacity). It is the number of cells lost over the network which has a direct and measurable effect on the network performance and QoS. The ultimate goal of the CAM is to maximise network utilisation while maintaining an agreed-upon level of QoS for the network users.

The majority of the CAM methods considered in this paper are based on the concept of an acceptance region. An acceptance region is defined as a set of traffic loads for which the QoS on a given link is acceptable. (For the sake of simplicity, we equate QoS with the cell-loss rate.) Thus the definition of an acceptance region partitions the set of possible traffic loads into two : inside the acceptance region we find loads which produce acceptable cell-loss rates; outside the acceptance region we find loads which produce unacceptable cell-loss rates. If an easily-defined acceptance region exists, then the CAF function is greatly simplified, since it need only determine whether or not adding the new call will cause each of the links to operate within its acceptance region. Since acceptance regions are, for most practical purposes, impossible to compute analytically (based on inexact data and imperfect models), research is aimed at deriving a successful approximation of such a region (see [10], [11], [12]).

3.2 Linear Weights

Methods using linear weights are based on the assumption that the network resources needed for a given CoS can be expressed by a weight. This weight is a fraction representing what portion of the link capacity is required by the service. This is also referred to as the normalised effective bandwidth and is dependent on the link speed. A new call request is accepted only if the sum of weights on the link (or virtual path) is smaller than 1.0. Given n CoSs, every traffic load on a link defines a point in an n-dimensional space. It is the function of the CAM module to set the boundary in the space Rn and to pass those boundary parameters to the CAF. In the linear weight methods, the acceptance region is assumed to be a hyperplane in the space Rn. Assuming that the acceptance region is a hyperplane defined by the set of coordinates W=(w1, w2,...wn) with wi, i=1,2...n being the normalised bandwidth of the ith class of service, ($0<=w_i<1$) and a traffic configuration T=(t1, t2,....tn) with ($0<=t_j$), j=1,2,...n, then this configuration will only be allowed by the CAF if the following holds true,

$$T.W = \sum_{i=1}^{n} t_i w_i < 1 \qquad (1)$$

i.e. the traffic configuration lies within the currently defined acceptance region. Clearly an important consideration is the maximisation of the acceptance region which can be measured in terms of the quantity

$$\prod_{i=1}^{n} \left(\frac{1}{w_i}\right) \qquad (2)$$

which we refer to as the volume of calls. Another constraint can be given by requiring that the wi >= pi, i= 1,2...n, where the pi's are normalised weights based on peak bandwidth PBi, with pi = PBi/LB and LB being the link bandwidth. Once the weights are found, these are passed to the network simulator who will use these in the CAF process. For a call to be accepted, equation (1) must be satisfied over all the links on the connection path.

Heuristic Method

While there are many avenues for a heuristic approach to a CAM design, the one considered here has the obvious merit of simplicity. Given a base QoS traffic configuration T, we allow the coordinate wi to vary so as to satisfy equation (1) with equality. The volume of calls as computed by (2) is then determined. The coordinate giving the largest volume defines the new hyperplane (H-Plane) acceptance region. Should the coordinate which maximises (2) be such that it falls within the P-Plane region, then the coordinates of the hyperplane parallel to the P-Plane and passing through the bad QoS point will define the acceptance region. Performance of this algorithm is measured in terms of the volume ratio of the H-Plane as well as the measured link utilisation.

Using two classes of service, we see in figure 4 preliminary results indicating a straight line approximation to the acceptance region in 2-dimensional space.

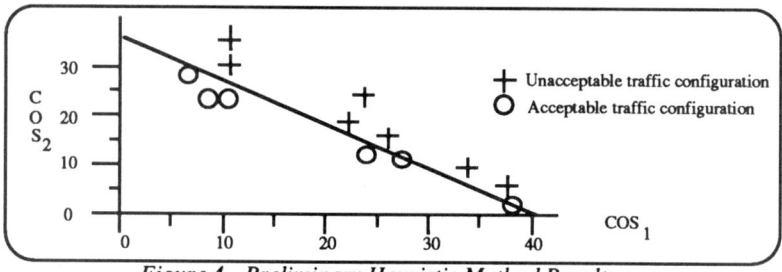

Figure 4 - Preliminary Heuristic Method Results

Using Constraint Logic Programming (CLP)

The use of a Constraint Logic Programming language such as CLP(R) [13] allows for a concise formulation and solution of a class of problems which may be described by a set of rules and constraints. In the context of the CAM problem a first solution for the weights which rejects all the bad configurations is found. In the second phase, the CLP attempts to find a better solution, i.e. one which increases the volume of calls. This process is repeated until successive increases in volume are deemed negligible or the constraints cannot be satisfied.

3.3 Two-Moment Allocation

Description of the Technique

The two-moment allocation scheme attempts to approximate the acceptance region discussed previously, but without considering CoS. It is a simple scheme in that only two traffic descriptors need to be provided by a new call, and the CAF requires only two parameters from the CAM in order to evaluate whether or not it will accept the new call.

For reasonably large buffers, analysis suggests (see [14]) that there exists an acceptance region approximately of the form:

$$m + a_j * s_2 < b_j$$

where m is the mean bitrate of the traffic and s_2 is the variance of the bitrate. The constants a_j and b_j depend on the switch characteristics (buffer size, capacity) and the acceptable cell-loss rate. The calculation of a (a_j, b_j) pair should be done for each different link descriptor. (A link descriptor consists of a distinct set of values for the three parameters: link capacity, buffer size, and acceptable cell-loss rate.) The CAM receives information in the form of configurations (different traffic combinations). For each configuration, the m, s_2, link descriptor and quality (good or bad) is provided. Once a sufficient number of configurations has been accumulated in the CAM, a discriminant function is applied to calculate the equation of the line separating the good configurations from the bad configurations. Since a configuration is simply a point in the m, s_2 plane, a two-dimensional line can be found separating the two regions. This line defines the acceptance region for the given link descriptor, and the a_j and b_j are easily derived from the equation of this line. The a and b are then passed on to the CAF for making its call acceptance decision.

There are certain limitations that apply to the two-moment CAM, although in the context of an ATM network, they should be accommodated easily by the inherent nature of the network and the traffic. The first limitation is that the existence of the two-moment acceptance region is only valid in an asymptotic sense. That is, the theory applies when the buffer size is reasonably large. Experiments also suggest that the capacity of the link must be large in comparison to the bitrate of the individual sources (thus, the input traffic to the link is composed of a large number of small sources).

The second limitation is due to the fact that the aj and bj coefficients are dependent upon three factors: the capacity of the link, the size of the buffer feeding this link, and the acceptable cell-loss rate on the link. This does not allow for calls with differing cell-loss rate requirements to share the same link. Therefore the two-moment CAF requires that any given link be shared only by calls having the same QoS (i.e. cell-loss rate) requirement. This limitation, in fact, is justifiable because we are only able to control the total cell-loss rate of a link, without distinguishing between different Virtual Circuits or Virtual Paths. If cell priorities were introduced, such that cells having different QoS requirements could be recognised and handled in different ways, then it would be possible to control cell-loss rates at a finer level.

Experiments

The experiments used a burst simulator which models a single buffer-link pair of an ATM switch. The buffer is fed a stream of ATM cells aggregated from a number of different traffic sources. These cells are then serviced by the outgoing link and cell losses are observed when the buffer overflows. The switch is an internally non-blocking, output-buffered fast packet switch. Thus, the switching of the cells has occurred prior to their arrival at the buffer (i.e., we do not model the actual switching of the traffic streams). We assume that all cells sharing this buffer and link have the same cell-loss requirement. The traffic is modelled as 53-byte ATM cells. The buffer size is 128 cells and the link capacity is 600 Mbit/s (1.415 x 106 cells/s). A 2-state traffic model is used for all of the traffic sources. During state 1 (state 2) cells are produced at a constant bitrate of B1 (B2) Mbit/s. The length of time spent in a state is exponentially distributed with a mean of S1 for state 1 and a mean of S2 for the state 2. We define our traffic sources as shown in table 1.

	B1(Mbit/s)	B2 (Mbit/s)	S1 (Mbit/s)	S2 (Mbit/s)	Mean (Mbit/s)	Variance	Peak/Mean
#1	1.7	1.3	10.0	10.0	1.5	0.04	1.3
#2	6.0	0.0	4.0	60.0	0.3775	2.1	16.0
#3	30.0	0.0	1.0	9.0	3.0	81.0	10.0

Table 1 - Definition of Traffic Sources

As the simulation runs, the amount of traffic offered to the buffer and the mixture of this traffic is dynamically changed. Each different combination of traffic produces a distinct (m, s2) point and an associated cell-loss rate. Thus "good" and "bad" configurations are produced and sent to the CAM. When a sufficient number of configurations has accumulated, the line defining the acceptance region is calculated. With each subsequent new configuration sent to the CAM, possible adjustments are made to this line.

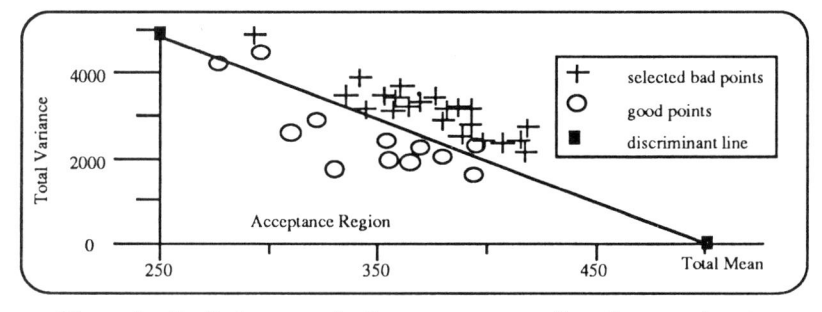

Figure 5 -. Preliminary results from two moment allocation experiments

Using 1E-5 as an acceptable cell-loss rate, figure 5 shows that the acceptance region is quite well approximated by the two-moment CAM for a buffer size of 128 cells and mixes of these traffic types. This figure also shows the good configurations produced, the bad configurations produced, and the line calculated by the CAM after the final refinement. Since the CAM algorithm does not use all accumulated points for calculating the line, only the points which were selected in the calculation process during the final refinement are shown.

3.4 Neural Networks (NN)

The CAM problem may be viewed as a learning problem, and as such, Neural Networks(NN) may be considered for QoS management in ATM networks. In an earlier phase of the project [15] a Kohonen self organising map for defining the boundary of the acceptance region was considered, but proved expensive in computational resources and in the time required to adapt to the network. Hence a neural network utilising supervisory learning was considered. Following [16], good and bad traffic configurations are presented to the input layer of a three layer back propagation network. After a number of traffic configurations has been received the network is set to learn the input pattern. This process continues until the network is able to delineate the boundary region. The CAF is carried out by applying to the neural network the new traffic pattern. The output layer of the neural network consists of one neuron having an output between 1.0 and 0.0. According to a predefined threshold the network decides to accept or reject the call.

Neural Network Technical Description

The NN CAM is based on the concept of a feasible region which is different from an acceptable region. A feasible region is defined as a set of traffic configurations (usually a vector of network services). A feasible region is bounded by a set of traffic configuration that have the same probability of containing at least one faulty service, and contains traffic configurations that have a bad QoS probability lower than a predefined threshold. The zero feasible region can be determined by not accepting any traffic configuration whose cumulative peak rate is higher than the bandwidth of the media porting it. We believe that the zero feasible region is a waste of network resources that does not take into account the random nature of traffic. The aim of the NN CAM is to teach a NN the probability that a connection is bad, thus enabling the user of the CAF to chose his accepted feasible region. The NN CAM is composed of two parts: a statistical module that aggregates network reports and a NN trainer. The statistical module receives reports and stores the configuration. If a similar configuration already exists, a new mean is calculated using the previous configuration and the last mean times the number of previous similar configurations that existed. If the number of past patterns stored for a given traffic configuration exceeds a given amount it is not increased when the mean is calculated. Thus our statistical module takes only into account a limited past history for each traffic configuration. After a number of aggregations, the learning module is launched. The learning module core is a back propagation training algorithm. If the learning is successful (i.e. if the new parameters of the neural network give better results than the old ones) the new parameters of the NN are sent to the CAF of the simulator. The learning module trains the NN on a training base and controls the results on the generalisation base (which contains all the configurations known by the statistical module). If the results of the training offer no improvement, a training phase is initiated with a new random NN. Only the best NN is saved for later training. With this method, we try to prevent the NN from getting blocked in a local minima (i.e. reaching a non optimal position from which training is useless).

Preliminary runs with this method have shown a good convergence. After a two hour run of the simulator generating 300 different traffic configurations, the Total Squared Summed error (TSS) of the resulting neural network was 3.8E-3. This means an error of 6.1 percent over 300 patterns which is 0.02 percent by pattern.

The advantages of this approach are many fold. First, the back propagation algorithm provides a significant distinction between good configurations - call configurations which should be accepted on our transport media (i.e. links or virtual paths) - and bad configurations. As the CAF algorithm is not tied to a theoretical traffic model there is no need to be concerned with the detail of the traffic characteristics of the connection types. Another advantage is the possibility of expanding the parameters of the NN to include any desired parameter, e.g. time of day, geographical location. Last but not least, as the NN converges toward the probability of a particular configuration being good or bad, we have a good grasp on the important ratio between QoS and network load. This allows the call acceptance algorithm to be tuned to business and service level decisions (e.g. adaptation to customer requirements, market segmentation, etc.) made higher up in the management hierarchy.

This method also has its disadvantages. First, it is difficult to design a completely automated learning scheme. Also, due to the random nature of traffic, a neural network needs to be defined for each physical or logical media capacity - accepting a single video conference connection onto a 40 Mb/s virtual path is not the same thing as accepting 15 video conference connections onto a 600 Mb/s link. In the second case a large number law could be applied with some success. Another problem is the time needed to train the NN - the back propagation algorithm is somewhat slow to converge. On the whole we believe there is an advantage in using NN for CAF, particularly its direct application to the QoS/network performance ratio, making it valuable for ATM networks.

3.5 Peak and Mean Method

The peak and mean method is a CAF with no associated CAM. Thus the method is interesting as it allows an entire management function to be removed. The theory on which the peak and mean method is based can be found in [17], briefly, the method works as follows. When a new call is requested the cell loss probability for the mix of all current calls plus the requested call is calculated, based on the peak and mean bitrates. If the cell loss probability is less than a pre-assigned limiting value the call is accepted. This method is more suited to bursty traffic than is a method which allocates bandwidth based solely on the peak bitrate of a call. In this paper we define the burstiness as the ratio of the peak bitrate to the mean bitrate, and so the burstiness is a real number greater than or equal to unity. Several runs have been planned with this method-runs which aim at quantifying the following two items:

- The additional traffic volume that is accepted with the peak and mean method compared to the peak allocation method.
- The traffic volume that is rejected but which could have been accepted before the observed cell loss rate equalled the limiting value.

However, the runs have not been completed, and so only preliminary results will be presented. Two such results are the following.

The first result is that the peak and mean method works best for bursty traffic. In the implementation of the peak and mean method a certain number grows infinitely large as the burstiness approaches the value of unity Therefore traffic with burstiness close to unity (e.g. telephone traffic) must be dealt with separately, for example by using a hybrid of the peak and mean method and the peak allocation method.

The second result is that the peak and mean method cannot collaborate with the VPM described in section 2. The reason is that the peak and mean method is a non-linear method in the sense that for a given combination of ongoing calls the peak and mean does not calculate an equivalent effective bandwidth (which is the bandwidth that would be sufficient to retain the level of quality of service for the calls). Had it done this, the effective bandwidth could be used as a measure of not just the link utilisation but also to work out the utilisation of virtual paths.

4. ACKNOWLEDGEMENTS

The work described in this paper was carried out in the RACE NEMESYS project as part of the investigation and evaluation of the role of Advanced Information Processing (AIP) techniques in Traffic and Quality of Service (QoS) management for IBC networks. The specific experiment described in this paper is the third "EMMANUEL" experiment.

5. REFERENCES:

[1] Patel, P. et al : "Viewpoints on Traffic and Quality of Service Management in Telecommunication Management Networks", Sixth RACE TMN Conference, Madeira, September 1992.

[2] Harksen, U. et al : "Experience of Modelling and Implementing a Quality of Service Management System", Sixth RACE TMN Conference, Madeira, September 1992.

[3] Gentilhomme, A. et al : "The Use of AIP Techniques in Traffic and Quality of Service Management Systems", Sixth RACE TMN Conference, Madeira, September 1992.

[4] RACE project NEMESYS 1005, : "Specification of the EMMANUEL Experimental System", NEMESYS deliverable 9 "05/KT/AS/DS/B/021/b1, December 1991.

[5] RACE project NEMESYS 1005, ": Experiment 3 design", RACE deliverable 10 "05/DOW/SAR/DS/B/024/a1," May 1992.

[6] CCITT draft Rec. I.150, "B-ISDN ATM Functional Characteristics", Study Group XVIII, Geneva 1990.

[7] Sato, K. I., Ohta, S., Tokizawa, I. :"Broadband ATM Network Architecture based on Virtual Paths", in IEEE trans. on Communications, vol. 38, no. 8, Aug. 1990.

[8] Griffin, D., Patel, P. : "Traffic Management for IBC Networks," in Proc. Fifth RACE TMN Conference., Church House, London 1991.

[9] GUIDELINE Deliverable ME8 : "TMN Implementation Architecture", 03/DOW/SAR/DS/B/012/b3, RACE Project R1003 GUIDELINE, March 1992.

[10] Appleton, J. : "Modelling a Connection Acceptance Strategy for Asynchronous Transfer Mode," Proceedings of the 7th Specialists Seminar, Morristown, NJ, 1990, Session 5.1.

[11] Hui, J. : "Resource Allocation for Broadband Networks," IEEE Journal on Selected Areas on Communications, Vol. 6, No. 9, Dec. 1988, pp. 1598-1608.

[12] RACE Project R1022, : "Technology for ATD," RACE Document TG_123_0006_FD_CC.

[13] Jaffer, J., Lassez, J. L. : "Constraint Logic Programming" IBM Yorktown Heights, N.Y. 10598

[14] Courcoubetis, C., Walrand, J. : "A note on effective bandwidth," (to be published).

[15] RACE NEMESYS Project 1005, "Nikesh Experiment 2", deliverable 6, June 1991

[16] Hiramatsu, A. : "ATM Communications Network Control by Neural Networks," IEEE Trans. Neural Networks, pp 122-130, March 1990.

[17] Rasmussen, C., Sorensen, J. H., Kvols, K. S., Jacobsen, S. B. : "Source-independent call acceptance procedures in ATM networks", IEEE Journal on selected areas in communication, Vol 9, No 3, pp 351-358, APRIL 1991.

The Use of AIP Techniques in Traffic and Quality of Service Management Systems

Alain Gentilhomme, Paul-Eric Stern (GSI ERLI, France),
Andrew Carr, Nicholas Mandich (Dowty, UK)

ABSTRACT

The aim of the RACE NEMESYS project is to evaluate the use of Advanced Information Processing (AIP) technologies in Traffic and Quality of Service Management for TMN. This paper presents a summary of this evaluation performed in the context of three experimental network management emulators developed during the lifetime of the project and whose purpose was to provide a representative testbed for AIP evaluation. We discuss the methodologies we have used during the project design phases, the experiment platforms used to facilitate AIP integration, and the evaluation of the AIP techniques we have used for solving various decision support and resource allocation problems.

1. INTRODUCTION

The primary objective of RACE project NEMESYS is to demonstrate and evaluate the use of AIP (Advanced Information Processing) techniques for the implementation of Traffic and Quality of Service Management for Integrated Broadband Communication (IBC) Networks [1]. This evaluation has been performed in the context of three experimental network management emulators, each designed to test a number of AIP techniques applied to selected aspects of Traffic and Quality of Service (QoS) Management in Telecommunications Management Networks (TMN) [2]. The goal of the third and largest experiment, Emmanuel [3], is to maximise the utilisation of the network resources whilst maintaining an agreed level of QoS to service users.

In this paper, we focus on the application of AIP techniques used in the project as a whole. Although the evaluation was performed in the context of Traffic and QoS management, some results of the evaluation are more general. The problems that were tackled using the various AIP techniques can be grouped into three main categories. The first deals with the problem of the analysis and design. In this part, we emphasise the benefits of using methodologies. The second category deals with the distributed architecture offering common services to the different Functional Blocks. Finally, the last category deals with various decision support and resource allocation problems investigated. For reasons of space, this paper does not cover all the AIP techniques used in NEMESYS. We draw the reader's attention to a paper on HCI considerations in TMN [4] in this volume as it includes material drawn in part from work in project NEMESYS.

2. DESIGN AND ANALYSIS

In the latter phases of the project, the Open Distributed Processing (ODP) framework [5] was used extensively in the analysis and design of the experimental system. ODP was complemented with other methodologies, such as Object Modelling Technique (OMT) [6] and ObjectOry [7], which were also used.

2.1 ODP Viewpoints

The ODP Viewpoints provide a framework for separating the various concerns when analysing a system. Each Viewpoint allows a perception of a system with emphasis on a particular concern. These viewpoints are addressed in a companion paper [8]. The third experiment (Emmanuel) was designed in part using ODP. It lived up to our expectations and allowed us to classify and analyse the various concerns in Emmanuel very clearly and effectively. The ODP-based analysis was supplemented with GDMO [8] used for describing the information model used in the system and its associated managed objects.

2.2 Object Oriented Methodologies

Based on a survey of state-of-the-art object-oriented methodologies (OOMs), two of the emerging OOMs, OMT [6] and ObjectOry [7], were investigated. They have been used for the design of components where the dynamic and functional points of view are of prime importance.

OMT

The Object Modelling Techniques (OMT) methodology provides for incremental development through *iterative* cycles between the analysis and design activities. The advantage is that the analyst, the designer and the implementer are not forced to "get it right first time". It is based on a description of the problem and its software solution using three orthogonal views on objects: the object model (OM), the dynamic model (DM) and the functional model (FM) (see figure 1).

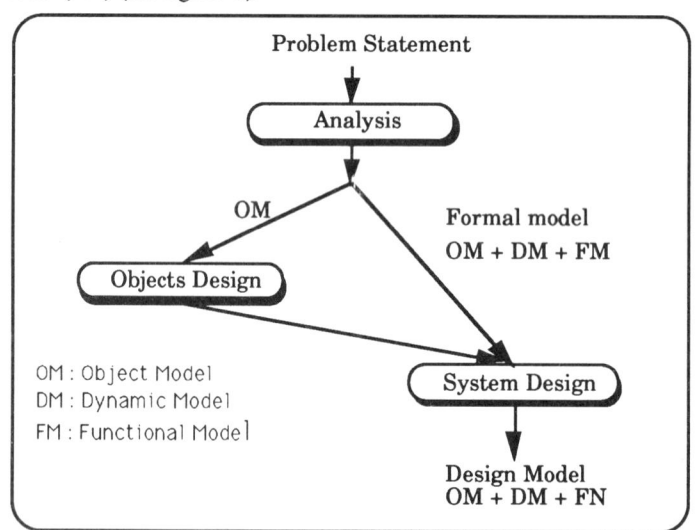

Figure 1 - The OMT phases.

Object Model

This is the framework for other models. During the analysis, it describes problem space objects. During the design, it describes problem space objects and objects of the software solution (general architecture objects as well as computer objects)

Dynamic Model

This takes into account time aspects, treatment of sequentiality, and control aspects. It is represented by a collection of state diagrams interacting with each other through shared events

Functional Model

This describes computational values and functional dependencies. It explains the meaning of operations of the object model and actions of the dynamic model. Some iterations have to be made among the three models so that, for instance, key operations can be discovered in the preparation of the functional model and added to the object model. OMT covers the whole software life-cycle (except testing and maintenance).The same notation and models are used for analysis and design.

We have benefited from the use of this methodology in the design of the Service Manager History function (see section 4.3). It has permitted us to describe precisely the functionality of this history and has made the implementation easier. However, it has not been possible to test fully its maintainability because the design was not changed after the first implementation.

ObjectOry

ObjectOry is a methodology for analysing and designing object-oriented systems. It is based upon a combination of conceptual modelling, block design and object-oriented techniques and is particularly intended for large systems. The overall approach is to model, and consequently design, the system based upon how it is going to be used rather than how its functions are decomposed. Subsequent modifications are then isolated as much as possible. It identifies two main phases - System Analysis and System Design. System Analysis consists of the creation of the following:

- *Entity Model* : Entities are things or events of interest to the system
- *Use Case Model* : A Use Case is a sequence of transactions performed by a user and a system in dialogue. A transaction is performed by either the user or the system and is in response to a stimulus. When no more stimuli can be generated, all the transactions complete and the Use Case ends
- *Service Model* : A Service is a behaviourally related set of functions which could be offered as a package to a customer. A Use Case may use several Services.

These models collectively capture the system specification. The System Design phase involves taking this specification and modelling functional blocks which can then be implemented.

ObjectOry was used for the analysis and design of the Virtual Path Manager (VPM) of the third experiment. The use of this technique produced many diagrams (thirty-seven specifically and this was not using all the aspects of the methodology). Since many of them were similar, a drawing package, with its ability to cut and paste, was extremely useful. No ObjectOry tool was available to the project. Organising a clear set of requirements is important since they are fundamental to the selection of entities, relationships, functional blocks and Use Cases.

The approach proved useful particularly in the analysis phase where it results in a thorough specification. The way the specification is structured then eases the transition to functional block design. The methodology is completely user-oriented. This is evident from the derivation of Use Cases and Services. It results in a design which should minimise the effect that maintenance of one functional block has on other functional blocks. Because of the short duration of the experiment an evaluation of the maintainability was not possible.

Conclusion

Both Object-Oriented methodologies have proved their usefulness for the design and analysis of complex system components. Each of them provides a model that defines precisely the behavioural aspect of objects (Dynamic model in OMT and Use Case model in ObjectOry).

The design and analysis of management application with these methodologies are not in conflict with the use of GDMO [9]. The latter permits the designer to describe the structure and a part of the behaviour of Managed Objects (MO), whereas the former helps in fully designing a management application. However, these kinds of methodologies are still in their infancy and require further development to ensure their success. In both cases, it is difficult to check the consistency of the different models. In addition, CASE tools that fully implement these methodologies and make them easier to use need further development to ensure their success.

3. DISTRIBUTED PLATFORMS

In the effort to investigate distributed TMN functionality, we have developed the experimental system around a distributed experiment control platform which provides common services to the various distributed management applications. This platform is required to provide:
- Transparent communications between applications
- Easy access to Managed Objects in the Management Information Base (MIB)
- Support for modelling and managing distributed problem solving
- A framework for integrating various AIP techniques.

The first two are Communication Platform-related, whereas the last two are Management Platform-related. A detailed description of both the Communication Platform and the Management Platform can be found in [9]. Some aspects of both are described below.

3.1 Management Platform

The Management Platform used in Emmanuel consists of two separate but interworking management platforms: the NEMESYS-developed and the OSI-based platforms. The NEMESYS-developed platform was developed at the start of the project in the absence of an agreed standard reference framework to which the experimental system architecture could conform so that experimental work could proceed. The OSI-based platform uses OSI management principles. Below, we describe one aspect of the NEMESYS-developed platform, the MU/MUIB architecture, as well as OSIMIS, the OSI-based platform.

The MU / MUIB architecture

This architecture has been used in traffic management and is based on multiple blackboard-like units running concurrently [10]. A management application at a particular layer of the TMN may have be implemented as one or more Management Units (MUs). A MU provides two kinds of services: problem solving services and data retrieval services. Notable aspects of a MU are as follows.

Knowledge organisation

A MU contains one or more Knowledge Sources (KS) representing network management problem solving knowledge. Each KS can be implemented with different problem solving techniques.

Enable cooperative problem solving

All KSs in a MU are able to communicate with each other and share common information. If the problem cannot be solved locally, a KS may communicate with a peer MU or another management application at higher level.

Common Management Information Base

A MU has a view of the MIB (Management Information Base) which is that MU's Management Unit Information Base (MUIB) (Fig 2). The overall MIB of the management application may be seen as the union of all MUIBs. A MU can contain several KSs that share the same MUIB This MUIB contains managed objects of two types: those owned by the MU itself and copies of MOs owned by other MUs which this MU needs to have access to. Data in all the MUIBs is accessed through (and only through) the use of MUIB services to ensure consistency. When a MUIB is modified by a function in a MU, the modification is passed on to all other dependent MUIBs.

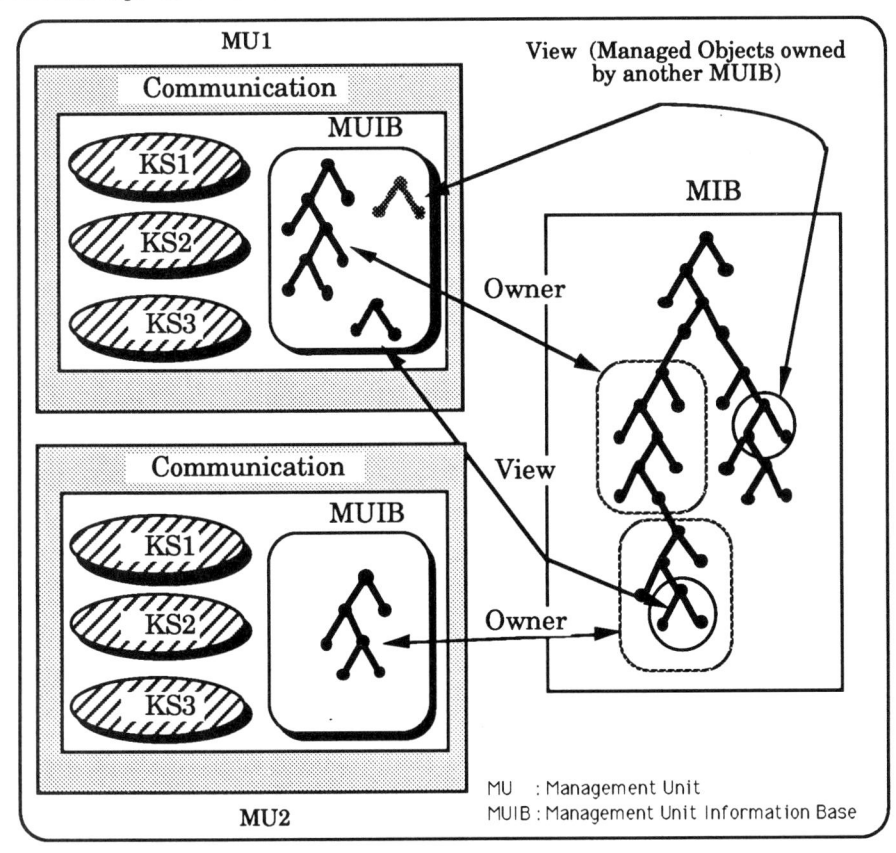

Figure 2 - MU / MUIB architecture.

Handling communications

A MU offers services to the KSs allowing them to send and receive events. A MU can also dispatch an incoming event to the appropriate KS. The communications are based on the low-level UNIX RPC and XDR facilities.

The MU/MUIB architecture models distributed decision making as a collection of cooperating Knowledge Sources as decision making components. The actual decision making within a particular knowledge source can be implemented using different AIP problem solving approaches. Integration of different AIP techniques is achieved by organising management functionality in line with this architecture and implementing different KSs using different AIP approaches. The MU/MUIB service is implemented using Objective-C. The existence of the Management Platform and the MU/MIB architecture makes it easier to use new AIP approaches within KSs. The MU/MUIB architecture has been validated by prototypes in a number of case studies. In particular, in the area of Traffic Management, we have integrated five different techniques to solve the same problem. Two of them, constraint programming and a neural network approach are presented in 4.1.

OSIMIS

With the emergence of a standard communication protocol between objects, i.e. CMIS/CMIP, it was considered useful to investigate this new protocol in the experimental work. At the beginning of the third experiment, it was decided by the consortium to build a new platform supporting this standard and which would interwork with the MU/MUIB architecture.

This platform, named OSIMIS [11], is object-oriented and based on real OSI management services. It is built on top of the OSI upper layers known as the ISO Development Environment (ISODE) [12].The OSI management model is based on the manager/agent relationship. All the management capabilities are expressed through managed objects which are handled by management applications in an agent role. The managed objects are accessed by applications in a manager role through CMIS operations. The distinction between managers and agents is not strong in engineering terms. An application may be in both roles, especially in a hierarchical layered management system. The managed objects capabilities are specified in an abstract form using an Abstract Syntax Notation One (ASN.1) [9] template formal description language known as GDMO.

The OSIMIS platform has a number of ideas in common with those of the MU/MUIB architecture. Each layer of the TMN can contain a generic management application. This unit provides the following features:
- It maintains one Management Information Base (MIB)
- It may comprise one or more Shadow Agents and their possibly associated Shadow MIB (SMIB). A SMIB represents a view of objects geographically distant and that are managed by another agent. The access to these objects is then transparent and they can be used as if they were in memory. The Shadow agent associated with one SMIB can access peer or lower level management units
- Several knowledge sources can be defined for a managed application at each TMN layer. Each knowledge source must specify the incoming events (notifications from the Managed Objects) in which it is interested.

In addition, during the third experiment, a Generic Management Information Base browser has been implemented. This MIB browser is a managing application which enables a human user to browse graphically through a MIB, making changes to Managed Objects to assist him with management decisions. This browser can be considered as generic since it can connect to any agent and access its MIB without having prior knowledge of the structure of the MIB, and cope with changes within an MIB. This flexibility is very important as the OSI management standards will evolve over a long period of time, possibly resulting in discrepancies between the managed object classes understood by managing and managed systems.

Conclusion

Although one architecture is proprietary whereas the second is based on a new standard, they fully interwork in the experiment. Both architectures enable management applications to be developed quickly and efficiently, allowing implementers to focus attention on management capabilities rather than distributed access methods. They are based on an object-oriented model which complies with TMN architectural principles.

3.2 Communication platform

Apart from using a management-oriented platform such as those mentioned previously, we investigated the use, in the experiment, of a general purpose commercially available communication platform. This would show whether the design of the experiment allows the

easy use of such a platform and give an indication of the likely performance. The platform chosen was ANSAware[13]. This is an open distributed platform providing some of the ODP distribution features. The requirements of its use were to:
- Effectively provide the functions of RPC and XDR as used in the second experiment
- Keep the interfaces unchanged
- Be able to run the experiment with or without ANSAware.

Using ANSAware it was possible to run two experiments simultaneously. Furthermore, it was possible to start up all the UNIX processes without having to resort to using site-dependent shell-scripts. However, there was insufficient time to optimise the use of ANSAware and consequently it runs more slowly than the original system. From a programmers' perspective it was considerably easier to use ANSAware for inter-process communication in a distributed system than it is to use the low-level UNIX RPC and XDR facilities.

4 SPECIFIC MANAGEMENT PROBLEMS

In addition to methodology and architecture, we have tackled more specific problems related to decision support and resource allocation at the level of Traffic and QoS Management applications. We have chosen in this paper to present three of them, each one implementing one or two different AIP techniques.

4.1 Call Acceptance Management (CAM)

In a telecommunication network, a number of connections share the same physical or logical transmission path. Traffic Management in IBC involves Call Acceptance Management (CAM) at link level [14]. CAM controls the Call Acceptance Function (CAF) in the network. The burstiness of some traffic types introduces complexity in managing the CAF. A number of CAM techniques have been implemented and evaluated in NEMESYS [8]. The two based on AIP techniques are described below. The first is a constraint programming-based approach using the Clp/R tool in which we try to optimise a linear function. The second uses a neural network which learns bad and good traffic configurations.

Constraint Programming Approach

The Clp/R based CAM optimises the accepted call volume while rejecting all the calls that have induced a bad quality of service in the past. The declarative nature of Clp/R has enabled us to formulate our constraints in a few simple forms. For instance, the simplest expression of our problem consists in the declaration of each bad configuration as a false predicate:

$$\text{Conf}([x1, x2, .., xn]) :-!.$$

One of the main advantages of the Clp/R approach is its ability to consider a large number of parameters at the same time (as opposed to an iterative method where only one parameter is considered at a given time). The main drawback in our implementation is its inability to revise past experience in order to find a better compromise between quality of service and network load. For example, once a configuration is classified as bad, it will stay so even though the CAF would be capable of improving this situation if this constraint was relaxed.

Neural Network Approach

In the CAM, a Neural network is trained. The input it receives is a call configuration (a 15 dimension vector) and it outputs the probability of the configuration being bad (i.e. inducing an unacceptable QoS). When the CAF is called a threshold is applied to this result. This threshold can be used to tune the network load versus quality of service ratio. A peculiarity of neural networks is their fault tolerance. An exception will be processed, but in the long run, its consequences will be ignored. On the other hand, if a new type of traffic is introduced, the neural network will at first ignore it, and then consider it by reducing slowly the knowledge acquired from the old traffic model. As this approach uses statistics, one of the drawbacks is the large number of samples needed to train the neural network properly. The learning time can become excessive if very precise results are expected. However, in the case of a long run, the neural network converges to a stable configuration and, if no new traffic is introduced, the amount of training required tails off, see figure 3.

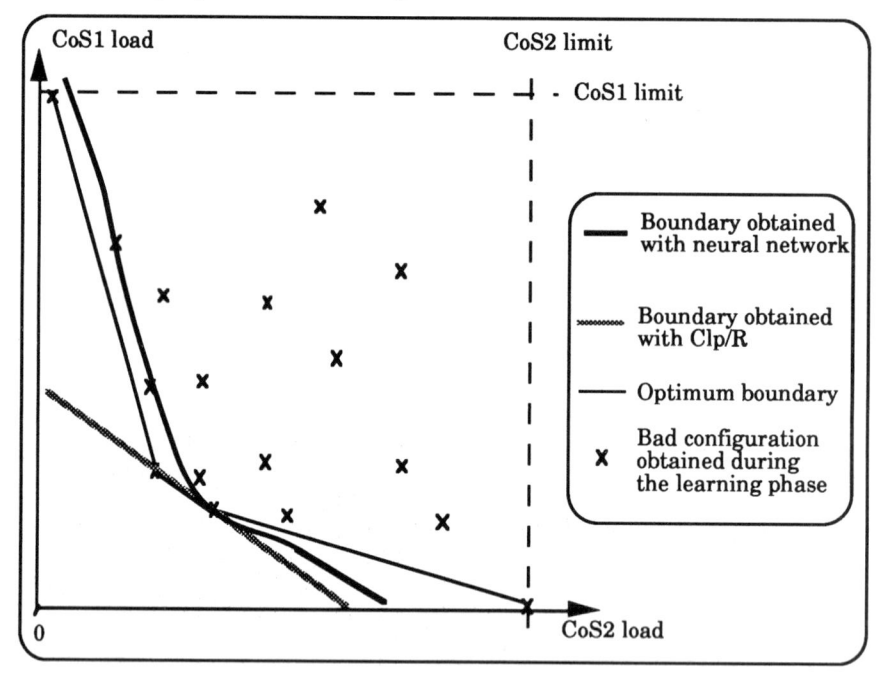

Figure 3 - Boundaries obtained with Clp/R and neural network for two classes of service (CoS).

Conclusion

Both algorithms are trained during a different run from that in which they are used. Thus, the generally long time needed to train the Call Acceptance Function has no impact on the speed of convergence of both of the above CAM techniques. The Clp/R CAM has the advantage of reacting faster than the Neural Network one. However the lack of fault tolerance of the former (due to the lack of constraints relaxation mechanism) tends to transform this initial advantage into a drawback. An interesting development would be to try to integrate statistics in the Clp/R CAM and thus gain some fault tolerance.

4.2 Virtual Path Bandwidth Management

The aim of the Virtual Path Bandwidth Manager (VPBM), as used in the second experiment, is to maximise the connection acceptance probability for all network users as user behaviour changes over time. This is done while minimising the amount of reserved, but unused, bandwidth allocated to Virtual Paths (VPs). The VPBM manipulates the bandwidth allocated to VPs based on load predictions from the Service Manager together with historical information.

The VPM has been implemented using rule based Production Systems (OPS83) and Object Oriented Systems (Objective-C). This allows the combination of a declarative approach (rules) without high performance degradation. The Production Systems technique was used to implement the Virtual Path Monitoring function of the VPBM. The tool chosen was OPS83 [15]. (Refer to [3] for details of the design.)

One of the general requirements of the Traffic Manager of which VPBM is a part, was to produce an integrated infrastructure containing object oriented programming, production systems and near real-time performance.

In order to produce a clean interface between OPS83 and Objective-C it was desirable to represent all relevant Objective-C objects with just one type of OPS83 working memory element (WME). It was also desirable to convert an OPS83 symbol into an Objective-C method selector at run time. A total of sixteen Traffic Manager objects was represented by one type of WME. These contain several fields copied from the corresponding object.

The results were encouraging for the future use of hybrid rule-based/object-oriented in real-time event handling. The suitability of rule-based systems in other real-time applications depends upon the definition of "real-time", i.e. the frequency at which input stimuli occur. However, it should be noted that OPS83 is a tool which generates fast, lightweight code. Other rule-based tools may not produce code suitable for this environment. Further work is required to optimise the conversion of OPS83 symbols into the selectors of Objective-C methods.

4.3 Service Manager History Function

The role of the history function, located at the Service Manager layer, is to store large amounts of information on Quality of Service. This information is used to send network load predictions to the VPM. These predictions need to be retrieved using various criteria such as Class of Service, User Group, Source node, Destination node and other more temporal criteria.

To implement this function, an Object Oriented database (ObjectStore [16]) has been used. This approach facilitates a direct mapping between objects used in the application and those stored in the database. Once the objects have been designed, there is no extra work to put them in the database. In addition, the advantages of having a database are retained. These include search, retrieval facilities, indexes to increase search speed, and transaction features such as roll back.

The performances obtained in the different tests we have made for our application have been very encouraging. However, for various reasons it has not been possible to compare in the same context Object Oriented to Relational Database implementations.

5. CONCLUSION

All AIP techniques described here have been implemented successfully during the project. Their evaluation have been done in context of a large integrated Traffic and QoS Management emulator. All experimental runs have been performed with realistic simulated users, services and networks. Therefore, assessment of the AIP techniques real-time performances has not been realised. It must be pointed out that a scientific and rigourous evaluation of the AIP techniques used is extremely difficult to perform. We have tried during the third experiment to use at least two different techniques to realise the same functionality.

The methodologies and management platforms used in this project have proved their efficiency and they will certainly be useful for building management applications in TMN.

6. ACKNOWLEDGEMENTS

This work has been conducted within the NEMESYS project which is part of the RACE program sponsored by the EEC.

7. REFERENCES

[1] Kühn, P. J. : "From ISDN to IBCN (Integrated Broadband Communication Network)", Information Processing 89, G.X. Ritter (ed.), Elsevier Science Publishers B.V. (North-Holland), IFIP, 1989, pp. 479-486.

[2] CCITT Recommendation M.3010: Principles for a Telecommunications Management Network (AP IX 31- E).

[3] Experiment 3 Design, RACE Project NEMESYS (R1005) Deliverable 10, May 1992, ref. 05/DOW/SAR/DS/B/024/a1

[4] Mandich, N., Belleli, T. : "HCI Considerations in TMN Systems", Sixth RACE TMN Conference, Madeira, September 1992.

[5] ISO: Working documents for the Basic Reference Model of Open Distributed Processing, ISP/IEC JTC1/SC21/WG7 (1990).

[6] J. Rumbaugh, M. Blaha, W. Premerlani, F.Eddy, W. Lorensen: Object-Oriented Modelling and Design, 1991, Prentice-Hall, Inc.

[7] Jacobson, I. : "Object Oriented Development in an Industrial Environment", OOPSLA '87 Proceedings.

[8] Patel, P. et al : "Viewpoints on Traffic and Quality of Service Management in Telecommunication Management Networks", Sixth RACE TMN Conference, Madeira, September 1992.

[9] ISO/IS 10165-4: "Information Technology - Structure of Management Information - Part4: Guidelines for the definition of Managed Objects", August 1991.

[10] Lebouc, P.,. Stern, P.-E: "Distributed problem solving in broadband telecommunication network management", Proc. Expert Systems and their Applications, Avignon, May 1991

[11] Knight, G., Pavlou, G., Walton, S. : "Experience of Implementing OSI Management Facilities", Proceedings of the IFIP Second International Symposium on Integrated Network Management, Washington, April 1991.

[12] Rose, M. T. , Onions, J. P., Robbins, C. J. : "The ISO Development Environement User's Manual Version 7.0", PSI Inc/X-tel Services Ltd, July 1991.

[13] ANSAware 3.0 Implementation Manual, February 1991, ref. RM.097.01

[14] Nikesh - Experiment 2 in NEMESYS, RACE Project NEMESYS (R1005) Deliverable 6, 28th June 1991.

[15] The OPS83 User's Manual, System Version3.0, April 1989.

[16] Object Store Reference Manual, Object Design Inc, 1991.

A Generic Maintenance System for Telecommunication Networks

J. Bigham, Dominic Pang, Tim Chau (Queen Mary & Westfield College, UK)
Walter Kehl (Alcatel SEL - AG, Germany), Carla Neumann (Danet, Germany)

ABSTRACT

This paper describes the application of model-based reasoning in the area of maintenance of telecommunication networks and specifies in outline form the architecture of a modelling and reasoning tool kit which has been built to perform diagnosis and maintenance. The contribution of model-based reasoning for containing the complexity of the network management task and feedback from the application to enhance the applicability of model-based reasoning techniques is discussed.

1. INTRODUCTION

This paper discusses the architecture and application of expert system techniques to the maintenance of telecommunication systems. The maintenance system described here is being developed for future IBCNs (Integrated Broadband Communication Networks) by the RACE project AIM. The maintenance problem described is the on-line corrective maintenance of hardware faults. Maintenance means more than solving the diagnostic problem for electronic components. It means coping with a vast number of fault reports, formulating and verifying hypotheses and carrying out repairs together with the necessary preventive actions. To solve all these different tasks, it was decided to use model based expert system techniques. These techniques were:

- modelling the telecom network and its behaviour explicitly
- using a model based inference engine which performs abductive and deductive reasoning on this model.

As practical experiments, AIM has built two prototypical applications of the developed automatic maintenance system. One application is for BT's System X Digital Subscriber Switching Subsystem (DSSS), and the other one is for the BERKOM network, a Broadband-ISDN field trial network built by Alcatel SEL. The primary source of information about the BERKOM and DSSS networks was design documentation with some additional expert interviews. Another prototype, which is still ongoing, uses the model of a Metropolitan Area Network (MAN). The MAN knowledge base was built using knowledge about a working MAN field trial in Stuttgart and deals more with errors occurring between whole network elements. Whereas the BERKOM application is nearer to the "classical" applications of model-based reasoning (like diagnosing electrical circuits and other kinds of hardware diagnosis), the MAN prototype brings in more telecom specific aspects. The MAN prototype is described more fully in [1].

The capability to have the use of design information is important as IBCN systems in the field will have many different variants of a much smaller number of basic designs. They will also be subject to design modifications and enhancements at regular intervals. To ensure that the reasoning mechanism is invariant to these changes the underlying design knowledge is represented in a declarative, modular and easily changed form which is then interpreted by the reasoning mechanism.

2. GENERAL ARCHITECTURE

The overall goal of the AIM project is to develop a Generic Maintenance System (GMS) for the IBCN. Such a highly generic system which has to be independent of vendor and technology-specific implementations can only be realized by a very clear separation between the generic procedural knowledge (i.e. the inference mechanisms and tools) and the specific declarative knowledge (i.e. the specific and explicit models of the different network elements of an IBCN). The implemented GMS is an integrated collection of tools to support the maintenance of a network and is described below. Within the AIM project the use of cooperation between GMSs, each working on different network elements, to provide additional information to assist in the diagnosis of problems in the network is also being investigated. This is not described here.

2.1 The Maintenance Cycle

The maintenance cycle is shown in figure 1. The different parts of the cycle are the modules which act like independent processes "piping" their results to the next module.

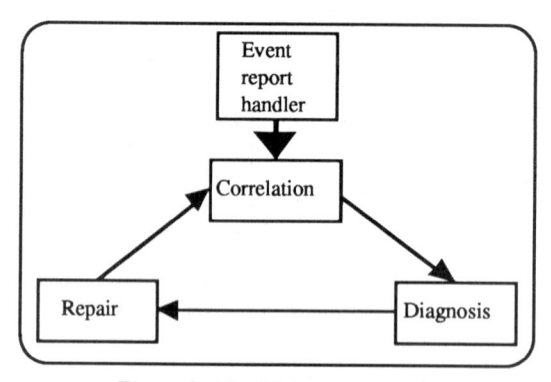

Figure 1 - The Maintenance cycle

The maintenance cycle covers the complete maintenance task from the first error report to the successful repair of the suspected component. The modules in the maintenance cycle are responsible for separate sub tasks:

Event Report Handler

The event report handler accepts the observed symptoms from the telecommunication network, processes a simple form of time correlation and forwards the symptoms to the correlation module.

Correlation

The main task of this module is the formulation of fault hypotheses. A specific feature of telecom systems is that one fault can result in a huge set of similar symptoms. These symptoms must be correlated and associated with a small set of possible explanations. As there is no single fault assumption built into the reasoning process, each possible explanation can be a conjunction of single causes. The output of correlation is therefore a disjunction of explanations. An explanation is defined more fully later in this paper, when a functional description of the reasoning process is given.

Diagnosis

The term "diagnosis" stands for the verification of the hypotheses supplied by correlation. Verification is carried out by applying tests to the suspected components. These tests can be executed either automatically or with the help of a human operator. The output is the physical component which has to be replaced, or if no hypothesis could be verified, a message to correlation.

Repair

This means the actual replacement of the faulty component. Several protective actions have to be carried out in order to do the repair with only a minimal disturbance to the subscriber traffic. After the replacement, the new component is tested again and a "success" message is sent out.

2.2 The Knowledge Base

The knowledge base used for maintenance should eventually become an integral part of a Management Information Base (MIB) within the TMN. The MIB is the conceptual information store for all management aspects of the TMN, with maintenance being one important part of management. The knowledge base is the part of the maintenance system which supplies detailed information describing the structure and behaviour of a particular telecommunication network. The knowledge base is divided into the following parts:

- *the functional model,* which includes structural and behavioural knowledge
- *diagnostic information:* information about available tests, test behaviour of components, test planning rules. Test behaviour is also represented by a functional model, but is separated from the main functional model because the use is different
- *the physical model,* describing the physical structure of the network: boards, cables etc. and their position and interconnections
- *repair information.*

2.3 The Knowledge Acquisition and Representation Tool

The Graphical Model Acquisition Tool (GMAT) and Graphical Acquisition, Browsing and Reasoning of Intelligent ELements (GABRIEL) toolsets, which have been developed to support the users in acquiring and representing telecommunication knowledge and are presently being integrated. They will be used as the GMS development environment, and runtime MMI, see figure 2. The use of user profiles will examine the requirement to enhance such tools or the provision of new tools or windows. The tool functionalities are:

The Graphics Editor allows an easy population of network element representations within the knowledge base. This population process is based on actions involving the graphical representation of these network elements and on the graphical representation of their connectivities. For this task, correspondence of concepts of abstract classes and abstract instances with generic graphical objects and graphical instances are performed, respectively. Thus changes in the graphical view of a knowledge base will update the knowledge base and other access and integrated MMI facilities will be provided.

The Object Editor is used to specify the structural representation of the network components. Once network elements are specified and stored in the knowledge base, it is possible to browse through the knowledge base by looking at all classes or slots or instances. Basic or generic telecommunication elements, e.g. multiplexer, can be stored in a library or generic knowledge base, so that these elements do not have to be defined several times, but can be duplicated or copied from the library. The Object Editor provides a management on the defined objects, so that add-on, delete and other transactions of knowledge within the knowledge base are performed in order to keep the knowledge base consistent.

The Rule Editor appears in the same layout as the Object Editor. This editor provides facilities to describe the functional behaviour of the network elements. This behaviour will be described in the form of if-then-rules. These rules can be attached to the defined classes. Rules can be clustered to rulesets. A complete management of links between rules and rulesets and classes and rules/rulesets is provided. The modification facilities of if-parts and then-parts of rules is supported by syntax checking.

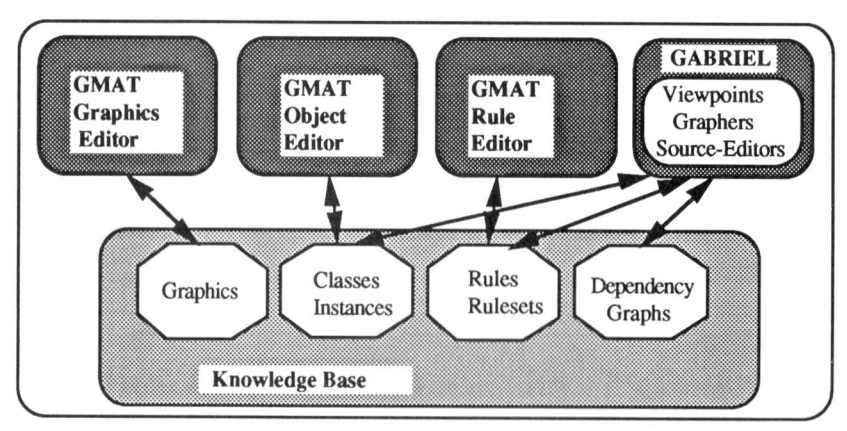

Figure 2 - Knowledge Acquisition & Representation Elements

The Reasoning Requestor enables users to interact with the reasoning system to test rule behaviour and to perform simulations and inferences on the network model in the MIB.

The Viewpoint Requestor enables users to view any part of the model from predefined viewpoints, which define how much information about a model is made available to the user, in terms of size and complexity. Users can also define new viewpoints. Such viewpoints use Computer Aided Design (CAD) layout Graphs, Containment Graphs and Manual layout Graphs. Viewpoints are also programmable, and can be called from application code.

The Filter Requestor enables users to select (via a menu or a graph) the ports that will be used to define the layout of a new viewpoint or a generic viewpoint of the automatic layout graphs. This functionality can be enhanced to enable viewpoints to be created using the current filter settings. The filter graph allows users to create new filter classes and to map them onto existing unit ports, providing user control of the filtering of ports.

All editors and graphs allow units to be selected and can call other editors or graphs to display that unit (and surrounding units) in a different way. Zooming can be achieved by users selecting a unit from a viewpoint and choosing a more detailed viewpoint of that unit. All editors and graphs have knowledge acquisition capability in the creation / deletion / modification of unit classes, unit instances, port and states, and connections between units. A multi-window environment has been chosen so that every editor is presented in its own window.

The knowledge base generated by the toolset contains a model description of networks and network elements and their corresponding graphical representation. This stored information will then be used by other modules of the Generic Maintenance System, e.g. the reasoning module or the simulator module. The toolset provides support to knowledge engineers and telecommunication experts towards the construction and modifications of a model of telecommunication concepts in an interactive and user friendly manner.

3. THE FUNCTIONAL MODEL

The functional model is built out of functional entities (FEs) which correspond to specific functionalities of the modelled telecommunication network. A functional entity is for example a multiplexing functionality or the functionality of transmitting data from one FE to another FE, or the behaviour of a test and the corresponding test results. There is a mapping between the FEs of the functional model and the elements of the physical model. This mapping is not necessarily a one-to-one mapping, it can be also a many-to-one or one-to-many mapping. This separation into a functional and a physical model allows the functional model to be independent of the physical configuration of the telecommunication network.

3.1 Knowledge Representation Language for Functional Models

A functional entity consists of internal attributes and ports. The ports are connected to the ports of other functional entities, e.g. the power-out port of a modelled converter is connected with the power-in port of a modelled multiplexer-group. The functional entities together with these port-to-port connections are the functional model.

A knowledge representation language (GOOD : GMS Object Oriented Design language) has been implemented to support the construction of functional models out of functional entities. Physical models are built using the object oriented base language CLOS which is part of Common Lisp. GOOD enhances the normal facilities provided in an object oriented language to allow the representation of behaviour in a modular and declarative manner. Behaviour is associated with functional classes (called FE classes), and more complex forms of behaviour can be constructed by combining the behaviour of parent classes. This is illustrated later.

This representation of behaviour reflects the following principles:
- Only local behaviour is described, modelling with rules which go from cause to effect
- Working and fault behaviour can be represented using the same formalism. If a FE can fail in a number of ways and knowledge about the failure models is known, then this can also be encoded in the knowledge base to be used by the reasoning system
- The rules are formulated in an abstract way. A rule is instantiated only when it is required by an application.

An input port can have only one input connection from another FE, though an output port can be connected to any number of input ports of other FEs. This is not a limiting restriction, it is simply to ensure that the behaviour is explicitly written as rules and not implied by the connectivity.

Linked to the knowledge representation language is a graphical tool which allows pictorial descriptions and construction of the functional and physical models. This maps to the same underlying knowledge representation languages as the functional and physical models. The tool is an experiment to determine the requirements of graphical front ends to MIBs supporting object oriented model based reasoning systems.

Network representations have very complex semantic interactions and must represent large numbers of resources. Managed Objects, i.e. software representations of real network resources, have many functionalities (power, clocking, subscriber information, etc.), and these functionalities are usually visible in the way the managed object semantically interacts with its peers. Because of the complexity of the MIB and managed objects, users must be given the ability *not* to view information which is not directly relevant to the type of information they wish to browse.

Ports can be used as a means of filtering out unwanted knowledge, as users can specify which ports to use as filters when a graph is shown. The graph can recursively show all children of a managed object which are connected using one of the selected ports. For example, selecting the 'power-out' and 'drain' ports will show the power hierarchy of the MIB, if a graph was started from the main power supply. All power providing and power receiving managed objects (with their associated power lines) will be shown.

Graphical creation or deletion of classes, ports, attributes, instances and behaviours is allowed. For example in order to create a class the user would first graphically select one or more superclasses from existing viewpoints (like the power class hierarchy and the clock class hierarchy), and would then ask the user for the name of the class to be created.

From this information the system will generate the class definition, which would be displayed in the appropriate viewpoint. This allows the user to rapid prototype the MIB models under construction by being able to immediately see the effect of his/her actions. For example creating a new semantic connection between managed objects, or testing new managed object behaviours. A fuller description of the graphical tool can be found in [2].

Behaviour defines the function of a managed object for the purposes of reasoning by a model based reasoning system. Such behaviour defines how a managed object works, and why it does not work. Behaviour can be created and immediately tested by the graphical tool using either abduction (inference) or deduction (simulation). For example, if power behaviour was added to a voltage convertor managed object, it can be tested deductively by turning off a connected power source and watching the effect on the convertor (i.e. it should go out of service and change colour). It can also be tested abductively by asking the reasoning system why it is not sending out power, which should reply that an upstream power source, or the convertor, is faulty.

Simple inheritance of rulesets has been augmented by a facility which allows the user to define how rulesets are to be merged. In the knowledge representation language combined rulesets can be constructed if we make the consequents of the rules (which belong to the rulesets to be merged) refer to intermediate states, rather than output ports. The user then defines an *additional* rule which combines the rules for the particular FE-class inheriting those rules. This additional rule defines the state of the output ports as a function of internal and intermediate states. Note that an intermediate state can only have one input rule. This is for reasons similar to the restriction of one input connection to an input port.

Several functional entities can be seen together as a higher-order functionality. This concept has been incorporated into the model as the organization of the functional entities on different levels of granularity which are connected via a has-part/is-part-of relation. At the moment there are four levels in the model which reflect (matched against the physical structure) the module, sub module, board and sub-board level. Most of the reasoning so far is conducted on the fourth level which has the finest granularity.

3.2 Reasoning with the Functional Models

A reasoning tool (called GMS^2, to be read as GMS-squared) has been implemented which uses a functional model built using the GOOD language described earlier, and identifies faulty functional entities. The reasoning tool is itself modular which will allow adaptation and enhancement to particular requirements. Its key components include a *control monitor*, a *suggestion interpreter*, a *model interpreter*, and a *simulator*. The reasoning tool can be applied to different functional models within the maintenance task, such as correlation of error reports, and test management. The underlying architecture is therefore designed to give flexibility but still presents a consistent set of concepts to the user and maintenance system builder. A GMS may have more than one instance of a GMS^2. *Each* GMS^2 instance reasons on a *functional model*. Functional models can exist at different levels. The correlation process takes the fault symptoms (in the form of fault reports) and uses the functional model (created using the concept of FEs) to produce a minimal set of suspect FEs. In the "diagnosis" step of the maintenance cycle, suspect FEs are first mapped to other FEs which have their behaviours under tests modelled. Secondly, the GMS² (possibly another instance) is applied to the new FEs in order to produce new explanations involving FEs. Thirdly, Substitutable Entities (SEs) for repair are produced by mapping to the physical model.

4. THE GMS²

4.1 Functional Specification of the GMS²

Following the notation of Konolige [3] a GMS² assumes a model which can be represented as a simple causal theory <C, E, I, S> where:

C, the set of *causes*, is a set of positive literals of the form (predicate-value (FE-name attribute)) e.g.

(not-working (FE-power-supply-1 working-status))

(working (FE-power-supply-1 working-status)).

Causes are associated with possible internal states of FEs in the FE modelling. In the case of internal states with binary domains then a cause may be a negative literal, e.g. if the internal state working-status has domain {working, not-working} then (not (not-working (FE-power-supply-1 working-status))), and (not (working (FE-power-supply-1 working-status))) are also valid as the meaning is unambiguous. However for non binary domains the meaning of a negative literal is ambiguous. For example if the internal state "setting" has domain {high, medium, low}, then (not (high (FE-name setting))) can be two causes.

E, the set of possible *effects*, is a set of positive literals, e.g. (out-of-service (line-card-1 output2)). The elements of E are output ports in the FE modelling. Ports with binary domains can have effects represented by literals, e.g. (not (in-service (mux-1 output1))), as these are unambiguous.

I, the set of possible *intelligent state values*, is a set of positive literals, e.g. (out-of-service (line-card-1 intelligent-state2)). The elements of I are operational state variables in the FE modelling. They are used to model the states in intelligent components which are observable in the MIB on request, though not known initially. They do not include intermediate states, as these are only states to allow the modeller to inherit and combine rulesets from functional entity superclasses. Intermediate states have no physical meaning. Intelligent states with binary domains can be represented by literals.

S is a domain theory, which is the union of the rulesets defined for each FE. An important restriction in the rule language is that no FE's ruleset can contain a rule which has non-deterministic output , e.g. of the form if <condition> then <consequent1> or <consequent2>. Although the rule language has been designed to give convenience (e.g. by allowing the antecedent to have nested "and"s and "or"s, and economy of representation using intermediate states and rules which can represent several rules by having variables for the attribute values), each rule in the FE's rule language can be mapped to a set of Horn Clauses. S implicitly contains additional constraints. Each internal state, port and operational state belongs to a set of mutually exclusive values. The domain theory assumed is therefore that which is representable by a set of Horn Clauses.

The reasoning process produces abductive explanations for symptoms individually or (more useful generally) an *abductive explanation* for a set of observations. An abductive explanation for a set of observations O which is a subset of E ∪ I, is a set A which is a subset of the set of causes C such that

A is consistent with S

S ∪ A ⊢ O

A is a subset-minimal cover over sets satisfying the first two conditions

A is read as a conjunction of elements taken from the set of atomic causes. A cautious explanation for O is a disjunction of all the abductive explanations for O i.e. $\vee\, A_i$ where each A_i is an abductive explanation. An option available is to produce a cautious explanation, though for non trivial problems only a subset is generated. As explained above the process of generating explanations can be continued when more information is available, e.g. a substitutable entity has been replaced but still cannot be returned to service.

An alternative provided is to produce a conjunction of the abductive explanations for each symptom. This can be of use where, for example, the disjunction relates to possible faults in a line, and loop testing can be performed on the line in say an incremental manner.

4.2 The Components of the GMS^2

The main components of the GMS^2 are as follows.

Suggestions Interpreter

The suggestions interpreter produces a set of abductive explanations S_{ij} for a symptom s_j where s_j belongs to those observations which are a subset of E ; i.e. the suggestions interpreter produces a set $\{S_{ij}\}$ where each S_{ij} is a subset of the set of causes C such that S_{ij} is consistent with S, $S \cup S_{ij} \vdash \{s_j\}$. S_{ij} is a subset-minimal cover over sets satisfying these conditions. In some of the implementations of the suggestion interpreter $\vee_i S_{ij}$ is a cautious explanation for s_j.

There are at least two kinds of suggestions. The first is based on the knowledge in the model. They are derived using a rule-generator which effectively inverts the behavioural rules in the model. For instance, the rule 'if ME1 is out of service then ME2 should also be out of service' can be re-interpreted by the suggestions interpreter to mean 'if ME2 is observed to be out of service, then it is likely that ME1 is out of service. The suggestions interpreter reasons with the inverted rules and outputs its conclusions as explanations to the control monitor.

Versions of the suggestion interpreter based on creating suggestions using the contrapositive of the model rules have been implemented also. When the assumption that the modelled behaviour represents all the ways the system can fail then the approaches are essentially equivalent. For intelligent components the suggestion interpreter can inspect the MIB and depending on the state (disabled or enabled) of the unit inspected, the suggestion interpreter controller can use this information to limit the work required to produce the explanations of the symptom.

A second kind of suggestion would encode heuristic and experiential knowledge and be used directly to generate explanations. These forms of suggestions have not been implemented. The suggestion interpreter could also be guided by the strategic control heuristics in the knowledge base, but this has not been implemented.

Control Monitor

The control monitor is the core of the inference architecture. It controls the invocation of the other components of the GMS^2 and synthesizes their results to produce the explanations. It constructs the explanations for O from the explanations generated by the suggestion interpreter from each symptom in E and consistency information available from using the model interpreter and the operational state values in I.

Two different control strategies have been tested. One motivated by the heuristic that faults are likely to show themselves by symptoms near the actual cause. Possible explanations are generated by the suggestion interpreter in the order that functional entities were encountered traversing upstream causally from the symptom in E. The controller computes the explanations for all the symptoms $\vee A_k$ as

$$\vee A_k = \wedge_j (\vee_i S_{ij})$$

In another strategy each FE can have a probability associated with its working-status internal state (if it has one). Any probability calculations relating to ideas like "It has been a long time since this item has been replaced and I now think it could be more likely to fail and so the probability of failure has increased to ..." are not done. Such adjustments to the prior probabilities would have to be made at the time of running the system.

The control monitor performs a best first search through the set of possible explanations constructed in a similar way to the first case but sorted by probability. The search tries to maintain the fact that every node has n children at any time - if possible. So there are n^d nodes, where d denoted the number of levels in the tree. Because of the scale and speed requirements in the application, in future work the search is to be controlled so as to limit the tree size more drastically.

Model Interpreter

The model interpreter is the component that has to perform deductions from a hypothetical explanation A_k (context explanation). It fires only those rules which can fire in the context. Intelligent components are queried for their values in the MIB and the model interpreter determines if the context explanation is consistent. The controller invokes the model interpreter with partial explanations, i.e. those which account for those symptoms incorporated to date. If the context is found to be inconsistent no more rules are fired and A_k is removed from the search by the controller.

The model interpreter consists of a *rule generator* and a *rule interpreter*. The former takes the rules as written for the FEs, which only refer to internal states, operational states, intermediate states and ports and uses the connectivity information to generate rules that explicitly refer to adjacent functional entities. This rule generator is also used by the suggestion interpreter for a similar purpose. Once a rule set has been generated it is saved so that it need not be done again. The rule interpreter then fires these rules and passes the deductions together with their justifications to PIE (see below).

The full set O of observations is not known at the start of the reasoning process. When the reasoning mechanism is checking a possible explanation A_k for consistency and the MIB is queried for the status of an operational state, then the corresponding value becomes known and added to the set of observations O. If A_k is inconsistent with O then A_k and any supersets of A_k should be rejected as explanations.

Simulator

The simulator is a stand alone version of the model interpreter which is of use in the initial construction of the model as it assists in ensuring that the knowledge is consistent.

Propositional Inference Engine Module (PIE)

The PIE which acts as a cache for the deductions generated by the model interpreter. As the model interpreter picks out the part of the model to apply deduction on, the PIE builds up a network of nodes to record all the supporting assumptions of the deduced propositions. It also maintains a list of inconsistent combinations of assumptions which are used by the controller to prune the task trees by deleting those nodes with an inconsistent focus and so avoid wasting resources following useless lines of reasoning.

By recording the dependencies between all the propositions, PIE maintains simultaneously multiple worlds. This capability allows the control monitor to switch between contexts fairly efficiently. The ideas in PIE build on the ATMS described in [4] and [5]. The nodes are connected by clauses and the propositional deduction performed is a form of unit clause resolution.

Repair Dialogue Monitor

The repair dialogue monitor which interacts with the human repair engineer to replace substitutional entities (SE - physical entities which can be replaced, as against the functional entities which are abstract) that are verified by the diagnosis module to be faulty. The repair monitor has to

- provide on-line assistance on repair steps to make
- allow the repair engineer to report observations that are not directly obtainable by the maintenance system
- react to such observations accordingly
- advise the repair engineer to perform tests to verify that the repair is successful and the symptom is cleared, and
- report back to the reasoning system of any test failures.

Figure 3 shows the important components of the repair module and their interaction.

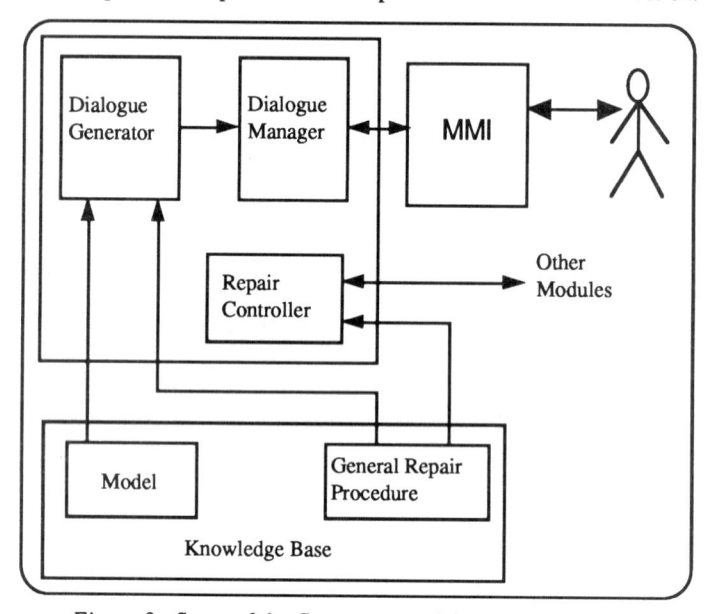

Figure 3 - Some of the Components of the Repair Module

The dialogue generator takes as inputs,
- an SE to be replaced
- repair knowledge stored in the physical model and devises a plan in a presentation-independent language of repair steps and control flow between these steps. This forms the repair dialogue and can be in the form of a state transition diagram. This is then passed on to the dialogue manager.

The dialogue manager executes the repair plan by interpreting the dialogue generated by the dialogue generator. When instructions are needed to be given to the repair engineer or questions are to be asked, they are passed on to the MMI. When data is required from other modules of the maintenance system or information is available as a result of performing the repair that would be useful to these modules, the dialogue generator invokes the repair controller to handle the inter-module communication. An example of why the latter is needed is when a repair test fails, the diagnosis module may have to be informed that its diagnosis is wrong.

The Dialogue Design Tool (DDT) is used to define repair dialogues (and dialogues to support manual tests) between the operator and the maintenance system. GMS^2 has been implemented in CLOS, using the Harlequin LispWorks development environment on SUN workstations, and the infrastructure in C on UNIX.

5. GMS CONTROL

The overall maintenance cycle was described in section 2.1. To perform the complete maintenance function the communication between the key components, event report management, correlation, diagnosis and repair has to be coordinated. Controllers have been developed for the prototype applications. These are not generic in the sense that they can be invoked by providing parameter values but do allow the application builder to see how to use the interface specifications of the component tools to construct intra-GMS control.

As part of its function an intra-GMS controller coordinates the input and output of instances of GMS^2. Instances of a GMS^2 with a specified identity can be created and invoked on a set of symptoms or facts and a set of explanations produced. If this instance of the GMS^2 is invoked again with another set of symptoms or facts then these symptoms are taken to be additional symptoms to be accounted for. The instance of the GMS^2 can also be made to start afresh. There may be as many GMS^2 instances as required, as they do not interfere with each other. For example if two GMS^2 "processes" are needed, one called correlation and one called diagnosis, then correlation may be run using a set of symptoms, then diagnosis on (typically) another set of symptoms, then correlation on further symptoms reported etc. Some of the symptoms for the second run of the correlation process can be "symptoms" output by the diagnostic process.

The implementation allows the instances of GMS^2 to be run concurrently. This concurrency is not used by current intra-GMS controllers. For example a "correlator" GMS^2 would be idle when the "diagnose" GMS^2 is running and vice versa. A communicating processes model which allowed interruptions would need to be developed to use such a feature effectively. Communications between different GMSs on different workstations are supported by the infrastructure tools, which provide transparent communication between processes and an object-oriented interface to some storage mechanism such as an Information Base (see [6]).

6. MODEL BASED REASONING FOR NETWORK MANAGEMENT

Much work in model-based reasoning has been in the area of circuit diagnosis and simple physical systems. Applying model-based reasoning to the management of telecommunication systems is different in two major aspects: modelling telecom systems leads to much more complex and dynamic models and the reasoning system will not be stand-alone, but integrated into an existing system with a large number of hard constraints. A good overview over the whole area is given in [7] which also gives many pointers to the literature. This section only discusses issues related to the experience gained from prototype construction and conceptual work.

6.1 Modelling Telecommunication Networks

Coarse Granularity Models

Technical systems like telecom networks can be looked at and modelled at any level of granularity, from the circuit level to the level of complete networks. But to cope with the complexity of telecom networks one has to start modelling at the highest possible level. So even in the (hardware-oriented) BERKOM model the elements on the lowest level of granularity were highly integrated chips with a complex internal functionality. For the MAN model typical units in the model are network elements like routers, gateways etc. which consist of a number of boards; the modelling here will not go below the board level. That means that by its very nature the modelling of telecom networks is an abstraction process and is started at the highest possible level of abstraction. From this modelling at high levels a lot of the other specific modelling features arise, like hierarchical modelling or dynamic behaviour.

Modelling Non-physical Entities

The "classical" model-based reasoning concentrates more on physical entities like electronic circuits, printed boards or different sorts of mechanical machinery. However in telecom systems, only (in terms of representing development efforts) a minor part is hardware. The larger part, and the part causing the hardest management problems, is software. Therefore software modules, services, subscribers etc. need to be modelled.

To have an entrance into the field, the AIM project started with the modelling of hardware equipment, but is now considering experiments with modelling and diagnosing software modules too. The strong demand for network management lies in the software area, hence research must be done for modelling software.

Hierarchical Modelling

Telecom systems - hardware as well as software - and the already existing management functions are designed and implemented in a hierarchical way in order to cope with their complexity. Therefore the modelling also has to follow this hierarchical approach. This allows for different viewpoints on the model (a "zooming in" on areas of interest) and has effects on the inference engine and the tools browsing the knowledge base.

One important issue is to maintain consistency between the information on the different granularity levels. This is done by identifying "layer border ports" which are the outer ports of a cluster of functional entities which are parts of the same higher level element. With these layer border ports and the appropriate mappings between the layers even an automatic update of the model can be achieved, i.e. if the model on one hierarchy level is changed, the effects of this change on the other levels have to follow automatically.

Management Domains

Management information for telecom systems is not always to be found at one single - logical or physical - location. Normally the management of a large network is distributed over various managers which manage (arbitrary) parts of the network. This means that the model of the overall network is cut into pieces and stored at different managers. In the area of maintenance, AIM has developed techniques for the cooperation of different GMSs which have each one part of the overall model they are responsible for. These techniques follow a hierarchical approach, i.e. wherever managers need information beyond their model knowledge they ask higher level managers which in turn have the right to request information from all managers which are inferior to them. With this cooperation and the necessary interfaces between the model parts, boundaries between management domains can be introduced at arbitrary positions in the overall model.

Dynamic Behaviour

As the models are at a high level of abstraction, the behaviour will not be as static as the behaviour of low-level physical entities. The behaviour of a network element can depend on the status of the environment, on administrative actions put on it or on a specific internal status. This means that the modelling language must allow the formulation of conditional statements which enable different types of reasoning according to the current status or must even allow the modelling of behaviour which is specific to only one instance.

Such enhancements of the behaviour modelling language have been implemented and these help to formulate the different kinds of behaviour components can have at different times: normal and fault behaviour, test behaviour, behaviour in active or stand by mode, behaviour dependent on a specific configuration etc.

Knowledge Acquisition and Consistency

Knowledge acquisition for the implemented prototypes was not too difficult because all the information was already available in the form of technical specifications and documents. The specifics of the telecom application is that - as a lot of effort is put into conformance to standards - there is already a good deal of generic knowledge which need not be acquired each time. Therefore the GMS does not contain only the procedural kernel but additionally includes the generic partition of the knowledge base. For a discussion of how knowledge acquisition could be carried out in more real-life environments, see section 7.3. A crucial problem is the consistency between the real world and the knowledge base. Not only does the status of some components change frequently, but also the configuration of telecom systems (which has to be mapped to the structural model) have a dynamic component. These changes can be caused by faults as well as by administrative actions of various kinds. The only way to solve this problem is to have the MIB as the single point where all the management-relevant information has to pass through. The concept of a managed object (see section 3.1) is considered to be helpful for this purpose. Especially in the case of administrative actions concepts similar to database transactions: "management transactions" which have locks, acknowledgements, roll-backs, etc. must be introduced.

6.2 Using the Inference Engine

The details of how telecom networks are modelled will have consequences on the reasoning mechanisms.

Hierarchical Reasoning

If the models are structured in a hierarchical manner the reasoning must make use of it. The reasoning switching back and forth between different granularity levels brings in very naturally the advantage of focussing the search for a fault reason. If, for example, a symptom occurs on a low level functional entity, the reasoning goes upwards to higher levels, searches there until it has found the higher-level element in which the cause of the fault is located and then "zooms in" to the detailed modelling of this element. This allows detailed statements to be made without having to do an ineffective search on a wide range of low level of granularity. It should also be possible to have symptoms on a more abstract level (maybe from another part of the TMN, e.g. reports on performance decreases in a whole network) which can then be explained with detailed causes (faults in a specific component). Similar ideas have been formulated in [8], but there is no distinction between structure and behaviour, only a hierarchical model in terms of states. The approach adopted in AIM is more flexible because the reasoning process can go up and down the hierarchy levels whenever this is indicated by behaviour rules. This is feasible because the subfunction/superfunction links between the levels are represented as normal port-to-port connections; therefore specialized behaviour rules can make use of them.

Efficiency

Efficiency is a very important topic for reasoning about telecom systems: on the one hand the models and the management tasks are very large and complex, on the other hand real-time (or almost real-time) responses are necessary to meet the availability requirements for telecom systems. Efficiency is a problem area also according to the telecoms domain. Whereas due to the techniques mentioned above no major efficiency decrease for large hierarchical models will occur, efficiency problems in the BERKOM prototype with its non-hierarchical model have been already encountered. These problems occur only when calling the model interpreter where they are due to the fact that the model interpreter is used in "full simulation mode". That means that all port states are computed which can be deduced from an assumption. However when using the model interpreter in the consistency checking in correlation, it could stop immediately after the first inconsistency has been found. With this and other efficiency improvements currently being implemented it is expected that the inherent exponential complexity of the ATMS (e.g. see [9]) can be controlled. To give an impression of the dimensions used in the models, approximate numbers are: for simulating typical events with only local consequences the model interpreter has to compute about 20 port states and needs about 1.5 seconds; for the computation of 200 port states it needs 50 seconds. But if the size of the simulation increases again to about 700 port states, the time goes up only to 60 seconds. Another answer to the efficiency problem arising from large knowledge bases could be the technique of knowledge compilation (see [10]). In some very primitive forms this has found entrance into the GMS[2], but it is subject of further study into how knowledge compilation and especially hierarchical modelling can be integrated.

Combination with other Types of Reasoning

Model-based reasoning depends on the accurateness of the model. One limitation of this accurateness is whether it is technically and economically possible to reach a 100 percent completeness of the model. Even if every available bit of knowledge could be put into the model there are faults (especially transient faults) which cannot be explained on the basis of the available knowledge. They simply happen and nobody knows why! To live with these limitations model-based reasoning should be complemented by a reasoning system which also makes use of experiential knowledge. Case-based reasoning or even machine learning approaches would fit into the existing framework. Based on event logs and history files (which are already present in the GMS) such techniques could be integrated with model based reasoning and could improve the efficiency of model based reasoning and expand the range of explanations. One could, for example, think of choosing the most probable fault candidate for (expensive) consistency checking or testing on the basis of experiential knowledge.

7. TMN:HOW CAN IT BENEFIT FROM MODEL BASED REASONING

7.1 The TMN Situation

The current situation in the area of managing existing telecommunication networks is characterized by the following facts:

- There are already a lot of good management functionalities available, due to the fact that telecom systems have always had to conform to very high reliability constraints (e.g. the German "DBP Telekom" requires an overall availability for digital exchanges of 0.9997)
- However there is no common or generic paradigm unifying all these good partial solutions. All the existing management functions are more or less specialized and proprietary
- There is the impression that future management tasks will be of a much higher complexity (like the management of IBCNs consisting of a large number of different types of equipment) and can no longer be handled with the current techniques. This is supported by the fact that managing telecom software seems already at its frontiers. The use of model-based reasoning techniques can be an answer to these challenges.

The benefit of model-based reasoning is twofold. The first benefit comes from the knowledge representation. Whereas in terms of functionality for specialized areas there can always be conventional solutions which are as good as or better than model-based reasoning, the main advantage lies in having a powerful, yet very clear, declarative and easy-to-understand representation of the management knowledge. This is especially important because of the following reasons:

- Telecommunication networks are usually quite large. To represent this large amount of complex knowledge a representation form is necessary which combines power with clarity. This can be achieved by building models in the way described above where the represented models correspond directly in an intuitive way to the real world units. The

importance of clarity and simpleness of the representation cannot be overestimated, as this knowledge must be maintained and worked on by human operators
- Telecommunication networks are quite often installed in variants of a given basic system. Modelling of these variants is very easy with a deep model based approach and with the strict distinction between generic and specific knowledge. The same holds for changes to the system.

The second benefit of model based reasoning is that it is a common approach which can be applied to several different management tasks. AIM has been able to experiment with model based reasoning only for fault management (maintenance), but cooperation with other RACE projects in the TMN area indicates that this technique could be applied as well to other management tasks, like performance management and configuration management. Further study in this area would be useful. The following advantages are identified:
- The knowledge base (the MIB) is accessible to all parts of management, i.e. the unification of the management functions starts with the common knowledge representation. This leads to the advantage that there must be only one information base which can reflect always the current state of the telecommunication network. A management function can use all the up-to-date knowledge which is available without worrying about consistency
- Another advantage of using deep knowledge is robustness, i.e. the ability to handle faults and events which are not explicitly foreseen
- As there is a simulation capability present with model based reasoning, it is possible to run certain scenarios with all management aspects included
- The knowledge base can in principle be constructed automatically from design data etc. which are available in a formalized electronic format. It is not possible nor necessary in practice to obtain all the required knowledge from expert interviews.

These two main benefits of using model based reasoning, namely, the power and clarity of the knowledge representation and its common usability, should be key factors in coping with the future complexity of telecommunication systems. Functionality can always be added or improved, but if the overview and general control over a managed system are lost then no partial solution will help.

Trends in Standards. The views expressed here are supported by the current trends in standardization. In the telecommunication area standards are very important, and much effort from telecom equipment providers and operators go into the development of new standards. The emerging standards for network management (see [11] and [12]) focus around the concept of a managed object. The communication between the management application functions (e.g. fault management) and the real piece of equipment goes always via a managed object (see figure 4).

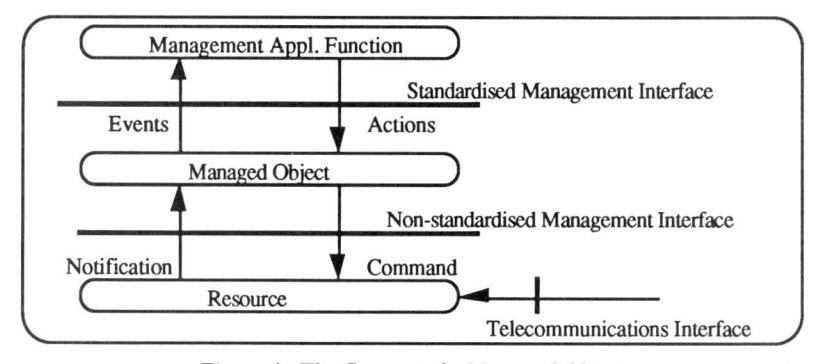

Figure 4 - The Concept of a Managed Object

These managed objects are subject to classifications according to the different types of network resources they are representing. Although it would be too much to say that using model based reasoning is a direct consequence of having managed objects, it seems very natural and intuitive to combine the standards bodies managed objects and model based reasoning. So for example in the experimental models built during the project it was very easy to integrate these two concepts: the instances in the structural model were implemented as managed objects, communicating with the network resources and management application functions via actions and events. The classification from the standards was not sufficient for the purposes of the project, but in principle the models are built using a generic class hierarchy, which can be adapted.

7.2 The Intelligent MIB

The integration of all the major topics discussed so far leads to the concept of an Intelligent MIB. The Intelligent MIB integrates the following features:

- *Object-oriented modelling.* There is a simple and intuitive way to represent complex knowledge about the telecom system using the principles discussed in sections 3 and 6
- *Model-based reasoning.* This accounts for the "intelligence" in that reasoning capabilities are integrated with the MIB
- *Common core of network management.* By having the MIB as the central institution through which all management actions must pass and by having the intelligent services (realized with model based reasoning) as the main functionalities of network management, a common management core where consistency is guaranteed and double or contradictory actions are avoided can be achieved
- *Conformance to standards.* As the intelligent MIB uses the concept of managed objects and standard interface protocols it can cooperate with any network resource or any other manager which conforms to the standards. The standards protocol could be used to integrate existing management functionalities by accessing them via these protocols. Although the intelligent MIB can support standards concepts it is not restricted to them.

It is not possible to describe in detail principles and construction of an intelligent MIB here; for an extensive discussion see [13]. However it is worth noting that some additional functionalities for the MIB are needed:

- The automatic maintenance of consistency within the knowledge base, especially during knowledge acquisition
- An environment which supports knowledge acquisition and knowledge maintenance
- A uniform and state-of-the-art man-machine-interface for all aspects of network management. This comprises an operator interface through which all operator management actions have to be achieved. Work has been done in this area in the AIM project (see [2]). Although the Intelligent MIB is supposed to be a common core for network management, this does not mean the MIB is realized in a centralized way. This is almost impossible due to the distributed nature of telecom networks. Therefore the Intelligent MIB can be distributed logically and physically, and interaction can take place between different managers which are responsible for parts of the whole model, i.e. for their respective management domains (see 6.1).

7.3 OPEN ISSUES

Economic Background

Network Management is becoming a major issue in the telecommunication business. In such an important area people tend to avoid risks as much as possible, and of course, using a very new technology like model based reasoning is a risk. There is also an environment of existing software and hardware which represents thousands of manyears of effort, even for one system.

With this background, model based reasoning must prove its capabilities; but it has a very strong support in comparison with existing systems.

Integration into product life cycle

To receive most benefits from using model based reasoning it must be fully integrated into the product life cycle. For telecom systems this should be feasible as all the design knowledge is available in electronic form as CAD or CASE data. From these the model could be built, with the additional advantage that the behaviour of the designed system can be simulated and tested even during the design phase. Integration with existing systems is also possible, especially in cases where the knowledge base can be derived from existing databases.

Scaling up

The usage of model based reasoning in a real-time environment would lead to several questions of scaling up and interfacing to existing software. All efficient sources for the reasoning process must be used to obtain real-time responses; speed and efficiency must become a criterion in developing model based reasoning systems. Another question is the use of a database; as the MIB is realized as an object-oriented model one should think of an object-oriented database as the first choice for achieving data persistency, but object-oriented databases are not yet mature.

8. CONCLUSIONS

The main benefit which network management can gain from model based reasoning is a common conceptual view on all aspects of network management and an easy-to-understand knowledge representation. This can help to cope with the challenge of the increasing complexity of future telecom networks. The work done in the AIM project has shown for the area of fault management that the model based reasoning approach is feasible for network management and has demonstrated some of the advantages of model based reasoning. As a consequence of this work the concept of an Intelligent MIB has been developed which can serve as an integration platform for knowledge representation and reasoning and as well for the different management tasks. On the other hand though, there are a lot of telecom specific issues which have to be solved in model based reasoning, because of the complex and dynamic nature of the models which must be built for telecom systems. Apart from these issues model based reasoning must prove its strength in terms of speed and efficiency in order to meet the high reliability constraints in the telecom area. Also the other management tasks should be tackled with model based reasoning methods to establish its applicability, and research needs to be done on the automatic construction of models. Experience within the project indicates that model based reasoning is potentially a technology that is powerful enough to solve a broad range of problems in TMN. A lot of work needs to be done but there is much mutual benefit to be gained from these two areas and which could eventually arrive at a model-based network management for telecommunication systems.

9. ACKNOWLEDGEMENTS

Several people have contributed to the development of the GMS. Particular thanks go to Gottfried Schapeler, Mark Newstead (Alcatel SEL - AG), Nader Azarmi, Steve Corley and Manooch Azmoodeh (BT), Yann Le Hegarat (DANET), Kevin Riley (UNIPRO), and Spyros Patatoukakis (QMW).

10. REFERENCES

[1] Hopfmüller, H. et al : "An Interconnected MANs Maintenance Protoype", Sixth RACE TMN Conference, Madeira, September 1992.

[2] Newstead, M. and B. Stahl, G. Schapeler :"Advanced Information Modelling for Integrated Network Management Applications" Proceedings of the Fifth Intern. Conf. on Industrial & Engineering Applications of Artificial Intelligence and Expert Systems, June 1992

[3] Konolige, K., "Abduction versus closure in causal theories", *Artificial Intelligence* 53, (1992), pp255-272

[4] de Kleer, J. : "An assumption based TMS, "Artificial Intelligence 1986, Vol 28, 127 - 162

[5] Forbus, K.D. and J. de Kleer, "Focussing the ATMS," *Proceedings of the AAAI*, August 1988, 193 - 198

[6] Riley, K. and T. Tin, "Infrastructure for the GMS," RACE AIM document, 06/UNI/000/DR/C/067/a1, June 1992

[7] Riese, M., : Model-based Diagnosis of Networks: Problem Characterisation and Survey, 1991

[8] Mozetic, I., Hierarchical model-based Diagnosis, International Journal of Man-Machine Studies, 1991, vol. 35, 329-362

[9] Selman, B. and H.J. Levesque, Abductive and Default Reasoning: A computational core, *Proceedings of the AAAI*, 1990, 343 - 348

[10] Karp, P.D. and D.C. Wilkins, An Analysis of the Distinctions between Deep and Shallow Expert Systems, International Journal of Expert Systems, Vol 2, 1989

[11] CCITT Recommendation M.3010 (4 - 8 November 1991), Principles for a telecommunications management network. Version R5, Temporary Document No 40 (REV. 1)-E, Geneva.

[12] OSI-NM Forum, OSI-NM Forum Architecture A&S 078 Issue 1 (working draft 11) 1989.

[13] Newstead, M. and H. Hopfmüller, G. Schapeler, A Design of the Operation, Maintenance and Construction of an Intelligent MIB, Proceedings of Fifth RACE TMN Conference, November 91

The Management of Telecommunications Networks
R. Smith, E. H. Mamdani, J. G. Callaghan (Editors)
© Ellis Horwood 1992

An Interconnected-MANs Maintenance Prototype

Heiner Hopfmüller, Walter Kehl, (Alcatel SEL - AG, Germany)
Gottfried Schapeler (Alcatel SEL - AG, Germany),
Nader Azarmi, Steve Corley (BT Laboratories, UK)

ABSTRACT

This paper describes a maintenance application prototype, the Interconnected-MANs Maintenance Prototype (**IMP**) which will demonstrate the applicability and appropriateness of Advanced Information Processing technology and an AIP-based Generic Maintenance System for the maintenance of Integrated Broadband Communication Networks.

1. INTRODUCTION

This paper describes the work done and results achieved in implementing a maintenance application prototype. The prototype (called the IMP) addresses the management of interconnected Metropolitan Area Network (MAN) switching systems. The choice of this switching system was due to : satisfaction of the requirements of the IMP experiment to test the applicability and performance of Advanced Information Processing (AIP) technologies and an AIP-based Generic Maintenance System (GMS) against realistic Integrated Broadband Communications (IBC) networks, and the top-level implementation information on MANs had already been published and was readily available.

The main technical interest in the IMP experiment is the study of concepts for distribution and cooperation of GMSs [1] which had been discussed theoretically and partially implemented in previous prototypes [2], [3]. These concepts have been more closely specified and implemented for the IMP. The IMP addresses the interesting central issues of problem solving across management domain boundaries and Hierarchical Maintenance Management. In order to support the implementation of a hierarchical maintenance system which is capable of hierarchical diagnostic reasoning a novel approach has been devised for the modelling of the underlying managed network at different levels of abstraction. A suitable cooperation model has also been developed which allows cooperative diagnostic problem solving among a number of GMSs while their hierarchy is preserved. The work has been influenced by the RACE-GUIDELINE Integration Task Force (ITF) studies and has adopted a similar network topology and management hierarchy to the ITF network scenario [4]. In addition, the IMP experiment has, as far as possible, attempted to apply evolving standards for Telecommunication Management Networks (TMN) and to verify these standards within the scope of maintenance. Section 2 of this paper gives a description of the IMP's underlying network. In section 3, functional and design specifications of the IMP are discussed. Realisation of the IMP with the use of the GMS tools is described in section 4. Section 5 describes the implementation details and the results of the IMP testing on a number of fault scenarios. Finally, section 6 presents a number of future work directions.

2. NETWORK DESCRIPTION

The QPSX MAN [5] technology has been chosen as the basis for the IMP experiment. The IMP network is based on a MAN-island consisting of three subnetworks, which are connected together via sub-network routers (see figure 1 for the network topology).

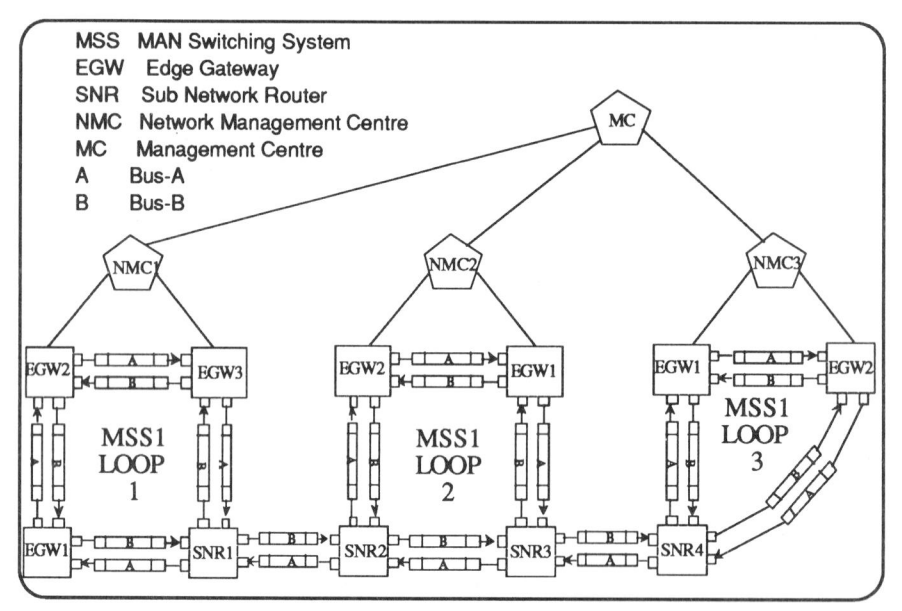

```
MSS   MAN Switching System
EGW   Edge Gateway
SNR   Sub Network Router
NMC   Network Management Centre
MC    Management Centre
A     Bus-A
B     Bus-B
```

Figure 1 - The IMP Network Topology

A layered management structure exists which involves management centres (NMC1 to NMC3) for the MAN Subnetworks and an additional central management centre (MC) at which service management and customer administration also reside. The QPSX MAN network management consists of the following classes of managers :

- Module Level Manager
- Cluster Level Manager
- System Level Manager.

Module Level Manager

The Module Level Manager works at a granularity level which roughly corresponds to the functionality implemented on a printed circuit board.

Cluster Level Manager

The Cluster Level Manager works at a granularity level which corresponds to what is shown in figure 1 as Edge Gateway (EGW) and Subnetwork Router (SNR). The Cluster Level Manager can be seen as a network element manager.

System Level Manager

The System Level Manager works at a granularity level which corresponds to what is shown in figure 1 as a MSS and its subnetworks. The System Level Manager is considered to be a network manager.

A Network Management Centre (NMC) which is used in the context of the QPSX MAN, see figure 1, can be regarded as an Operation System Function (OSF, see [6]). The OSFs are able to handle events from the resources, to trigger actions on the resources (e.g. change of operational states, execute tests, information requests) and to cooperate with other (non-maintenance) OSFs as well as human operators. Resources in this context are represented by a network simulation model which is able to create appropriate responses to any input, e.g. to perform what-if-scenarios.

3. FUNCTIONAL AND DESIGN SPECIFICATION

3.1 Functional Specification

The IMP consists of a simple Network Scenario Simulator (NSS) and a Maintenance System (MS), see figure 2. The functional definition of the IMP will be given in terms of the functional definitions of its sub-functions (i.e., NSS and MS). Since the IMP is divided into the MS and the NSS modules, the specification of the functionality of these modules and their interaction is crucial. The input to the IMP will be a number of pre-determined fault scenarios which can be either direct hardware faults or can come from other TMN functions, such as traffic management or provisioning functions [4]. The output of the IMP will be the diagnosis of faulty resources and the production of corrective actions for the repair of these resources.

The Maintenance System (MS)

The function of the MS is to carry out maintenance diagnostic reasoning in order to identify faulty resources and to initiate test or repair procedures. It will also be the responsibility of the MS to update incrementally its Management Information Base (MIB) network model for it to be consistent with the underlying network at all times.

As illustrated in figure 2, the MS will consist of a number of cooperating maintenance managers (NM-GMS). Each maintenance manager will be in charge of an area of the network. It is assumed that the network will be divided into a set of mutually exclusive areas (as far as the network elements are concerned). That is to say that an over-lapping management domain situation is not being considered in this prototyping activity. The point being made is that the MS must demonstrate the capability of cooperation among a number of maintenance managers for the task of total network fault diagnosis.

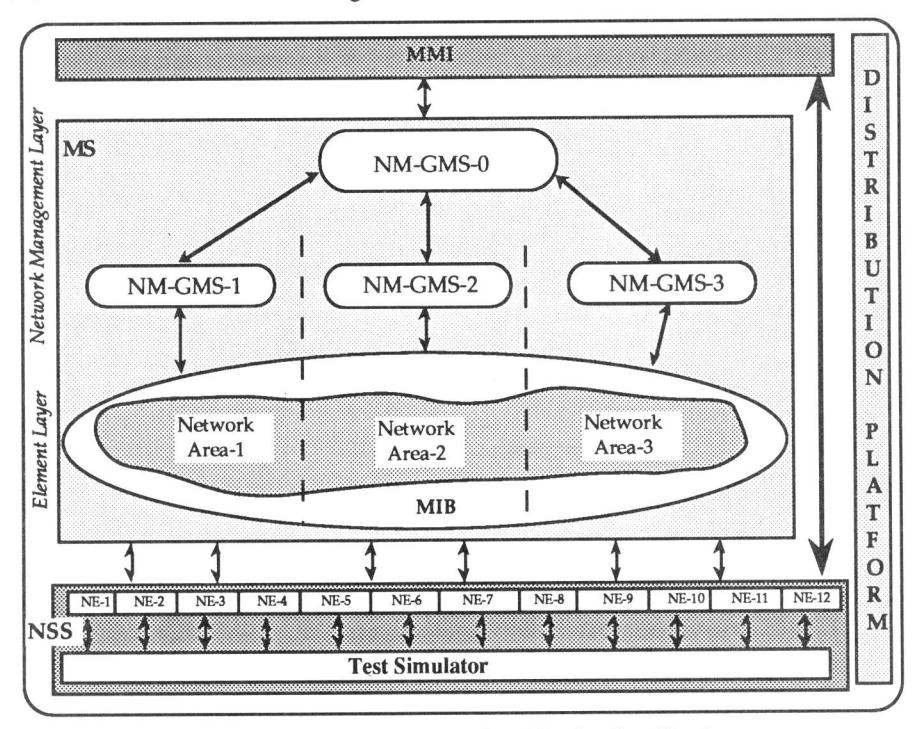

Figure 2 - The IMP Functional and Design Specification

The input to the MS will be provided by the NSS and the output from the MS will be a list of faulty resources and (for some cases) repair procedures for repairing each faulty resource.

The Network Scenario Simulator (NSS)

The function of the NSS is to simulate the behaviour of a real IBC network. The input to the NSS will be a set of management queries/actions from the MS or the user, and the output of the NSS will be the appropriate responses/messages, i.e. event reports (for example, clear/fault alarms, NE state reports, put-out-of-service confirmed, etc.) to the MS. The NSS event reports correspond to the messages/responses which will be received from a real IBC network had it been in existence. For example, putting a resource out of service would result in all of its functionally dependant resources going out of service. The response/message generation will be achieved by simulating the network behaviour. It is expected that the model of the network (in the MS's Management Information Base, the "MIB") would provide sufficient information to enable simulation functions to be developed for the NSS. Furthermore, the NSS will provide a test simulator which simulates the testing of the NEs and uses the test models and test information stored in the MIB for this function.

3.2 Logical Design

The logical design definition of the IMP will be given in terms of logical design definitions of its sub-functions (i.e. the NSS and the MS).

Logical Design of the NSS

Figure 2 illustrates a high-level design specification of the NSS. It consists of two modules. The first module is the Test Simulator which simulates the running of the diagnostic tests on the resources modelled in the MIB. The Test Simulator uses the diagnostic test model in the functional model of the network. The second module is an MIB front-end which allows specification of the IMP's fault scenario, by setting internal states of NEs within the MIB.

Logical Design of the MS

Figure 2 illustrates a high-level design specification of the MS. In essence the MS will consist of a number of GMSs and a conceptual MIB. This paper does not describe the GMS tools. These are described in [1]. In the following sections, two important issues for the design of the MS, i.e. modelling (mainly from the cooperation perspective) and cooperation are discussed

3.3 MODELLING APPROACH

The AIM approach to modelling allows for a flexible deep object-oriented model of the underlying telecom network. Levels of abstraction appear in the network models. In the earlier prototypes [2], [3] different levels of abstraction were used to set up the class hierarchy, but no use was made of this abstraction in the context of diagnostic correlation. This resulted in the existence of a non-hierarchical model (a flat model) which was sufficient for the initial experiments. For large scale networks flat models appear not to be suitable, because:
- They do not reflect the hierarchical structure of the real world they represent
- In a non-hierarchical model there is generally a 1:1 relationship between resources to be managed and managed objects
- A non-hierarchical model would cause a high load on the management system for the forwarding and administration of information..

Using a layered modelling concept, reasoning can be performed at different levels of abstraction. These levels of abstraction allow focusing during problem solving by transferring control to a higher level where a more abstract model will provide a set of 'high-level fault candidates'. More detailed explanations are generated by using less abstract models at lower levels.

The MIB of the MS will contain functional and physical models of the underlying network. The partitioning of a network model into a number of sub-networks has been based on the IMP network topology and its non-overlapping maintenance management domains. The partitioned network model will contain valuable information which is needed for the cooperation aspects of the GMSs. In order to realize a maintenance management system which is capable of fault diagnosis in a hierarchical fashion the underlying network needs to be modelled at different levels of abstraction. This important issue is discussed in the next section.

Hierarchical Modelling Technique

The use of levels of abstraction should not be interpreted as having different models at different levels but to have different views on the model. To provide these more abstract views it is necessary to have additional higher level objects and as such the collection of these objects can be considered as a model (that is, an abstract model).

Abstraction could be used differently for
- "global" assertions about the respective area
- "global" operation on the respective area
- description of relationships between areas of the same level of abstraction.

One technique for representing abstraction is the Has-Parts/Is-Part-Of hierarchy. This hierarchy is well known and has been used in previous prototypes to model the containment of subcomponents within a component [2], [3]. In the IMP model there are four levels of abstraction : network, subnetwork, network element and component level. The component level model was essential for carrying out realistic fault scenarios. Moreover, it was found that only modelling components/hardware (which corresponds to the OSI physical layer) was not enough, since the consequences of a fault can affect the operation of upper OSI layers. Therefore, modelling of the data link layer is important if a complete fault management system is to be achieved. For the IMP prototype, modelling of the data link layer was partially realised.

Although the containment relationship can be used to model physical or functional abstraction, a further aspect of abstraction layering is a logical mapping between more and less detailed information. In terms of cooperating GMSs this means that the lower abstraction level information must not be interpreted (used for reasoning) at the higher level - it is purely syntactical and only used for forwarding information to a cooperating GMS. On the other hand the higher abstraction level uses the lower level as a service, comparable to the OSI definitions.

Unit-Port Hierarchical Modelling

An object oriented modelling language has been developed which represents network components in terms of units and ports [7]. A unit is an object which specifies a component's structure and behaviour. Ports describe a unit's relationships with other units. In order to model the hierarchy and separate management domains in terms of units and ports, specialized ports are introduced which allow both layer (i.e. a level of abstraction) and domain borders to be bridged. This approach provides a uniform means for modelling an arbitrary number of levels of abstraction and individual domains. It must be possible to model and allow exchange of information across levels of abstraction and domain borders independently. In other words, functions mapped on one layer have to operate whether or not there is a domain border across elements in the lower layer(s). Although in the IMP model hierarchy borders and management domain borders are the same, this is not always true for the general case. Hierarchy borders come from modelling and reflect different levels of granularity, whereas management domain borders come from arbitrary organisational requirements of the real world, e.g. country borders or ownership of equipment. Therefore these two kinds of borders should be kept conceptually separate. For a function to be performed across a domain border, cooperation will be required. As shown in figure 3 three types of ports are required (i.e. Domain Border Ports, Layer Border Ports and Containment Ports) for this purpose.

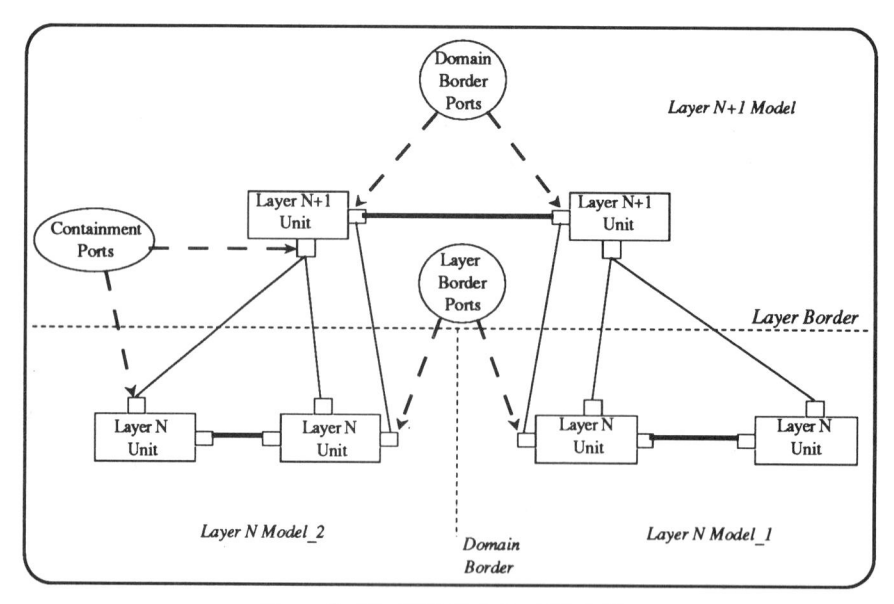

Figure 3 - The IMP Unit-Port Modelling

It is assumed that domain borders can appear at arbitrary points in a network. Therefore no restrictions should come from the modelling and cooperation approach used. The Unit-Port language also had to be enhanced to allow the specification of dynamic behaviour inherent in high level network components. The behaviour of a network element can depend on the status of the environment, on administrative actions applied to it or on a specific internal status. This means that the modelling language must allow the formulation of conditional statements which enable different types of reasoning according to the current status or must even allow the modelling of behaviour which is specific to only one instance. Such enhancements to the behaviour modelling language, which has been used in the previous prototypes (see [2], [3]), have been implemented. This helps the formulation of different kinds of behaviour that components can have, for example normal and fault behaviour, test behaviour, behaviour in active or standby mode, behaviour dependent on a specific configuration, etc.

One of the functions of maintenance at the Subnetwork Layer is to perform loop-back-tests, either upon the request of the network manager or autonomously. The execution and coordination of the test has to be modular. This can be assumed because tests are part of the structure and behaviour of the resources and as such are captured within the corresponding units. This also holds for objects at different levels of abstraction.

3.4 GMS Cooperation

The IMP has been designed to incorporate cooperating GMSs in performing faults diagnosis across the network. The network is assumed to be partitioned into a number of mutually exclusive domains, each containing a number of NEs which are managed by one and only one GMS.

There will be real network resources connected together across management domain borders. Since it is assumed that the knowledge of a domain manager (i.e. a GMS) is limited to its domain, such outside domains connections are not known to the corresponding domain managers. This situation results in a knowledge gap for the purpose of network maintenance management and cooperation among sub-network GMSs becomes necessary.

In the context of hierarchical maintenance management and hierarchical modelling of the network, the role of each GMS is to manage a number of NEs at a particular level of abstraction and it only deals with this level of detail. This framework allows a unified approach to fault diagnosis, since the GMSs at level (n) can perform model-based diagnosis in the same manner as the GMSs at level (n+1). The resulting mechanism is a hierarchical diagnosis process as the hypothesis suspecting an element/unit at level (n+1) can be refined and focused at level (n).

Each GMS in the IMP, upon receiving a set of event reports, will generate (correlate) a number of internal and external hypotheses. An internal hypothesis of a GMS indicates a network element belonging to the domain/model of that GMS is faulty, whereas an external hypothesis identifies a faulty managed object outside the domain of that GMS. For the case of an external hypothesis, the adopted cooperation mechanism allows GMSs to send requests to their parent GMSs in order to identify the probable source of the problem. Through higher-level reasoning, the parent GMSs which have knowledge of the neighbouring domains of lower-level GMSs, identify the domain in which the problem can be solved for the received requests. The parent GMSs would then send appropriate requests/messages to the corresponding GMS of that domain to carry out reasoning in its domain.

In summary, the above cooperation mechanism works in an ascending and descending fashion. The ascending process is used when the GMS's reasoning generates external hypotheses. The descending process is used to refine a diagnosis.

4. REALISATION WITH GMS TOOLS

The GMS tools have been used extensively in all aspects of the IMP implementation. In the following sections, the applicability of these tools for the realisation of the IMP is described.

4.1 Knowledge Representation / Acquisition Tools

In the course of the construction of the IMP model, the Knowledge Representation / Knowledge Acquisition tools developed by AIM, GABRIEL and GMAT have been used mainly for knowledge representation. For a detailed description of these tools see [1] and [8].

4.2 The Inference Engine

The GMS-GMS [6] reasoning tool has proved sufficiently generic for the implementation of the IMP without major changes. This has been possible due to the Unit-Port modelling of the underlying MANs network of the IMP.

Application of the GMS-GMS to the IMP's hierarchical diagnostic reasoning was easily achieved due to the hierarchically structured models of its network. All the relations between the hierarchy levels (e.g. parts-of, subfunction-of) are formulated as normal ports which can be used by the inference engine, not different from normal functional-dependency ports. With these ports it is then possible to trigger the reasoning - with specialised behaviour rules - to switch back and forth between the different levels of granularity.

4.3 GMS Infrastructure Tool for Distribution and Communication

The IMP experiment requires a number of distributed cooperating GMSs. The GMS Infrastructure tool has been used for Realization of the IMP's distributed MIB and the communication among the GMSs. The distribution details of the IMP which describes the logical location and communication aspects of the GMSs are kept in the Distribution model provided by the infrastructure tool. The communication between the Management Application

Functions (like correlation and diagnosis), the Managed Objects in the MIB and the network resources is the task of the MIB Handler. The MIB handler is the collection of all methods and procedures which make use of the MIB, formalised as actions and events in a protocol-like way. This protocol approach has the following advantages:
- Interaction with the MIB is independent of the physical implementation of the MIB
- Information hiding, i.e. the Management Application Functions get only that part of the information which is relevant for them
- The MIB is independent of the Management Application Functions. It offers a set of services to the outside world, but is not concerned who makes use of theses services.

4.4 The Maintenance Cycle

The basic concept of the maintenance cycle of the maintenance system is the same as described in previous publications [1], [2]. There are four major modules, event report handler/manager, correlation, diagnosis and repair. What is different now is that one fault might not be handled by one maintenance cycle of one GMS but has to be passed through various GMSs each of which performs a part of a complete maintenance cycle, as described in the following scenario.

The GMS-1 receives an event report and starts correlation1. Correlation1 suggests that the possible fault cause lies outside the management domain of GMS-1. Therefore a message is sent to the superior manager, GMS-0 (see figures 1 and 2, GMS-1 to GMS-3 correspond to the NMCs, GMS-0 corresponds to MC), requesting that the fault treatment for this particular situation is to be finished. After correlating at a higher level, GMS-0 asks the responsible GMS (here: GMS-2) to execute correlation, diagnosis and repair. Also a message must be sent back to GMS1 that the original event has now been cleared, this message additionally going via the superior manager GMS-0. Note that there might be cases where local symptoms (that is, symptoms which occur in the management domain of the faulty resource) and distributed symptoms (that is, symptoms which occur in another management domain) exist at the same time. In these cases, fault management can work locally with the local symptoms and in addition, explain the distributed symptoms (via the cooperation and distribution techniques). In other cases, for example, if there is not enough local fault detection equipment, there may only be distributed symptoms and it is then necessary to reason backwards from these symptoms.

5. THE FAULT SCENARIO

A basic fault scenario was constructed to outline the overall goal of the demonstration as well as to guide the development of the prototype. The fault scenario should help to define the necessary units, ports, and behaviour rules for the purpose of modelling and to identify the necessary and possible cooperation of the different maintenance managers/agents aimed at.

From this three constraints follow:
- Cover the overall network
- Provide sufficient detail of information on the building blocks
- Support the definition of the management architecture.

The last constraint was satisfied by the decision to apply hierarchical management and by the standardised structure of MANs to consist of sub-networks. A sub-network is mapped into a high level object class and instantiated as a unit.

To cover the overall network and to achieve cooperation of distributed agents, a local fault with impact on a distant data transmission was assumed. It was necessary to consider data traffic because a limitation to simple (on-line corrective) maintenance cases would not allow for the investigation of a real world situation. For this reason the model had to be extended to contain sufficient functionality modelling data traffic and event reporting. These kinds of functionality are already on the OSI layer 2, the data link layer. To provide sufficient detail of information on the building blocks, the alarm reporting capability of the Network Elements has been investigated and some of the alarms have been selected to be incorporated into the scenario. Only a rough outline of the scenario will be given here to omit the details of MAN-internal functions.

The initial situation
Consider the three MAN subnetworks as in figure 1. A user starts to transfer data from EGW2 of Loop1 to EGW2 of Loop3 via EGW3 of Loop1, SNR1, SNR2, SNR3 and SNR4.

The fault
Both line sending functions of EGW3 of Loop1 fail (are pulled out), i.e. neither Bus A nor Bus B can be served by EGW3 of loop1.

The Events
The following alarms are received by the management system:
- EGW2 of Loop1 detects Signal Level Below Threshold on Bus A
- EGW2 of Loop1 detects No Frames on Bus A
- SNR1 of Loop1 detects Signal Level Below Threshold on Bus B
- SNR1 of Loop1 detects No Frames on Bus B
- EGW2 of Loop1 forwards a Healing Event
- SNR1 of Loop1 forwards a Healing Event
- EGW2 of Loop3 detects Packet Pipe Timer Time-out.

A *Healing Event* means a special reconfiguration from a loop as in figure 1 to a single sequence of NEs, an "open bus configuration". In this configuration it is still possible for all nodes to transmit data despite the break in the connection. *Packet Pipe Timer Time-out* means that the reassembling of a packet was not possible within an expected time.

The events inside loop1 can be explained locally by GMS-1. Reasoning backwards from the given alarms it finds that the reason was the faults in the line sending function of EGW3 of loop1. For the event in loop3 no causal explanation according to other events within the loop is possible. Because GMS-3 can use only the model of loop3, it has to call GMS-0 for further treatment of the event which in turn determines that the fault cause lies in the responsibility of GMS-1, i.e. in loop1. GMS1 is informed and can then give the information about the fault which explains as well the symptoms in loop3.

6. CONCLUSIONS
The application of the GMS to the MAN model has brought several new aspects which have not been tackled before; the network aspect, i.e. having a model where one fault can produce not only local symptoms but symptoms which are distributed over the network, the cooperation aspect, i.e. doing fault management as a cooperative task of not only one, but several managers, and the aspect of modelling a different transmission technology. A DQDB loop differs considerably from a "classical" telecommunication network in its greater dynamics and flexibility. This was an additional challenge especially in modelling the behaviour. With the techniques of hierarchical structural modelling, of cooperation between distributed managers and with a more dynamic modelling of behaviour the GMS has shown that it is generic and flexible enough to also allow the modelling of such a DQDB network which differs in the aforementioned aspects from the prototypes which have been built so far in the course of applying and testing the GMS. But there are also new questions arising from this prototype which ask for further work in the following directions:

Temporal Reasoning

Hitherto, it was sufficient to reason with the exact time point of a symptom. However, for problems occurring at higher levels of functionality, temporal reasoning may be necessary.

Second Order Maintenance

This means the maintenance of the built-in maintenance functionalities of a telecom system, like alarm circuits, alarm lines etc. In the case of the MAN, the data links and the maintenance system are already partly overlapping, e.g. some maintenance information is transported on the same bus as normal user data. This is conformant to a trend in TMN where the TMN uses the resources of the managed network. Although some work has been done in this direction, it needs to be represented in a broader and more systematic way.

Empirical Testing

With the current model it should be very easy to do some empirical testing of the efficiency of the inference process for large models. This can be done by adding more high-level components to the model, by making it more distributed and by running it in different "reasoning modes" (i.e. "flat" against hierarchically focused reasoning).

Modelling

Our experience has shown that in order to tackle more complex faults it is necessary to model also functionality at higher levels of a communication protocol stack. This leads very naturally to an integrated management where fault management is only one task inside an integrated TMN.

7. ACKNOWLEDGEMENTS

The work described in this paper has been performed within the RACE project AIM. The IMP system builds upon previous work done by all partners of the AIM project. This work was supported by the Commission of the European Community under the RACE programme. Thanks are also due to the reviewers for their useful comments.

8. REFERENCES

[1] Bigham, J. et al : "A Generic Maintenance System for Telecommunication Networks", Sixth RACE TMN Conference, Madeira, September 1992.

[2] Final Report on Integrated Maintenance Application Prototype for BERKOM,,AIM Deliverable 06/SEL/FZS/DR/B/075/b1, Jan 1992.

[3] Deliverable on Maintenance Application Prototype - System X, AIM Deliverable 06/BTR/DNM/DR/B/073/b2, Jan 1992.

[4] Initial Report on the ITF Common Case Study, GUIDELINE ME7 Deliverable 03/BCM/ITF/DS/C/011/b1, Feb 1992.

[5] Alcatel MAN, The QPSX Connection, 1989

[6] CCITT Draft Rec. M.3010

[7] Maier, F. et al : "Recommendations for the Use of AIP Techniques for Maintenance in Telecommunication Systems", Sixth RACE TMN Conference, Madeira, September 1992.

[8] Final Documentation on the Generic Maintenance System, AIM Deliverable 06/QMC/INS/DR/B/081, forthcoming, 1992.

VI - Recommendations

The Management of Telecommunications Networks
R. Smith, E. H. Mamdani, J. G. Callaghan (Editors)
© Ellis Horwood 1992

Experience Designing TMN Computing Platforms for Contrasting TMN Management Applications

V. Wade (Trinity College Dublin, Ireland), W. Donnelly (Broadcom, Ireland),
S. Roberts, D. Harkness (ROKE Manor Research, UK),
K. Riley (UNIPRO, UK), A. Carr (Dowty, UK),
J. Celestino (UPMC Versailles, France)
R. Shomaly (BT Laboratories, UK),

ABSTRACT

TMN management applications are many and diverse, frequently employing different technologies and isolated solutions. Yet as telecommunications management becomes more sophisticated, there is a growing need for telecommunication management integration and interoperability. This paper presents a General TMN Computing Platform architecture flexible and powerful enough to support contrasting TMN management applications while facilitating integration and cooperation. To validate the architecture, three diverse management applications have been chosen. Profiles, based on selected components of the General TMN computing platform architecture, have been developed which would provide the necessary computing platform implementations for the applications. The paper also examines how emerging general purpose computing environments may be adapted and utilised to (partially) support TMN systems. The paper identifies and examines core infrastructure components required across a broad range of TMN applications. Finally technologies required specifically for telecommunication computing platforms are identified and discussed.

1. INTRODUCTION AND MOTIVATION

The last five years have seen a rapid growth in the capabilities and interoperability of general purpose computing platforms. Much of the sophistication which was originally built into telecommunication management systems is now being offered by these emergent platforms. Because of the difficulty and expense of re-developing Computing Platforms (CP), the TMN community must utilize and adapt general purpose CP systems to their specific needs. In order to achieve this saving in development cost and effort, a common understanding of the capabilities of modern platforms needs to be achieved and TMN management systems have to be designed in such a way as to avail themselves of these CP systems. However such general purpose systems will never provide a complete solution for telecommunication management. The telecommunications community must therefore identify TMN specific technologies and develop these within a general TMN computing platform infrastructure.

This paper describes the experiences of four RACE TMN projects in developing a General TMN Computing Platform. It describes the general architecture for TMN platforms and develops three profiles, based on the architecture, capable of supporting contrasting management systems. These profiles are used to illustrate the flexibility of the General TMN CP Architecture and validate it's applicability across a wide range of management applications. This paper also describes how some TMN CP profiles may be implemented (at least in part) using emergent CP systems. Conclusions are drawn concerning the core components of the General TMN CP. These core technologies can act as a common level(s) of support across a range of TMN platform implementations and would provide an effective means of achieving integration and cooperation of management applications. However some important technologies are specifically required by TMN platform implementations that are not supported by more general purpose platforms. The paper highlights these TMN specific technologies and discusses their significance across a range of management applications. Finally several open issues in the area of TMN computing platforms are discussed.

2. EXPERIENCE DEVELOPING A TMN CP INFRASTRUCTURE

A General TMN CP architecture, capable of supporting a wide variety of TMN management applications, has been developed over the last two years [1], [2]. This General TMN CP architecture was the result of prototyping experience of several independent RACE projects, namely ADVANCE, AIM, NEMESYS and ROSA. As well as this experience, the development of the General TMN CP architecture was influenced by TMN related standards [3], [4].

The General TMN Computing Platform for management applications consists of five layers. The goal of these layers is to abstract away the complexity and heterogeneity of the underlying host systems and communications protocols. The layers also provide transparencies for distribution, information access. and replication. These five layers are (i) CP Kernel, (ii) Distributed Processing Support (iii) Computing Platform Interface (iv) TMN Support Environment (v) User Generic Applications (see figure 1).

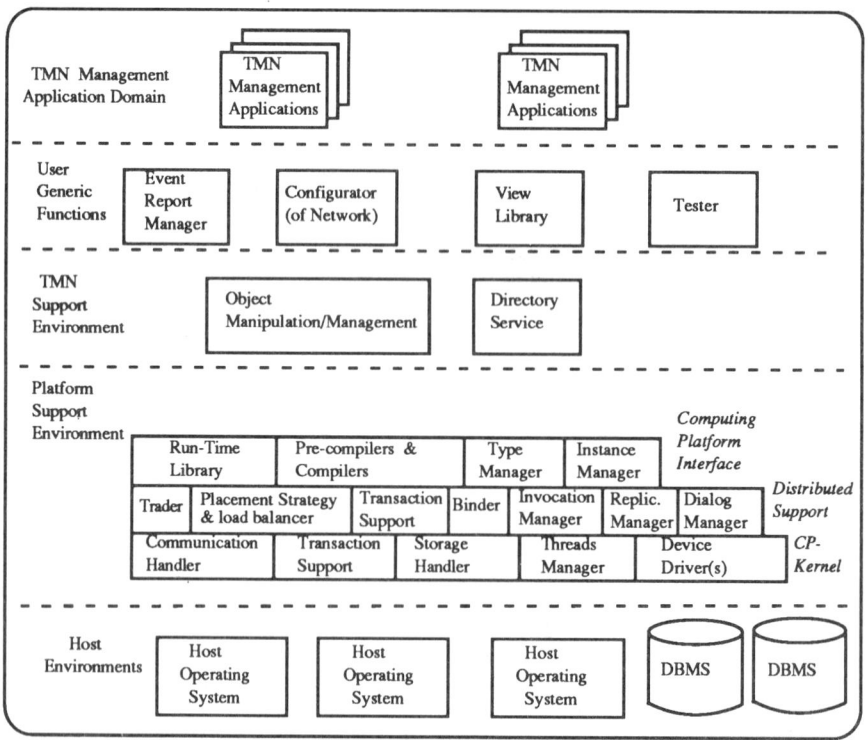

Figure 1 - General TMN Computing Platform Architecture

The CP-kernel consists of a number of components that mask heterogeneity, and provides a common view of the world. This decreases the knowledge that an application programmer requires, and increases portability, interoperability and vendor independence. As it provides abstraction from both host operating systems and native databases, the CP Kernel consists of a communication handler, transaction support module, storage handler (access to files and databases), threads manager and various device drivers for control of heterogeneous devices, for example, printers etc. The Distributed Processing Support layer essentially provides abstraction from the distribution of the applications. This involves trading and binding to locate and gain access to remote resources, placement management which controls application instances and load balancing, transaction management (across the distributed hosts/databases), invocation and replication management. The Computing Platform Interface provides the link between the programmes of the application developer and the underlying processing support. This link has off line and on line (run time) aspects. It does not necessarily encompass all the tools used to develop applications. It contains pre-compilers and compilers, runtime libraries and a type manager (which would facilitate cooperation of multilingual application implementations). The TMN Support Environment provides the platform services that are required by the TMN Management Applications and the User Generic Functions to fulfil their respective roles in managing the telecommunications network. As such, the services that the TMN Support Environment provides are specific to the requirements of telecommunications management. These include a Directory service containing network management specific information and Object Manipulation/Management which facilitates applications accessing TMN managed objects. User Generic Functions perform key tasks which are required across a range of TMN applications. These include Event Report Management, Configuration (for the network) and View Library (for abstracting and providing views of Management Information Bases). This layer may also contain some ISO defined generic functions.

3.   THREE TMN MANAGEMENT SYSTEMS

In order to test the general architecture three management applications from different functional areas of the TMN [5] were chosen. These management applications had been devised and developed by three RACE research projects prior to the specification of the General Computing Platform. Thus by adequately supporting each of these applications, the General Computing Platform could validate it's usefulness across TMN management domains. Three functional profiles (or cross sections through the General TMN CP architecture) were developed which were capable of supporting one of the TMN management systems. Each of these three TMN management applications is presented with the design of the computing platform profile needed to support it.

3.1   Customer Complaint Handler Management Application (CCH-MA)

The CCH operates within the context of a customer/provider relationship. Complaints originate from the customer, who perceives unsatisfactory behaviour in some equipment or service which is being received from the provider (e.g. no dial tone when making a call). On the provider side, the provider MA seeks to resolve all complaints received in order to satisfy an adequate level of service. Thus the provider will seek to establish the reason for any complaints (i.e correlate complaints with known fault reports, flag unresolved faults for further scrutiny by the fault repair service). There are two major areas of interest, namely complaint transmission from the customer to the provider, and complaint processing/resolution by the provider. This system involves a customer who has management capability (complaints can be as a result of using a service as well as a result of managing a service). The customer has the ability to log complaints through his own complaint management application on the customer network management system. There exist two TMN domains, TMN 'A' being the provider management system, and TMN 'B' being the customer management system. Communication for the purpose of complaint logging between the two TMNs (across the X interface [2]) is achieved through a wide area network.

Computing Platform Profile for Customer Complaint Handler

The CCH-MA resides at the Management Application layer of the general CP architecture [6]. The User Generic Functions layer and the TMN Support Environment layer facilitate the high-level operational requirements of the CCH-MA whereas the Platform Support Environment layer facilitates the low-level computing requirement of the Applications. The following functional blocks are required by the CCH-MA within each of the Computing Support layers:

Management Application layer

Management Applications which play a role within Customer Complaint Handling will be present at this layer:

TMN-A (Network Provider system)	TMN-B (Customer system)
Customer Complaint Handling MA	Customer Complaint Agent
Customer Complaint Logging Management	Complaint-Fault Correlation Mgt
	Operator Complaint Handling Mgt.

User Generic Functions layer

View library: This component provides the CCH-MA with host independent access to the platform. A specific requirement for this application is access to management information. These information stores may be distributed and multi-vendor. Examples of information bases include objects within a common information model shared between the two TMNs, private information kept within various relational databases, and Management Applications.

TMN Support Environment

Directory Service: The common information model provides a unified and logical representation of the information objects and management functions of the system. However, the actual resources of the system are widely distributed and heterogeneous. To achieve location transparency and to facilitate access transparency, logical naming is applied to them. Thus a Directory Service is needed to provide mappings of logical names to logical addresses for all the functions and applications of the system.

Object Manipulation/Management: Many entities are represented and accessed in the form of objects by the CCH system. This functional block is needed to abstract away the complexity of access and manipulation of these objects across the system.

Platform Support Environment

Computing Platform Interface: The applications associated with the CCH are written in a variety of languages and development environments. The implementers of the system require various platform facilities such as,

- Pre-compilers and Compilers which facilitate the reduction of runtime checking overheads
- Run-time Libraries which enable various platform functionality to be added to the system
- Instance Management which facilitates the creation and deletion of object instances within the system.

Distributed Processing Support: The support of this layer is required to provide transparency of distribution and access to information resources. The functionality of a Trader and Binder is needed to locate various system components and to establish logical communication channels across applications [7] . Other functions such as Transaction Management and Placement Management are used to enable creation of instances and provide coordinated access to information.

CP Kernel: The CCH is required to run on a variety of host environments. This layer is needed to abstract away from the complexity of the underlying host components and to increase interoperability and vendor independence. The use of an object-oriented structure provides the application programmers with a common view of the system resources and facilitates easier access to functional components. The functions of the Communication Handler, Processing and Storage Handler are required to provide common access to local and remote resources, to provide a uniform interface to processing and scheduling capabilities of the platform and to enable access to files and various native databases.

Host Environment

The CCH system may be implemented on a variety of technologies. There is expected to be a mixture of host technologies including operating systems, database management systems, communication protocols. The complexity of the components of the host environment has to be abstracted from the CCH system by the CP kernel.

3.2 Generic Maintenance System (GMS)

Maintenance consists of three primary functions: fault correlation, fault diagnosis and repair. These functions are implemented in a Management Application called the Generic Maintenance System (GMS) [8]. Faults are detected by the delivery of event reports to the GMS from the managed objects representing the resources being managed. These event reports are correlated resulting in a list of candidate faults. Diagnosis refines the suspected faults using deep model-based reasoning and testing of the suspect equipment. Finally, the repair of the equipment that has been established as actually being faulty, is scheduled.

The GMSs are anticipated as being widely distributed throughout the network element layer of the TMN, each having responsibility for a local domain in the telecommunications network [9]. These GMSs model the detailed functionality of the network elements that they manage. Links between domains are managed at the network layer of the TMN. GMSs at this level, model the connectivity of the domains managed at the network element layer. Thus at the network layer, the GMSs orchestrate the cooperation between GMSs at the network element layer. Given the frequency with which event reports will be delivered and the dependency of the services provided over the network on the reliable and timely operation of the network, maintenance is considered to be a time-critical application, particularly with regard to determining which equipment is faulty and their subsequent removal from service.

Computing Platform Profile for Generic Maintenance System

This section describes the CP profile selected for the Generic Maintenance management application.

User Generic Functions layer

The User Generic Functions of interest to maintenance are listed below. Standards for the interfaces between management applications and these functions are currently being specified by the standards bodies [10].

Event Report Manager : Maintenance relies heavily on the regular and reliable delivery of event reports from the network for it to correctly model the current status of the network. The Event Report Manager passes these messages to the MAs for which they are relevant.

Configurator : Maintenance frequently needs to isolate equipment so as to perform tests without degrading the performance of the network and also bring repaired equipment back into service. Requests to change the status of equipment must be made via the Configurator to avoid conflicts between MAs.

Tester : Maintenance needs to perform tests on equipment to establish whether the diagnosed cause of a fault is correct and also to determine whether or not a repair has been successful. However, other MAs may also wish to perform tests (e.g. customer complaint handling needs to correlate complaints to actual faults). Each MA wishing to perform a test makes a request to the Tester so that it can handle conflicting requests.

View Library : Maintenance uses this function to access other components of the TMN and also to query databases providing persistency of the data used.

TMN Support Environment

Object Manipulation/Management : When accessing common objects of the TMN, maintenance uses this function to avoid conflicts.

Directory Service : Maintenance assumes that addressing between modules is by logical name and so expects location to be handled by this service.

Platform Support Environment

Computing Platform Interface : The development of the Maintenance system requires the use of all of these components e.g. pre-processors, run time libraries etc.

Distributed Support : Maintenance has particular requirements on the Trader and Placement Strategy/Load Balancer since it is necessary to state the preferred location of certain modules for them to perform adequately. Replication of GMSs within the same domain (e.g. to provide fault tolerant processing) is not required, but automatic restart of modules executing on a host that has crashed is assumed. Binding is needed as well as the Dialogue Manager and Invocation Manager. Transaction processing is not used in the GMS but this will be needed for the smooth operation of the TMN.

CP-Kernel : Maintenance uses all of these features (with the exception of Transaction Support) but does not access them directly. Maintenance places particular demands on the Communications Handler. The GMS consists of a set of autonomous, co-operating agents. Co-operation between these agents is achieved by asynchronous message passing so as not to impose a fixed order on the maintenance cycle. Thus it is assumed that the Communications Handler is capable of supporting interactions of this type.

Host Environments

Maintenance needs to take advantage of a range of available hosts and databases that may be used by an organisation.

3.3 Traffic and Quality of Service Management

The goal of this management application is to maximise the utilisation of network resources whilst maintaining an agreed level of Quality of Service (QoS) to users of the network. The approach taken is two-fold. Firstly, to determine call acceptance criteria at Virtual Path (VP) entry points in order to maximise VP bandwidth utilisation whilst maintaining a predetermined QoS. Secondly, to maximise the proportion of successful connection attempts by optimising the bandwidth allocated to VPs and the routing of VPs. Both of these are addressed by the Traffic Manager, whereas the monitoring of QoS, generation of QoS warnings and provision of network load predictions for the Traffic Manager is undertaken by the Service Manager.The primary components of the Traffic Manager are a Call Acceptance Manager and a VP Manager. The Call Acceptance Manager adjusts the parameters of the network function which accepts calls onto VPs. The VP Manager modifies VP bandwidth and routing-tables in the network.

Computing Platform Profile for Traffic and Quality of Service Management

This section describes the profile, based on the CP-SIG Architecture, capable of supporting the Traffic and QoS Management system.

Management Application Layer

Both the Traffic Manager and the Service Manager [11] reside at the Management Application Layer of the CP architecture.

User Generic Functions Layer

Event Report Manager: The Traffic Manager receives VP events and network events, while the Service Manager receives events on Service Associations and QoS complaints. All of these are handled by the Event Report Manager.

Configurator: The Traffic Manager interacts with the network in order to set Routing and Configuration parameters which effect VP bandwidth changes, and to set Call Acceptance Function parameters. The Service Manager does not interact with the network directly and will therefore not make use of the Configurator.

TMN Support Environment

Directory Service: Traffic and Service Management are two distributed sets of applications which make use of location and access transparency and therefore will use the Directory Service.

Object Manipulation Management: This provides access to an OSI conformant Management Information Base (MIB) as used in the Service Manager as well as private MIBs. For example, the Traffic Manager makes use of the Management Unit Information Base (MUIB) which provides a model of the network for Performance Management purposes.

Platform Support Environment

The Traffic and QoS Management system is a set of distributed applications designed to run on a heterogeneous computing environment.

Computing Platform Interface: All off-line and run-time support facilities of the Computing Platform Interface are used in the development of Traffic and QoS Management applications.

Distributed Processing Support: The applications require the use of location transparency. It therefore makes use of the services of the Distributed Processing Support provided by the CP-Platform. Dialogue Management, and more specifically, the User Interaction Management Applications, identified in relation to the TMN Workstation Function [2], would be used in implementing the Human Interfaces.

CP-Kernel: The objects provided by the CP-Kernel which mask heterogeneity of the platform are of use to this system. Objects for transaction support will be of value in a multi-operator situation.

Host Environment

The support for mixed operating systems, hardware architectures and so on, provided by the Host Environment are essential to support this heterogeneity.

4. COMPUTING PLATFORM TECHNOLOGY

There are a number of initiatives and organisations that currently provide technology appropriate for a TMN Computing Platform. These may be roughly divided into two camps: Industrial Initiatives and European (research) initiatives. Although individual manufacturers have provided some proprietary support for distributed processing - such as Remote Procedure Call (RPC) and networked file systems - products which attempt to provide open distributed processing are only now beginning to emerge. The major initiatives in this area are as follows.

The Open Software Foundation's (OSF) *Distributed Computing Environment (DCE)* [12] and *Distributed Management Environment (DME)* [13] are sets of (industrially) standardised interworking components to support basic distributed processing and management across a number of heterogeneous hosts. DCE is currently commercially available and supported by a large number of information systems vendors.

Unix International's *Atlas.* [14] is described as a distributed processing architecture, and is intended to add to DCE rather than compete with it. It provides additional tools for distributed processing in a UNIX environment.

The *Object Management Group* (OMG) [15] is another industrial grouping, containing software vendors and users in addition to system vendors, which are concerned with defining concepts, such as the Object Request Broker (ORB). These concepts are being refined into a model and specification of the Object Management Architecture. This architecture is intended to provide a framework for the construction of general purpose open distributed processing systems.

There have also been a number of significant initiatives in the area of distributed processing within the framework of the European Commission's ESPRIT programme. Projects of interest include: Integrated Systems Architectures (ISA) [16], COMANDOS [17], HARNESS [18] and DOMAINS [19]. ISA is a continuation of the UK Alvey project *ANSA*, and their work is the development of the ANSA distributed processing architecture. This architecture is intended to support open distributed processing in a heterogeneous environment. The currently commercially available system derived from this work is called ANSAware [20]. Sections 4.2 and 4.3 below investigates the correspondence between OSF DCE/DME, ANSAware and the General TMN CP architecture.

4.2 Correspondence of DCE/DME technology to CP-SIG architecture

Because of the distributed nature of the TMN many of the computing platform requirements identified for the TMN are similar to those identified for OSF's DME. The scope of DME is focussed on two issues; a management framework constituted around a conceptual model of distributed systems, and the provision of tools and facilities to support the development of management applications. The technology to be utilised in DME has been identified, the system is currently under development. DME operates over DCE which already provides the basic core technologies such as remote procedure call, directory services and time services. In the following comparison only the General CP architecture functional blocks which have been identified as being supported by DME are discussed.

Mapping to the The CP-Kernel layer

The CP-Kernel consists of a number of components that mask heterogeneity and provides a common view of the world.

Threads : The DCE Thread Services which provide portable facilities that support concurrent programming. The threads service includes operations to create and control multiple threads of execution in a single process and to synchronise access to global data within an application.

Time Service: The DCE distributed Time Service is a software based service that synchronizes each computer to a widely recognised time standard. This provides precise fault tolerant synchronisataion for systems in both local and wide area networks.

Communication Handler : DME provides Application Programmer Interfaces (APIs) at two levels. At the lower level it provides access to the management services SNMP, CMIP and RPC. At the higher level two further APIs are provided which hide the complexity of the underlying communications protocols. These are based upon ANSI C and C++ respectively.

The DME management request broker is responsible for choosing the right communication protocol to route messages between applications and /or object servers.

Processing objects : The DCE Remote Procedure Call supports execution of processes on remote hosts and allows direct communication between them.

Storage objects: DME provides two possible ways of accessing file and database systems of the underlying hosts: RPC and a distributed file system. The RPC presentation service masks the difference between data representations on different machines, allowing applications to interoperate across heterogeneous systems. RPC supports large data processing applications by permitting unlimited argument size thereby efficiently handling bulk data. The OSF Distributed File System Service solves the problem of accessing remote files. The Distributed File System appears to the user as a local file system, providing access to files from anywhere in the network for any user.

Device Drivers : DCE provides a distributed print service and management of that service. This service includes reference implementations for print servers and print supervisors and API's and drivers for traditional print systems (i.e Unix and System V.4)

Mapping to the Distributed Processing Support layer

Trading: The trader provides facilities for the identification of system and application entities. The directory service used by OSF DCE is integrated into OSF DME. The DCE distributed system is designed to meet the needs of an integrated distributed computing environment. It is based on DEC's DECdns and Siemens DIRX. It provides the means for all applications and services to locate and share information about objects.

Binding: The binding function provides a logical connection between two application entities via the location mechanism of the platform. OSF's RPC provides programmers with the tools necessary to build client/server applications. Integration of the RPC with the DCE Threads Service allows clients to interact with multiple servers and servers to handle multiple clients simultaneously.

Location management: Location management provides facilities for the initial placement of newly created object instances on machines within the system. The OSF DME Request broker performs as a location manager.

Mapping to the Computing Platform Interface

Run Time libraries : Class libraries are provided by DME or third parties.

Compilers: The RPC provides a compiler which converts high level interface descriptions of the remote procedures into portable C language source code.The OSF DME supports ANSI-C and C++

Mapping to the TMN Support Environment

Directory Services: Provided via the DCE Directory Service

Object Manipulation Management: DME provides a simple form of object manipulation via CMIS and SNMP.

Event Report Management: DME provides a notification service.

4.3 Correspondence of ANSA technology to General CP architecture

This section is a comparison of the facilities provided by the current version of ANSAware and the facilities defined in the General CP Architecture for supporting distributed TMN Management Applications. Each layer of the General CP architecture is examined to reveal how ANSA may be used as a basis for implementing (part of) the architecture.

Mapping to Host Environment

ANSAware, like DCE, provides distributed platform support upon heterogeneous hosts e.g. Sun 3s and Sun 4s (UNIX based), Hewlett Packard 300 Series (UNIX), Vax VMS machines, DEC Ultrix machines, Acorn RISC, PC AT compatible machines running MS DOS. ANSAware is written in C, and does not prevent access to the native Operating Systems using C libraries.

Mapping to the CP Kernel

The Communication Handler: ANSAware supports communication over TCP/IP networks. This uses ANSA's own Message Passing System (MPS) protocol which in turn supports the higher level Remote Execution (REX) protocol. Neither of these protocols are conformant to any current standards.

Threads: ANSAware supports multiple threads. The number of threads within a capsule is defined by the application programmer. There is no theoretical limit to the number of threads within a capsule however these limits are determined by the resources and ability of the underlying operating system.

The current version of ANSAware does not provide any support for persistent storage or device drivers other than those available through the local OS.

Mapping to the Distributed Processing Layer

Trader: The ANSAware Trader is key to the overall ANSAware system operation. The ANSAware Trader is a form of directory service and uses the client server model. The entries in the directory relate to interface offers. An interface offer (or binding address) is exported to the Trader by an object which supports a particular interface on which it may be invoked. The instance that makes the offer is known as a server, and the invoker of that interface is known as a client. Note that this mechanism does not imply that interfaces *not* in the Trader cannot be invoked, it just means that they must be known about in some other way.

Placement Manager: ANSAware does provide a Node Manager which, as its name suggests, manages the services available on a particular node. These may be static (always running) or instances (created on demand). However, ANSAware provides no help in deciding which capsules, and therefore which instances, should be placed on which machines. This a manual task for the system developers. ANSAware does not support migration of objects or dynamic load balancing.

Transaction Support: The ANSAware v3.0 does not provide any support for transactions.

Binder: A client who possesses an interface (binding address) of a server object may invoke on that object. ANSAware sets up the message buffers and invocation management routines required for binding.

Invocation Management: The ANSAware capsule library provides support for Invocation Management. This takes a number of forms:
- ANSAware starts off a new thread when an incoming invocation is received by a server.
- ANSAware converts the passed and reply parameters in any Invocation from the native format to a host independent format described in the Interface Definition Language (IDL) which describes the interface to any object
- ANSAware supports asynchronous invocations without termination notification, synchronous invocations with termination notification, and asynchronous invocations with termination notification
- ANSAware manages the buffers for waiting incoming and outgoing invocations

- ANSAware provides some exception handling routines for errors that occur during invocations
- ANSAware does not provide explicit concurrency control
- ANSAware does not provide any form of access control.

The current version of ANSAware (v3.0) does not provide any support for Dialogue Management, Replication and Group Management and distributed Time Management..

Mapping to the Computing Platform Interface Layer

Compilers: ANSAware provides preprocessors for Interface Description Language and Distributed Process Language. They produce C code which is compiled with the appropriate libraries.

Run Time Library: Some components of ANSAware must be run as separate processes, i.e. the Trader, Node Manager, Factory and Notification Service, together with their associated client utilities.

Type Management: As mentioned briefly above in the Trader section, the ANSAware interface type tree is held within the Trader and clients of the Trader provide support for managing that type tree. New types must be added manually to the tree. The IDL files which describe the interface to a particular application describe the types of application components and their compatibility, but it is up to the application developer to ensure that those relationships are installed in the type tree and that the compatibility statements are true - ANSAware does not check claims that interfaces are compatible.

Instance Management: ANSAware provides support for the dynamic creation of capsules and thus interface instances though the Factory service. The Factory may be invoked with a request to create a particular capsule, in which it may create instances of objects and interfaces of particular types on request.

Mapping to the TMN Support Environment layer

ANSAware does not provide any TMN specific, or system management, components. The only component that may be mapped easily to ANSAware is the TMN Directory Service which may be considered similar to the ANSAware Trader.

5. CONCLUSIONS

The management applications were chosen to test the applicability and validity of the General TMN CP architecture. Each of these management applications were developed by different RACE TMN projects using divergent platforms and technologies. These systems reflected the specific requirements of the projects' domains and were not concerned with development of a common inter-project infrastructure. This paper has demonstrated that, by use of profiles, the General CP architecture is capable of supporting these diverse management systems. The paper has illustrated the necessary infrastructure from specific management systems down to underlying host environments. This demonstrates the flexibility and applicability of the General CP architecture.

A second important result of the profiling work is that the selection of functional blocks of the General TMN CP architecture required to support each of the management applications show a large degree of commonality. Thus the profiles provide explicit evidence for common infrastructure across TMN management systems. This suggest that a powerful means of achieving network management integration is via the General platforms functionality. This means of management integration is significantly preferable to integration at the management application level.

Because of the diversity of requirements, it is sometimes difficult for a system designer to balance the needs of various TMN management applications. However, by using profiles to focus the functionality of the TMN platform, these conflicts may be made explicit and therefore better addressed. In addition to this, by developing profiles for platform types, a taxonomy of TMN management application may be realised. This would provide significant benefits in configuring and distribution of TMN installations.

Taking a broad view of the RACE community, it is envisaged that the TMN will be widely distributed and multi-vendor. Identification of the functional and non functional requirements of the TMN CP is the first step in its realisation. The following requirements are commonly accepted by the RACE community:

- The non functional requirements include support for heterogeneous hosts, a high level of performance, reliability, interoperability and fault tolerance. Extensibility is also an important non functional requirement
- The functional requirements include storage, multilingual programming support and HCI support.

The degree to which these various requirements are to be supported will depend on the type of applications that are using the platform. Maintenance applications operate at near critical time scales and therefore a high level of performance is a priority while for non real time applications such as those related to accounting, reliability is a more important requirement.

Assessment of these functional and non functional requirements in conjunction with the experience gained by the individual projects from their prototyping work allows for identification of the core technologies which must be part of any implemented platform for TMN.

5.1 Description of Core Technologies

This section gives a description, based on a consensus of the authors and their research groups, of some core technologies which exist at each of the layers of the General CP architecture.

User Generic Functions

In this layer the view library was identified as very important for inter and intra management application communication as it provides for mapping from languages and protocols used by the applications to the interaction language and protocols used by the rest of the systems. It provides a very necessary abstraction of information and functionality.

TMN Support Environment

A key component of any TMN installation is the ability to access and manipulate management objects. This object manipulation and management is therefore considered as core technology for a TMN computing platform. Another core technology is the Directory service which is concerned with managing the name space of objects within the TMN.

Platform Support Environment

Because of the multi-vendor environment on which TMN operates on, Run Time libraries, Pre-compilers and Compilers are important services that the platform must provide.

The trader is considered essential to the General TMN CP because of the distributed nature of the TMN and the requirement for distribution transparencies support (such as access and location transparency). The Binder function which provides a logical connection between two application entities and the Invocation Manager which is concerned with ensuring that interactions between application entities conform to the common computation model are considered as core functionality of the platform. Because of its importance in providing access to the TMN by the user, some dialogue management functionality should be provided by the platform.

It is envisaged that the TMN CP will be supported by heterogeneous operating systems, databases and communication systems. Therefore the platform should incorporate communication handlers, storage handlers and device drivers to mask this heterogeneity.

Conclusion

It can be concluded that a large percentage of the functional blocks identified as part of the General TMN CP Architecture are also core components which must be incorporated into any realisation of the TMN CP. These conclusions vindicate the approach of using profiles to identify application specific requirements on the TMN CP.

5.2 Identification of TMN CP Specific Technologies

This paper has demonstrated that the General TMN CP architecture is both flexible and powerful enough to support a broad range of TMN management system. It was also highlighted the profiling technique for focusing the platform issues on specific TMN domains. The review of the emerging platforms has provided an insight into the type of support that such general purpose systems could be expected to provide. The two platforms chosen for examination are merely a representative sample of the currently emerging platforms. The paper has highlighted strong consistency in the fundamental abilities of the emerging platforms. Also highlighted is the strong evidence for the view that implementation of the General CP architecture would be supported to a large degree by these emerging platforms.

Although some core technologies have been identified for the General TMN CP, there still remains some *TMN specific platform functionality* which is not expected to be supported by these emergent general purpose systems. These key TMN specific platform functionality must be provided by the TMN community. The TMN specific functionality tends to be concentrated at the User Generic and TMN Support Layers of the General CP architecture. The functions identified as TMN specific include Event Report Management, View Library, Object Manipulation and Management and Directory Services (X.500) with specific TMN management objects. As the ISO/CCITT TMN management protocols evolve extra communication handling support may also need to be provided.

5.3 Open Issues

Interoperability : The standardisation of communications services and protocols and object interfaces in Network Management will provide the infrastructure for enabling TMNs to cooperate. From the ODP perspective, this interoperability will be made transparent by the Trader utilizing Directory Services. Thus some level of interoperability will be achieved. Difficulties will arise at the level of the policy differences between enterprises (a management action allowed in one TMN may not be provided, be illegal or produce a different result in another). Where this occurs the TMNs must be able to negotiate a satisfactory conclusion to a request. Thus interoperability is an issue that is not just addressed at the low levels of the TMN platform but occurs at different (higher) levels in the TMN CP architecture.

Heterogeneity -: The TMN is a complex system, being comprised of a number of applications with different requirements, particularly across the four management areas/layers of the TMN [5]. Each of these applications are suited to a different types of host (cf. customer billing and maintenance). Thus the platform must be able to support different distribution profiles for the various applications, particularly when required to perform dynamic placement of processes.

Time-Critical Performance : This is a low-priority issue in TMN and ODP [21] at the moment although some management applications have been identified which require time critical responses. The architecture allows TMN management applications to make explicit access to (possibly high performance) specializations if available in the host environment. However, such close interaction between the host and management application has to be traded off against the need for integration and cooperation with other TMN applications. Thus achieving time critical performance of management applications will incur a tradeoff in the ability to integrate these applications via the platform infrastructure.

6. ACKNOWLEDGEMENTS

This paper is the result of the research of the TMN Computing Platform Special Interest Group (CP-SIG). The special interest group comprises representatives drawn from four Telecommunication Management RACE projects GUIDELINE, ADVANCE, AIM, NEMESYS. The authors would like to acknowledge the cooperation of the project managers for facilitating this research and the support of the CEC RACE programme. For further information concerning the research described here please contact Vincent Wade, CP-SIG Chairman, Trinity College Dublin, Ireland. EMail: VWade@Cs.Tcd.Ie

7. REFERENCES

[1] Wade, V., Donnelly, W., Harkness, D., Riley, K., Shomaly, R., Celestino, J., Chapman, M. : "A Framework for TMN Computing Platforms", proceedings of the 5th RACE TMN Conference, London Nov. 1991.

[2] GUIDELINE Deliverable ME8 : "TMN Implementation Architecture", 03/DOW/SAR/DS/B/012/b3, RACE Project R1003 GUIDELINE, March 1992.

[3] Open Systems Interconnection, "Common Management Information Protocol Specification", ISO/IS 9595, July 1991

[4] CCITT Recommendation M.3010 :"Principles for a Telecommunication Management Network - Version R5", C, temporary document no. 40 (Rev1) Geneva.

[5] RACE Project NETMAN (R1024), Deliverable 6 :"Telecommunications Management Specifications", document no. 24/BCM/RD2/DS/A/006/B2, 1991.

[6] RACE Project ADVANCE Deliverable,: "MA scenario Functional Requirements Capture (Complaint Handling MA)", p12-23 ADBC0187

[7] Hurley, C. B. : "An Architecture and other Key Results of Experimental Development of Network and Customer Administration Systems", Sixth RACE TMN Conference, Madeira, September 1992.

[8] Bigham, J. et al : "A Generic Maintenance System for Telecommunication Networks", Sixth RACE TMN Conference, Madeira, September 1992.

[9] Riley, K., : "Design of GMS Infrastructure", RACE Project AIM deliverable,1991.

[10] CCITT X.731, ISO 10164-2, "State Management Function".

[11] Gentilhomme, A. et al : "The Use of AIP Techniques in Traffic and Quality of Service Management Systems", Sixth RACE TMN Conference, Madeira, September 1992.

[12] Fauth, D. et al : "OSF Distributed Computing Environment Overview", Technical Report, Open Software Foundation, Cambridge, MA, May 1990

[13] Open Software Foundation, "OSF Distributed Management Environment Rationale", Cambridge, MA, September 1991

[14] Hubley, M. : "Distributed Open Environments" ,Byte,November 1991,pp 229 - 237.

[15] Object Management Group, R. Soley (ed), "Object Management Architecture Guide", OMG TC Document 90.9.1 November 1990.

[16] ISA, EC ESPRIT Programme Project 2267

[17] COMANDOS, EC ESPRIT Programme Project 2071

[18] ESPRIT Project 5279, "HARNESS Platform: Basic Specification. Platform Specification and Evaluation", HARNESS Consortium, June 1991

[19] ESPRIT 5165 Project Team : "DOMAINS Object Model and Object Machine Refinement", DOMAINS Task 2.3 Final Report, , November 1991

[20] Architecture Projects Management Ltd,"ANSAware 3.0 Implementation Manual", Document RM.097.00, January 1991

[21] ISO : "Recommendation X.9yy: Basic Reference Model of Open Distributed Processing, Part 2: Descriptive Model", ISO Proposal for Committee Draft, ISO/IEC JTCl/SC21 N6079, August 1991.

The Management of Telecommunications Networks
R. Smith, E. H. Mamdani, J. G. Callaghan (Editors)
© Ellis Horwood 1992

Recommendations for the Use of AIP Techniques for Maintenance in Telecommunication Systems

Franziska Maier, Carla Neumann, Yann Le Hegarat (Danet, Germany), Gottfried Schapeler (Alcatel SEL - AG, Germany), Nader Azarmi (BT Laboratories, UK)

ABSTRACT

This paper provides recommendations on the application of Advanced Information Processing (AIP) techniques for the maintenance of telecommunication networks. The recommendations are based on a number of studies and in particular the experience gained in developing the Generic Maintenance System (GMS). This approach of using the GMS as the focus for the AIP recommendations allows us to demonstrate the integrated utilisation of the techniques. These include : object oriented modelling, model based reasoning, man machine interface techniques and distribution and infrastructure.

The recommendations are based on the experiences with the Generic Maintenance System, but are given in a general way and hence are useful for everyone concerned with maintenance systems for telecommunication networks.

1. INTRODUCTION

This paper provides recommendations on the application of Advanced Information Processing (AIP) techniques for maintaining telecommunication networks. Its intention is to transfer the experiences gained within the RACE technology project AIM to those concerned with maintenance in telecommunication networks.

Telecommunication networks are getting larger and more complex: their equipment is provided by different vendors and based on different technologies. The demands of a highly reliable and easy to operate telecom network require standardised interfaces of the network to a powerful network management system: the Telecommunication Management Network (TMN). A TMN must offer all functions necessary to manage a telecom network: maintenance (fault management), configuration management, accounting, performance management and security management. The work presented focuses on research into appropriate techniques to support on-line corrective maintenance of hardware faults. Maintenance often has to cope with many fault symptoms from various sources. It also needs to formulate and verify hypotheses and to carry out repairs together with the necessary preventive actions. On the other hand the maintenance process must not affect the telecom network activities. The AIP techniques recommended in this paper help to realise an easy to handle and widely automated maintenance system.

1.1 Advanced Information Processing Techniques

Advanced Information Processing describes modern information processing technologies including Distributed Processing and Databases, Real Time and Distributed Knowledge Based Systems as well as Advanced Processor Architectures [1]. This paper presents some recommendations on AIP-techniques for the areas outlined below.

Network Modelling Techniques

The AIP-Technique recommended for modelling telecom networks for use in computer applications is *Object Oriented Modelling* (OOM) [2]. The OOM approach models telecom networks in a modular way. Objects are the primitive elements of this modelling approach. They comprise the behaviour of the entities they represent. Objects communicate via messages. To structure the overall domain, a taxonomy of classes can be built. The objects can be considered as instances of a class. There can be super- and subclasses, so that a hierarchical or heterarchical structure can be realised. Information can be inherited from superclasses to subclasses. Therefore only the local information has to be stored separately in each object.

Reasoning Techniques

There are different techniques for processing the modelled information. For knowledge based systems (KBS) the processing is handled by an inference engine using reasoning techniques such as Model Based Reasoning or Case Based Reasoning [3].

Telecom networks are characterised by their behaviour and structure. Both, behavioural and structural knowledge - also referred to as deep knowledge - is modelled and used by the *Model Based Reasoning* (MBR) approach [4]. By using deep knowledge MBR differs from Rule Based Reasoning, where rules contain shallow expert knowledge. MBR is either based on a model of the "working" system or the "not working" system. Within AIM both, the "working" and the "not working" system are modelled by a set of production rules. A detected symptom is matched against these production rules in order to find the possible faults. As the knowledge required for MBR is already available during the design and the specification of a telecom network, the maintenance system using MBR can be built in parallel to the telecom network.

The *Case Based Reasoning* (CBR) approach uses a knowledge base built of standard cases [3], [4]. Each case has to be coded as scripts based on the experience gained from the working system. The different cases represent a well-defined application field. Each problem handled by the reasoning mechanism is, if possible, mapped into an existing case stored in the knowledge base. Hence, this technique is suitable for applications, which can be reduced to a small set of already available and known cases. This means, that the development of the application, e.g. telecom networks, has to be completed in order for the (case) knowledge to be available.

Man Machine Interface Techniques

No matter, how extensive the results of a knowledge based system are, they will only reach the user's attention if they are presented in an adequate way. Hence, the Man Machine Interface (MMI) will contribute crucially to the acceptance of the system [5]. One common MMI technique is the use of windows to structure information. In some MMIs information is presented according to the different needs of the users.

Distributed Processing & Infrastructure Techniques

Telecommunication networks are complex and logically and physically distributed. To ease the use and the development of distributed systems different levels of transparency are offered to its users. Details can be hidden away from the user by providing an application program interface [3]. Maintenance, as one management function of a TMN, must be able to cope with distributed telecom networks.

1.2 GMS: A Maintenance System

This chapter gives a short introduction to the Generic Maintenance System (GMS) developed within the RACE project AIM [4]. The task of a maintenance system - in this context - is to maintain telecom networks. It can be seen as one part of the TMN. GMS consists of tools for

knowledge acquisition, reasoning, MMIs and for distribution and infrastructure. The GMS mainly fulfils three tasks:
- correlation of event reports
- diagnosis of possible faults
- support to repair faulty parts.

The three tasks correlation, diagnosis and repair are implemented in the GMS. They make use of a Knowledge Base (KB) and an Event Report Manager. The logical model and the physical telecommunication network are connected via mediation functions, whereas the communication between users and the generic maintenance system is handled by the Man Machine Interface.

Because of the complex structure of telecom networks GMS supports non-distributed networks as well as distributed networks. Hereby distribution can be logical or physical. In addition, GMS is designed in a generic way and therefore applicable to different networks and network elements. The maintenance system is portable to different domains and hence widely reusable.

A major advantage of the GMS is, that its models can be developed concurrently with the telecom network. This means a working maintenance system can be provided right at the beginning of the market introduction of the network or network element. Another advantage is its vendor independence.

2. AIP RECOMMENDATIONS FOR THE REALISATION OF A GENERIC MAINTENANCE SYSTEM

This chapter gives recommendations on AIP-techniques for the realisation of a maintenance system. The recommendations are based on experiences gained within AIM and are illustrated with examples taken from the GMS. For further studies see the final report on AIP Evaluation & Results [3]. The subsections align with the phases of a development system life cycle.

These are as follows:
- requirement specification
- design phase
- implementation phase
- test phase.

The AIP recommendations are mainly addressed in the design phase.

2.1 Requirement Specification

The first step in realising a maintenance system is to specify its requirements. To fulfil the overall objective of a maintenance system - the maintenance of telecom networks - there is a requirement of knowledge acquisition, ways to model and represent the network, reasoning mechanisms, reliable communication & control of the network components and an MMI for man machine interaction.

2.2 Design Phase

In the design phase the techniques and means are established to fulfil the above mentioned requirements. For each requirement a set of problems can occur. Here, typical problems are pointed out and recommendations to solve them are given. Both, problems and recommendations are based on the experiences gained during the design phase of the GMS and the maintenance application prototypes for System X (DSSS) and BERKOM (an experimental broadband network).

Knowledge Acquisition

The applicability of a maintenance system highly depends on its KB. The acquired knowledge needs to be correct and to be kept up to date. Knowledge acquisition (KA) is therefore an important task. Difficulties in KA mainly occur when complex scenarios, such as IBC networks, are newly acquired or updated. We therefore provide recommendations to guide knowledge acquisition engineers (KE) in order to prevent them from acquiring inaccurate network configurations and to enable them to verify the correctness of the acquired network structures. In what follows, we point out two ways how this can be achieved by using constraints:

- user driven constraints : Explanations are available for the KEs so that corrective modelling actions can then be taken. In this case, the user asks for consistency checks. This means, that the user has to take care not to violate the constraints, but he can ask for assistance from the knowledge acquisition tool
- tool driven constraints : Using tool driven constraints means that the tool will assist automatically in rectifying invalid structures defined by the KE. The KE is guided by the KA task in order to achieve completeness of the acquired knowledge, but also consistency checks are offered. This approach is more user friendly than the user driven constraints approach, but takes more effort to realise.

Supporting KEs can also be achieved by providing KA Tools. The experience with GMS show, that there should be at least two tools for this purpose. An object editor, which accepts formalised textual input is one of them. Objects proved to be suitable information units for KA tasks. A more sophisticated tool is the graphical editor, which accepts graphical input. A graphical representation gives a better overview of the knowledge and is nicer to handle by the user than mere textual objects.

Especially for complex scenarios, where the domains that are managed are distributed, it is essential to have a KA Tool that allows for a good overview of the whole management system. KA within several management domains differs from KA within a single management domain. New aspects arise, such as the need for a conceptual definition of management domains, the assignment of managed objects to domains and the need of access control. KA Tools, suitable for TMN applications, which are widely distributed, have to cover these aspects.

Complex systems, irrespective of their distribution, lead to a further major problem for KA: the acquisition of a KB containing the complete functional and physical structure of the network or network element. This will be a tedious and time consuming effort, if every single variant of a basic configuration needs to be implemented separately. We recommend to use network design, specification or configuration tools, which provide output that could be used - either directly or via data translation - to create automatically the KB. Such tools would be a great help for KA. The design and implementation of a KB could then be a part of the network design and configuration process. In this context it is also worthwhile following the improvements made in the area of text analysis systems. These systems select relevant information from a non-formalised text, such as network specification documents and automatically formalise it, for example, in a production rule structure. This, however, is still a research topic.

Modelling

The acquired knowledge of the structure and the behaviour of a telecom network has to be modelled in order to design a maintenance system. Multiple modelling techniques are used in the area of management systems. However, there is no unified approach to specify and utilise the information required by an integrated management system. We recommend to use object oriented modelling. Especially for very large and complex systems as telecom networks tend to be, this approach proves to be best suited. It allows for a modular structure and can be easily developed and expanded. Within GMS we use object oriented modelling and have various application experiences with this approach. The following example shows the object oriented modelling approach used for modelling the KB.

Figure 1 shows the overall organisation of the Knowledge Base (KB) representing the structure and behaviour of a telecom network. The physical and functional models capture information on physical and functional characteristics of the underlying telecom network. Repair information for replaceable units is represented in the repair model which is associated with the physical model. Detailed knowledge on fault reports and fault descriptors, diagnostic tests and their descriptors and heuristic diagnosis rules are represented in the diagnosis model. The maintainable and substitutional models are subclasses of functional and physical classes which inherit appropriate diagnosis or repair knowledge. Alongside these, the network generic structures are represented as constraints of these models. Object Orientation in this area is useful to capture structural semantics which cannot be adequately achieved using a relational model of data.

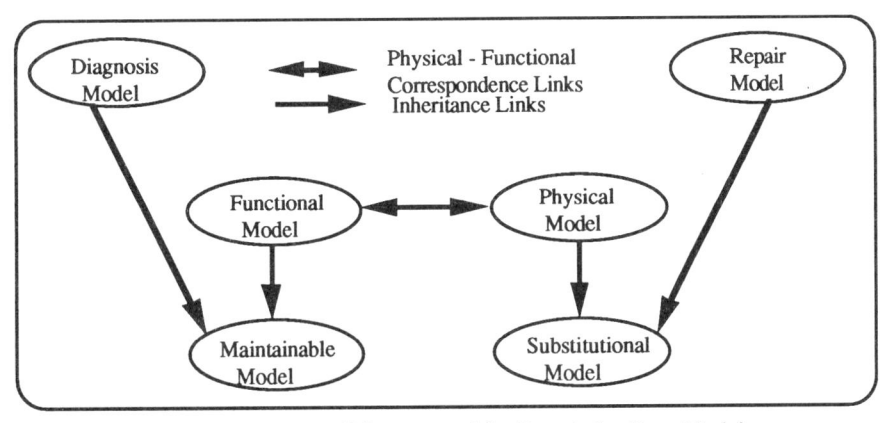

Figure 1 - The overall Structure of the Knowledge Base Models

The KB can be viewed as a set of abstract models containing real world facts. Each model comprises three components, namely structure (what there is), function (what is being done) and dynamics (when things happen). These components are integrated into the notion of an object, as is shown in figure 2. Application areas with specific focus and aims concentrate on one or more of the components. However, more complex applications have led to increased expectations and hence there is need to incorporate all three components.

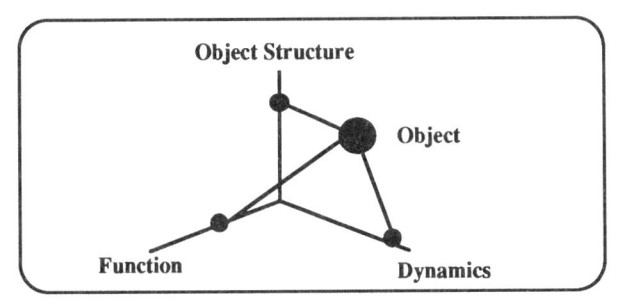

Figure 2 - The three components of an abstract model integrated into the notion of an object

An object oriented maintenance language allows users to define such objects and their interactions, using the Unit-Port Concept, see figure 3. A unit is viewed as an object which contains a description of its internal behaviour, a list of ports which defines its interactions with the outside world and a list of internal state variables which defines the internal status of the units. A unit is defined in terms of its class. Units are connected via ports. A port is defined in terms of its domain, which provides a clear interface to other ports. Ports relate to other ports via a connection. Semantics are attached to ports via the localised behaviour of the unit. The directional properties of ports are also defined using unit behaviour. The unit behaviour describes the unit functions. It also describes what is received and sent from its ports. Such an approach keeps all benefits of the object oriented paradigm, like information hiding, specialisation and inheritance.

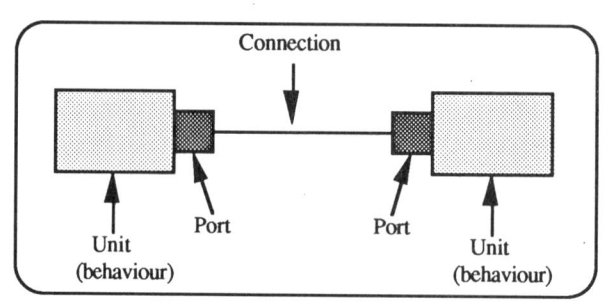

Figure 3 - The Unit-Port Concept

Given a Unit-Port model of network functionality, a reasoning system can use the behaviour and the port connection to make deductions between the units. What is demanded now, is to find appropriate reasoning techniques.

Reasoning

The central reasoning technique used within the GMS is Model Based Reasoning (MBR) [3, 4]. One reason why MBR is applied is the fact, that MBR uses deep knowledge for its reasoning process. Deep knowledge, i.e. behavioural and structural knowledge, is explicitly available for telecom networks and can therefore be modelled well. By using deep knowledge MBR differs from Rule Based Reasoning, which is restricted to shallow expert knowledge.

Because of the complexity of telecom networks a modular knowledge representation is desired. Within the GMS, MBR benefits from the object oriented knowledge representation. There is a second benefit MBR can gain from the structure of its knowledge base: invariance to telecom network changes. This is achieved within GMS by distinguishing between generic knowledge and specific knowledge. The generic knowledge represents the inference mechanisms, whereas the specific knowledge covers the underlying design knowledge of the telecom network. The specific knowledge is represented in a declarative, modular and easily changeable form, which is then interpreted by the reasoning mechanisms. An invariant inference mechanism in terms of network changes is especially valuable, as telecom networks tend to change and expand constantly.

There are not only the changes of the telecom networks which have to be considered during the development of a maintenance system, but also changes in the requirements of maintenance systems. Reasoning, being a major component of a maintenance system will therefore be affected by changes in requirements. The answer to this problem, not only within GMS, is the modular construction of the reasoning tool. This allows adaptation and enhancements to particular requirements.

Furthermore, the models for a generic maintenance system for telecom networks using MBR can be built currently with the telecom network. This is possible, as the required knowledge is available during the design and specification phases of a telecom network. Herein lies a major advantage of MBR over CBR, where cases are usually based on experiences gained from an already working system. However, further improvements are expected by combining Model Based Reasoning and Case Based Reasoning.

Applying MBR to the maintenance of telecom networks requires reasoning on complex and dynamic models. The shortcomings of the MBR approach, therefore, concern speed and efficiency. It can be recommended to consider Case Based Reasoning as a mechanism for providing heuristics for guiding the search of the suggestion interpreter. The suggestion interpreter is in charge of searching possible causes for detected fault symptoms in the telecom network [4]. This combined approach could speed up the process of finding a solution.

Within the GMS another approach has been successfully undertaken in making MBR more efficient. A propositional inference engine (PIE) module is used within GMS which acts as a cache for the deductions generated by the model interpreter. The model interpreter checks, whether the suggested causes of a symptom are consistent with the current status of the system [4]. The underlying concept of PIE are Assumption Based Truth Maintenance Systems (ATMS) [3]. An ATMS maintains global consistency, extends the cache idea, simplifies truth maintenance and avoids most dependency-directed backtracking. Progress has been made in realising a very fast implementation of ATMS, but there is still scope for additional improvements.

One further improvement may be achieved by using temporal reasoning; but, is it worthwhile to consider temporal aspects? In many telecommunication applications it is possible to abstract out the temporal aspects of the problem so that the reasoning system only needs to handle a static representation of the problem. In the current GMS notions of time are required for correlation, where the exact event time of the fault report and of subsequent state changes and automatic reconfigurations is necessary, and for physical changes when network components are replaced. However, time information is available with every symptom. This information is recorded into the time-stamped error event reports. It is recommended to use this time information in order to automatically group related symptoms together taking into account the temporal order of the symptom report.

The improvements expected from applying temporal reasoning mechanisms within the maintenance system are:
- better searching for a solution
- better fault localisation by performance of the reasoning systems, because of the use of additional information in the process of the improving the focus of the reasoning.

However advanced the reasoning mechanisms are, they will only reach the maintenance system user's attention if an adequate Man Machine Interface is provided.

Man Machine Interface

Man Machine Interfaces (MMIs) may vary for different tasks and different needs [5], but can there be an overall concept for developing MMIs? We think there can and recommend to base the development of MMIs on the Seeheim Model. The Seeheim Model [6] , see figure 4, separates out the functionality of user interactions into three logical components: presentation, dialogue management and application interface.

The presentation layer is concerned with the representation of the physical appearance of the system. It makes the lexical analysis of the interaction language and converts external world format into internal computational format. The presentation is the only component of the system which directly manipulates physical I/O devices. Other components of the system necessarily use the presentation as an intermediary to exchange information with the user. The dialogue manager is the medium between the presentation and the application interface. It is in charge of the syntactic analysis of the interaction language and constructs sentences from the syntactic units. Correct sentences correspond to queries and information the user wants to provide to represent specialised elements. The dialogue manager is also in charge of controlling the state of the interaction. All possible states, their relationships and sentence composition define the structure of the dialogue between the user and the application. The definition of a state can include the set of allowed queries; in this case the dialogue manager performs a part of the semantic verification before sending a query to the application interface. The application interface is a representation of the application from the viewpoint of the user interface. Separating the user interface and the application code makes it possible to change the interface without modifying the underlying functionality and vice versa. The Seeheim Model allows a step-by-step design of the MMI. Other advantages are the clear definition of a conceptual framework, as well as its genericity and generality.

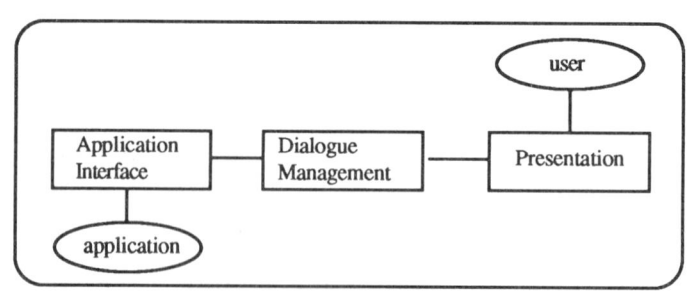

Figure 4 - The Seeheim Model

Based on an overall concept, like the Seeheim Model, MMIs still need to be realised for different application tasks and different needs of users. We recommend to use a MMI tool, which generates different MMIs. The idea is, to have a modular structured set of tools, where only the MMI tool is responsible for MMIs. This allows for global concepts, which may be used for additional MMIs. Apart from having a general concept on which the development of MMIs can be based, it is also desirable to have a User Interface Management System (UIMS) by which the development of MMIs could be automated.

There are several suitable ways for realising MMIs. A rudimentary, but quick and powerful way of demanding information is by using a command language. This, however, needs special training and is therefore suitable for more experienced users. People unfamiliar with the system may prefer natural language queries. Windows, menus, panels or colours are used to structure information. The most authentic presentation offers multi media, where, for example,. text, graphic and sound are combined. Within GMS different MMIs are realised. For example, a graphical MMI for knowledge acquisition. The focus here is on the needs of system developers. For future use of the GMS, however, other users also need to be considered. Different users are expected to communicate with the maintenance system, such as operators, telecom engineers or software engineers, at both beginner and expert levels. They communicate with the system for different tasks and with different skills. Therefore, we recommend to create user profiles to capture the different needs. If the MMI takes the user profile information into account, this results in user specific communication and increases the acceptance of the system.

Distribution & Infrastructure

Within a TMN parts of information are distributed. This information needs to be shared between different application functions. Especially for the purpose of constructing an Interconnected Metropolitan Area Networks Maintenance Prototype (IMP) [7], distribution and infrastructure have been addressed concentrating on the support required by the distributed GMSs.

Maintenance is one management application within a TMN that is time critical. One fundamental requirement of maintenance is to allow as much maintenance activities as possible to be performed in parallel. This can be achieved through distribution. Infrastructure techniques will have to ensure the necessary interoperability between maintenance applications.

Due to known constraints and requirements, the infrastructure should in principle include:
- platform support for process placement/balancing and software/hardware fault management
- security and control aspects like user access controls
- management aspects of the communications defined in OSI management using directory services and object manipulation.

Standards are now emerging in Open Distributed Processing that reflect a wide range of applicability [8]. Distribution is often discussed with reference to various forms of transparency that should be offered to the applications supported. Which of these transparencies an application takes advantage of should, however, be optional.

A trade-off needs to be made between the acceptable performance and the utilisation of the platform for interoperability with other applications. In the future it is anticipated that the TMN, and hence maintenance, will be built on top of general platforms [9], which are customised, by higher-level functions, to the telecommunication domain.

2.3 Implementation Phase

As a consequence of the requirements and the design phase, an object oriented programming language is recommended. Object oriented programming languages emphasise structuring and reuse of code. This can be a crucial advantage for the software implementation. Within the AIM project CLOS is used. CLOS is the object oriented extension of COMMON LISP and as such well suited for rapid prototyping. We use rapid prototyping for the realisation of the GMS. However, it must be understood that problems may occur with LISP when real time requirements need to be fulfilled, as in maintenance.

For the realisation of a generic maintenance system it is essential to use techniques based on market availability and standardised software. The GMS, for example, runs on commercially accepted platforms, such as SUNs under UNIX.

For more detailed information see [3].

2.4 Test Phase

Using an object oriented approach testing of the software, in this case the GMS, is a more modular task and is therefore expected to be easier. In addition, the GMS can be used to test the concept of the managed telecom network. The GMS will also help to test the specification of the telecom network entities. This can be done by simulating faults and by analysing the result of the correlation process, i.e. the set of possible fault causes. If too many fault causes are reported, more or different entities of the telecom network need to be enabled to report faults in order to allow a better focus on the possible fault causes.

3. CONCLUSIONS

The recommendations on AIP techniques given in this paper are based on experiences gained during the development of the GMS and the maintenance application prototypes for BERKOM and System X (DSSS). For the areas of knowledge acquisition, modelling, reasoning, man machine interfaces, distribution and infrastructure a number of AIP techniques are recommended, which help to realise a maintenance system.

For knowledge acquisition the benefit of KA Tools like graphical or object editors are pointed out. Recommendations are also given for automating the creation of the knowledge base by using design, specification or configuration tools. Modelling telecommunication networks is stated to be best done by Object Oriented Modelling. This allows for a well structured and easily maintainable representation of complex telecom network knowledge.

The central reasoning approach used by the generic maintenance system developed within project AIM is Model Based Reasoning. MBR has proved to be suitable for this task. It makes good use of the object oriented knowledge base. Furthermore, the models used by the MBR approach can be generated during the development of the telecom network that needs to be managed. Recommendations on how the shortcomings of MBR concerning speed and efficiency can be overcome are given as well. Assumption Based Truth Maintenance Systems, for example, are used for speeding up the phase where the suggestions for possible faults are checked for consistency with the actual system. Further recommended AIP techniques are Case Based Reasoning and Temporal Reasoning.

Recommendations are also given in respect of communication either between human and machine and between distributed applications. The Seeheim Model is suggested as a unified basis for Man Machine Interfaces. Furthermore the need of applying user profiles in order to achieve user specific communication is stated. For distributed systems infrastructure tools are required, which ease and maintain the handling of distributed applications.

Valuable AIP techniques have been identified and used by AIM, nevertheless there still remain challenging fields of study in the area of AIP techniques for use in maintenance systems.

4. ACKNOWLEDGEMENTS

The authors wish to thank all the AIM project members for helpful discussions, in particular T. Chau (QMW) and K. Riley (UNIPRO). Support for this work has been granted by the CEC RACE Programme.

5. REFERENCES

[1] GUIDELINE Deliverable ME6 : "The Application and Integration of AIP Techniques within the RACE TMN", 03/BTR/712/DS/B/009/b1, RACE Project R1003 GUIDELINE, April 1991.

[2] Azmoodeh, M., Enstone, C. : "Object Oriented Modelling in RACE TMN". Sixth RACE TMN Conference, Madeira, September 1992.

[3] AIM RACE Project R1006 deliverable, "Final Report on AIP Evaluations and Results", 06/DAN/WBS/DS/A/083 (to be published in December 1992)..

[4] Bigham, J. et al : "A Generic Maintenance System for Telecommunication Networks", Sixth RACE TMN Conference, Madeira, September 1992.

[5] Mandich, N., Belleli, T. : "HCI Considerations in TMN Systems", Sixth RACE TMN Conference, Madeira, September 1992.

[6] Pfaff, G.E., "User Interface Management Systems", Proceedings of the Workshop on UIMS held in Seeheim Germany, November 1-3 1983, Springer-Verlag, 1983.

[7] Hopfmüller, H. et al : "An Interconnected MANs Maintenance Prototype", Sixth RACE TMN Conference, Madeira, September 1992.

[8] ISO6083, Information Retrieval, Transfer and Management for OSI, "Working Document - Partial text for the ODP Reference Model - Part 1 and Part 4", 1991.

[9] Wade, V. et al : "Experience Designing TMN Computing Platforms for Contrasting TMN Management Applications", Sixth RACE TMN Conference, Madeira, September 1992.

The Management of Telecommunications Networks
R. Smith, E. H. Mamdani, J. G. Callaghan (Editors)
© Ellis Horwood 1992

HCI Considerations in TMN Systems

Nicholas Mandich (Dowty, UK), Thierry Belleli (CRI A/S, Denmark)

ABSTRACT

Human-Computer Interaction (HCI) is a topic of importance in the Management of Networks and as such it is one of the Advanced Information Processing technologies investigated by RACE TMN technology projects.

The TMN projects are technology-oriented and are therefore concerned mainly with systems, technology, and implementation issues. In the case of HCI, technology orientation is not at the expense of human-centred design issues, which are of key importance in HCI, but only in terms of primary investigation.

This paper concentrates on what are essentially HCI technology considerations. These are grouped under the ODP viewpoints headings. In the technology viewpoint in particular, a number of new and emerging technologies are covered and their relevance to TMN in the short and medium term are discussed. A discussion of the limitations on HCI work resulting from the scope of the RACE TMN projects is also included.

1. INTRODUCTION

Human Computer Interaction refers to the processes and mechanisms which enable a human user to communicate with a computer system and which allows the user to monitor the system's state and control the system's behaviour.

When the concept of the Human Computer Interface first began to emerge, it was commonly understood as the hardware and software through which a human and a computer could communicate. As it evolved, the concept has come to include the cognitive and emotional aspects of the user's experience as well [1].

It is therefore useful to talk about Interaction and what may be referred to as the User Interface (UI) platform as distinct aspects of Human-Computer Interfaces.

Interaction

Human users are part of the organisational system which includes the technological system, so that the design of a technological system must take account of human concerns and factors if it is to meet acceptability and usability goals. Furthermore, in order to meet the broader objectives of users the design of the HCI must be human-centred.

Interaction is the domain of HCI designers. It focuses on task analysis, human factors and the design of the human-computer dialogue and concerns itself with human aspects such as cognitive issues, mental models, metaphors, usability and so on.

UI Platform

The UI platform is the software and hardware that make interaction possible. It is the domain of engineering and systems designers and is technology, and implementation, centred. It encompasses the software and hardware needed to support the interface including both the development and the execution environments. Important concerns include technology issues such as tools, techniques and methods, standards, performance, reliability, security to name but a few.

A unified reference model within which both of these sets of concerns can be considered is the ODP framework [2]. Broadly speaking, the human-centred aspects are part of the enterprise viewpoint, while the implementation aspects are part of the information, computation, engineering and technology viewpoints.

In this paper we present and discuss a range of issues and recommendations pertaining to HCI for TMN systems, grouped as considerations under the five ODP viewpoint headings.

1.1 Scope

It should be noted that, collectively, work of the RACE I technology projects does not exhaustively cover the Network Management problem domain. Each project investigated AIP techniques by developing prototypes within its domain of TMN functionality. These were either stand-alone or were run in conjunction with relatively small scale real or simulated networks. Both of these facts impose limitations on scope. It should also be noted that the workplans of the projects did not include physical integration of the prototypes across projects, although project GUIDELINE, in its coordination role, studied conceptual integration across the three functional domains [3]. A discussion of the effect of this limitation in the context of this paper is discussed in section 7.1 below.

2. ENTERPRISE CONSIDERATIONS

2.1 Multi-Disciplinary Nature of HCI Design

The ability to handle complexity is essential in complex systems such as the increasingly sophisticated TMN. Tasks that human users need to perform in complex control systems are themselves complex. The increase in the complexity of tasks should not, however, mean that Human Interfaces themselves become more complex, but rather that the design of those interfaces becomes more complex. The reason for this is the limiting factor of human cognition. This can also be expressed by saying that the ratio of task and development complexity is a measure of the complexity of the interface which has an upper limit set by human cognitive limitations. This is summarised in the following expression :

Complexity (Human Interface) =
  Complexity (Tasks) / *Complexity* (Development of the Human Interface).

In the case of real operational systems, tackling the complexity of Human Interface development must involve, in addition to technologists and application domain experts, a range of specialists from a number of fields including : Human Factors, Ergonomics, Cognitive Psychology and Graphics Design. Issues of modelling, abstraction, effective presentation, appropriateness, cognition, and usability can only be addressed by such a multi-disciplinary team.

Furthermore, and more fundamentally, the importance of the Human Interface should be recognised and addressed a one of the central aspects of the design of a complex interactive control system.

2.2. Human Interface Types

The experimental work in RACE TMN has highlighted (at least) two distinct types of Human Interface, distinguished by the purpose which they serve.

First, is the experimenters' interface. This is a Human Interface to an experimental system where the interface serves the purpose of observing and interacting with the experimental system in order to demonstrate aspects of the solution or approach taken. The point is that the experimenter's interface may not necessarily be of relevance to a TMN operator of such a system, even if it made sense for the prototype to be part of an operational system. In many cases the scope and emphasis of the experiments is such that they are not directly part of an operational context. Consider, for instance, a performance management experiment in which we wish to demonstrate distributed problem solving.

We therefore have a situation where the purpose of the Human Interface is to allow the behaviour and other aspects of the experimental system to be shown, rather than providing in-service capabilities as required by operators in an operational context.

The second type of interface is the TMN Operators' interface which should be based on an analysis of the tasks which need to be performed, the type of users and the organisational context within which the system is being used. Although it is the case that TMN operators' requirements are being considered explicitly in some cases, it may arguably be unfeasible to address adequately operators' requirements meaningfully because of the new technology syndrome, which leads to a circular situation as far as tasks are concerned. (This is discussed further in section 6.1 below)

2.3 System Lifecycle Timeline and the on-line/off line Distinction

In addition to the the experimenters'/operators' interface dichotomy outlined above, another dimension for classifying Human Interfaces is the system lifecycle time line. There are interfaces to planning applications, system commissioning, in-service operation and maintenance and system decommissioning and modernisation interfaces. Each may in turn be either an off-line or an on-line interface. Examples include: Off-line planning applications, on-line monitoring and control, on-line fault management, off-line administration, off-line fault management, to name but a few. Identifying the various dimensions of classification of Human Interfaces is therefore important.

2.4 Consistency

A range of constraints will typically apply to Human Interface development within an enterprise and thus to Human Interfaces to TMN systems as well. These may come from both within and outside an enterprise and include:
- Standards
- Design Guidelines and Principles
- Style guides.

3. INFORMATION CONSIDERATIONS

3.1 Information Management Bandwidth

As the complexity of systems increases, so does the amount of information involved in controlling these systems. Automation of functions and abstraction and layering of information presented at the Human Interface are used to deal with an increase in system complexity. Nevertheless, complex systems do require the ability to manage, at the Human Interface, an amount of information proportional to the complexity of the underlying systems.

We refer to the volume of information a Human Interface is capable of managing as the Information Management Bandwidth (IMB) of that interface. A distinguishing characteristic of Windowing Interfaces is their relatively high IMB. This is because multiple windows are capable of supporting multiple output and input streams.

Graphical User Interfaces almost always exist in a Windowing environment. Furthermore, GUIs by definition involve symbolic representation and metaphors both of which allow information to be communicated in aggregated and abstracted form which can not only be simpler and more intuitive, but more significantly, can result in representations which involve less cognitive effort on the part of users. Visual interfaces were originally designed expressly so that they engage the symbolic, enactive and iconic mentalities [4] of human users, and in so doing, can tap into the human cognitive apparatus. It is therefore our contention that Graphical orientation per se is capable of contributing to the IMB, quite apart from the multi windowing aspect.

From the Information Engineering point of view, perhaps the most significant advantages of GUIs is the therefore the high IMB.

A stage beyond Windowing and GUIs in terms of increase in IMB is 3-D visualisation, as in the Information Visualiser [5] as used in Virtual Reality Systems [6][14][18]. (Also see section 6.2 below).

3.2 Requested and Unsolicited Information

A TMN Management System will have to handle two different types of information: requested and unsolicited.

Requested information is that which the user has directly requested, or indirectly caused. In either case the information results from some user action and it would not have occurred had the user not performed a previous action.

The second type is unsolicited event reporting generated by the system spontaneously. We will not enter into a philosophical discussion that an event may have resulted from a user action, but as the system is complex, the relationship is not obvious. There is arguably a grey area between these two types, but hopefully the distinction is meaningful.

Requested information should be more predictable, and from the HCI point of view, the major issues are: establishing relevance, structuring and navigation.

Unsolicited event reporting is potentially highly unpredictable in terms of timing, quantity and urgency. It can range from an indication of a relatively innocuous malfunction, to a condition which requires the user to suspend the current task and deal with the problem in real-time, right through to a catastrophic failure situation with guaranteed cognitive overload.

Cognitive Overload and Filtering

The problem is that it may not only be difficult to generate a problem space to base and test the solution against, but worse still, it may be impossible to even identify the whole of the problem space. One thing is certain, and that is that intelligent filtering and management of incoming information at the level of the Human Interface is required. An Artificial Intelligence (AI) approach is arguably the only promising solution available.

Autonomous behaviour and Explanatory capability

Another consideration in intelligent control systems capable of responding autonomously to incoming event information is understandability of actions taken by the systems so that an explanatory capability may be useful.

Online Support

Many TMN tasks involve complex interactions with human users and will require skill and experience on the part of users. Support for users with varying levels of expertise may therefore be useful and necessary. This support can potentially be passive and active. An example of a passive function is a user-driven help function, and an example of an active function is a proactive help function which may be triggered by poor performance or errors on the part of a user in performing a task.

Another aspect of online support is the use of the management system in conjunction with simulations as discussed in section 6.2 below.

3.3 Representation and Modelling

A Human Interface is not only a doorway to a system, but it is also a model of the system that a user sees beyond the door. Real challenges in representation and modelling are to be found in Human Interface design of complex systems. Physical analogies are appropriate some of the time since there is real hardware out there being controlled.

However, as we move away from the physical levels, we encounter a range of abstractions. Effective modelling of these abstractions within a TMN for different tasks may require innovative new metaphors and analogies.

Representation and modelling are a very important topic particularly considering the many abstractions that are possible and indeed necessary in systems as complex as a TMN. Please also see 3.1 above.

4. COMPUTATION CONSIDERATIONS

4.1 TMN Workstation Function

It is well known that Graphical Interfaces require significant computing power. Good performance, especially in colour, comes at a price. When we further consider the computation requirements of a User Interface Management System which may be used to build Human Interfaces to TMN we are looking at workstation class computing power just to run the UI Platform. Multi-media in the interface, and particularly three dimensional visualisation, push this much higher. The Workstation function of M.3010 [7] may therefore require a workstation in which to run.

To emphasise this point, a computation model of the TMN workstation function proposed by RACE TMN [8] is based on the Seeheim model of User Interface Management Systems [9]. The rationale for Seeheim is two fold: the ability to manage complex interfaces and the desire to achieve separation of the Human Interface from the application code.

5. ENGINEERING CONSIDERATIONS

5.1 Distribution

An operational TMN system will be a distributed system. We assume that the TMN computing platform [10] will have a number of transparencies, and in particular distribution transparency which will hide the distribution from the various applications including Human Interfaces.

In an IBC context, the bandwidth of the Data Communication Network [7] may well be of a different order compared to management systems today. Delays and cost of communications may therefore well be a lot lower, and consequently we may have a situation where it is faster and cheaper to access data across the distributed TMN, than fetching it from local storage.

In essence, the point we wish to bring out is that a quantitative change in network performance may result in a qualitative change in the way we design systems and we may therefore need to look at the communication/computation tradeoff in the light of this.

5.2 Human Interface Separation

Separating the Human Interface from the application code has a number of advantages:

- *Modularisation* : Separating the UI results in a more modular system which is more manageable
- *Modifiability* : The Human Interface can be changed independently of the application code thus reducing the dependency and increasing reliability
- *Focusing of Skills* : Separating the UI development task makes it possible to focus skills and develop Human Interfaces by Human Interface specialists in addition to application developers
- *Distribution* : A UI separate from the application code, provides the opportunity to better distribute and decentralize the overall system.

5.3 Support for Terminals of Differing Display Capability

In contrast to the above, it may be necessary to provide access to TMN functions using a variety of display terminals with different degrees of graphics handling and processing support. This may range from glass teletypes, X-terminals to graphics workstations and the spectrum between direct manipulation and command line interfaces. This has implications on:

- interaction style
- scope of interface
- modelling and representation
- location of the Human Interface subsystem.

5.4 Simulation

Model-based reasoning could in theory provide the basis for a simulation of parts of the TMN, which could in turn be used for training and familiarisation for operators in the same way as flight simulators are used in the aircraft industry for training pilots.

This is an area of great potential and formidable challenge given the complexity of the systems to be simulated.

It may not be as daunting as it may at first appear, considering that flight simulators exist, and it may be necessary to develop such simulators in order to investigate human responses to extraordinary network states which result from system failure conditions.

6. TECHNOLOGY CONSIDERATIONS

6.1 New Technology Syndrome: Technology push or Requirements pull?

All new technological systems suffer from the *new technology syndrome*. The less precedent for a system, the more acute this problem is. The challenge is illustrated by the following circular statement.

"Usability and suitability for purpose of a system are related to how closely the system meets the operational requirements of its users which, in turn, are related to the tasks performed, tasks which tend to be shaped by the system being used".

HCI practise promotes a requirements-driven, Task Analysis-based approach followed by usability testing and iteration. In the case of a genuinely new technology, with few precedents, there has to be, almost of necessity, a degree of technology push at the start. It is therefore inevitable that the only practical way forward for developing TMN Human Interfaces is to resort to rapid prototyping which by definition involves iteration.

6.2 New Technologies

There are a number of currently leading edge technologies which are relevant to Human Interfaces to TMN.

Interface Agents

User Interface agents [11] can be likened to software robots; they are task specific computer processes, that can act as guides, coaches or secretaries [4]. Trivial examples are an alarm clock and a backup agent. A more realistic example is an agent that sifts through event logs, searching for patterns, and draws inferences. In some instances Agents are a type of expert system. In some ways, agents are comparable macros, and may well be used for end-user programming of repetitive sequences and specific tasks. Seel [12] and Bass [13] provide a very worthwhile insight into Agents in the Human-Computer Interface.

Effective use of multiple media in the interface

In the context of Human Interfaces, Multi-media refers to a multiplicity of forms of communication between a human and a computer. A GUI today typically employs text, graphics and, to a limited degree, sound.

Animation, live video, CD quality sound, speech synthesis and to a modest degree speech recognition and handwriting recognition are additional media that are already available.

The real challenge in Multi-media is to design new ways of utilising the various media to broaden the communication channel non-trivially, and thus use the media effectively.

Multi-Sensory Interfaces

Somebody once pointed out that, if an alien civilization tried to infer what kind of beings humans are, based on contemporary workstation remains as artifacts, they may well conclude that we have no legs, no mouth, one eye, underdeveloped ears, two arms and an overdeveloped index finger.

This is a good reflection of how well we make use of our senses in the Human Interfaces we now have. In a high information content situation such as a complex Management System, it may be possible to spread the cognitive load by employing multi-sensory I/O devices. See [14].

A related area is that of Multi-Modal Interfaces [15] which involve the use of a combination of media to execute a single task.

The Information Visualizer

The Information Visualizer [5] is an experimental User Interface developed at the Xerox Palo Alto Research Centre as part of the Interactive Information Access project. Its aim is to help users better manage jobs that involve vast amounts of information. It uses 3 dimensional real-time animation to present information as 3-D interactive objects.

The Information Visualizer was designed in a different context than Virtual Reality (VR) systems described below and unlike VR systems does not require special head–mounted displays and datagloves to allow visualisation and interaction in 3-D space. Instead, it uses solid modelling on an "ordinary" CRT display and also uses the familiar mouse endowed with additional degrees of freedom to allow navigation in the 3-D space. The Information Visualizer incorporates the Rooms metaphor [16]; rooms are 3-D rooms.

As mentioned already, the key driver behind the Information Visualizer is the requirement to manage large volumes of information and it does so by providing an interface whose Information Management Bandwidth is considerably larger than that achievable with a "Standard" GUI available today.

Virtual Reality

Buxton [17] discusses new directions in Human Interfaces, among which one of the most exciting developments is that of Virtual Reality (VR) which was developed initially by NASA in the context of remote manual control of robots in hostile environments.

Virtual Reality is a technology that will take the user through the screen into the world "inside" the computer : a world in which the user can interact with three dimensional objects whose fidelity increases as display technology progresses [17]. Interactions in this virtual world [6][14][19] mimic interactions with real-world objects and direct manipulation acquires greater realism. This three dimensional virtual reality is known as Cyberspace, a term which denotes a three-dimensional domain in which cybernetic feedback and control occur.

In order to interact with this virtual world, a person has to wear a special head-mounted display which replaces the monitor, together with a special glove, known as dataglove, which has optic fibres running along its surface. These are used as sensors to detect hand movements which are then translated into movement of a simulated hand visible to the person in the virtual space much in the same way that the movement of a mouse pointer is controlled by physical movement of the mouse.

This technology is not very far off. In fact, games manufacturers are already seizing the opportunity and working on arcade games based on VR. As with the Information Visualizer discussed above, the Information Management Bandwidth which VR can handle is enormous and its potential in direct manipulation applications seems very significant indeed.

7. CONCLUSIONS

7.1 What RACE TMN will not cover

The three experimental RACE TMN technology projects AIM, ADVANCE and NEMESYS are each concentrating on subsets of Network Management functionality. Between them, they do not exhaust the problem domain at least from the HCI viewpoint and there are HCI areas which will remain unexplored, not through any omission on part of the projects, but simply because they do not fall within their scope. (See also section 1.1 above).

Examples are: unsolicited event handling, filtering and presentation; catastrophic event management; graphical representation of large networks; navigation in large data sets; cognitive issues resulting from combining a large number of management applications in one system, particularly when they interact or are used in a combined way.

Some of these will be difficult to address, either because of the scale and scope of experimental systems, or indeed emphasis and recognition, or because they are also extremely complex issues, as for instance catastrophic failure management.

This line of argument leads to an uncomfortable conclusion, namely that RACE TMN is not going to be in a position to address some hard HCI problems because, on the one hand there is not going to be a unified, physically integrated TMN prototype, and on the other, there is no IBCN in existence to test it against, nor is there, or is there going to be a large enough simulation of an IBC network, which could generate a realistic management problem. It therefore seems that RACE TMN must of necessity leave out chunks of the problem domain, at least as far as HCI is concerned. This is not to say that that useful experimentation is not possible, just that parts of the story may have to remain untold.

7.2 Evolutionary Issues

TMN systems will need to be developed on available, proven technologies. In the early to mid nineties, X-Windows looks like just such a choice, state-of-the-art commercially, and arguably technologically.

In the short to medium term, technologies that add value and build on the GUI paradigm will be of interest. In the medium term, new technologies better able to overcome some of the limitations of the present state-of-the-art windowing interfaces will consign GUIs (as we know them today) to the wasteland of history.

8. ACKNOWLEDGMENTS

This paper is based on work by the authors within the RACE Programme. The authors would like to acknowledge support by Projects GUIDELINE, NEMESYS, ADVANCE and AIM.

The paper presents an overview of HCI Considerations in TMN from the authors' perspective and does not necessarily represent a complete view of HCI Issues in TMN. Responsibility for omissions, and oversights therefore remains with the authors.

9. REFERENCES

[1] Laurel, B. : "Introduction" in The Art of Human-Computer Interface Design. Brenda Laurel, ed. Addison-Wesley, 1990.

[2] Recommendation X.9yy, Basic Reference model of Open Distributed Processing – Part 2: Descriptive Model. ISO/IEC JTC1/SC21. 21 Aug 1991.

[3] Turner, T., Callaghan, J. G. : "Towards Integrated TMNs - The Global Conceptual Schema", Sixth RACE TMN Conference, Madeira, September 1992.

[4] Kay, A. : "User Interface: A Personal View." in The Art of Human-Computer Interface Design. Brenda Laurel, ed. Addison-Wesley, 1990.

[5] Clarkson, M. A. : "An Easier Interface", BYTE, Feb 1991, pp. 277-282.

[6] Rheinghold, H. : Virtual Reality, Seeker and Warburg, 1991.

[7] CCITT Recommendation M.3010. Principles for a telecommunications management network. Draft revision of the Paris Editing meeting, Dec 1991.

[8] GUIDELINE Deliverable ME8 : "TMN Implementation Architecture", 03/DOW/SAR/DS/B/012/b3, RACE Project R1003 GUIDELINE, March 1992.

[9] Maier, F. et al : "Recommendations for the Use of AIP Techniques for Maintenance in Telecommunication Systems", Sixth RACE TMN Conference, Madeira, September 1992.

[10] Wade, V. et al : "Experience Designing TMN Computing Platforms for Contrasting TMN Management Applications", Sixth RACE TMN Conference, Madeira, September 1992.

[11] Laurel, B. : "Interface Agents: Metaphors with Character." in The Art of Human-Computer Interface Design. Brenda Laurel, ed. Addison-Wesley, 1990.

[12] Seel, N. : "Agents in the Human-Computer Interface: (Extended Abstract)" in Proceedings of the IEE Colloquium on "Intelligent Agents", Digest Number 1991/048. The Institution of Electrical Engineers, Computing and Control Division, Savoy Place, London WC2R 0BL.

[13] Bass, L., Coutaz, J. :, "Developing Software for the User Interface," Addison Wesley, 1991.

[14] Fisher, S. S. : "Virtual Interface Environments." in The Art of Human-Computer Interface Design. Brenda Laurel, ed. Addison-Wesley, 1990.

[15] Richard, A. Bolt, "Conversing with computers", in Readings in Human-Computer Interaction: A Multi-Disciplinary approach, Ronal Baecker and William Buxton, Eds., Morgan Kaufmann, Los Altos, CA, 1987.

[16] Henderson, A., S. Card, S. : "Rooms: "The use of multiple virtual workspaces to reduce space contention in a window-based graphical user interface", ACM Transactions on Graphics, Vol. ?, No. ?, 1986.

[17] Buxton., B. : "Smoke and Mirrors" BYTE, Special Feature on "Computing without Keyboards", July 1990, pp 215-221.

[18] Walker, J. : "Through the Looking Glass." in The Art of Human-Computer Interface Design. Brenda Laurel, ed. Addison-Wesley, 1990.

[19] Fisher, S. S., Tazelaar, J. M. : "Living in a Virtual World." BYTE, Special Feature on "Computing without Keyboards," July 1990, pp 215-221.